TEXTBOOK OF
BIOTECHNOLOGY

TEXTBOOK OF BIOTECHNOLOGY

S.C. Bhatia
B.E. (Chemical), MBA

ATLANTIC®
PUBLISHERS & DISTRIBUTORS

Published by
ATLANTIC®
PUBLISHERS & DISTRIBUTORS
B-2, Vishal Enclave, Opp. Rajouri Garden,
New Delhi-110027
Phones : 25413460, 25429987, 25466842

Sales Office
7/22, Ansari Road, Darya Ganj,
New Delhi-110002
Phones : 23273880, 23275880, 23280451
Fax : 91-11-23285873
web : www.atlanticbooks.com
e-mail : info@atlanticbooks.com

Printed in India
at Nice Printing Press, Delhi

Preface

Biotechnology impinges on every one's life. It is one of the major technologies of the twenty-first century. Its vast, wide range, multi-disciplinary activities include – microbiology, chemistry, biochemistry, chemical engineering, genetics and immunology. Today, the world is facing four major problems – malnutrition, diseases, energy scarcity and its high cost, and environmental pollution. Biotechnological developments are going on continuously to overcome these problems. Better understanding of the growth requirements, improvements in micro-organisms by gene manipulation techniques, control on large-scale growth and metabolic activities, immobilisation techniques, more efficient reactors are some of the important features responsible for the biotechnology's success in different fields.

Micro-organisms are the oldest inhabitants of earth. They are masters in versatility and adaptability to the changing environment. They will definitely prove to be most cost-effective partners in our efforts for sustainable development. So limitations faced by physical and chemical methods of pollution control will be overcome if we take the help of these environmental masters (micro-organisms).

This invaluable textbook gives a full account of the current state of Biotechnology, providing reader with insight, inspiration and instruction.

Biotechnology precisely applies engineering and other principles to add the abilities of organisms and thereby develops process useful to mankind's progress. So, it has a role in taking care of our health, in producing chemicals, in giving cheap abundant food, in improving plant productivity, in providing alternative energy and in pollution control (environment protection).

Each chapter covers an important aspect of Biotechnology with an accurate and up-to-date account of each topic. Chapter 1 is devoted to basic concepts of biotechnology and Chapter 2 deals with substrates for biotechnology. Genetic engineering is the fundamental basis for the modern biotechnology as it involves the ability to transfer specific genes from one kind of organism to another, thereby producing new biological properties. Chapter 3 focuses on genetics and biotechnology. Chapter 4 describes in detail the fermentation technology and industrial microbiology. Various types of fermentation process are discussed. Chapter 5 is devoted to downstream processing in biotechnology such as solid-liquid separation, concentration and formulation etc. Enzymes are obtained from plants, animals and micro-organisms by extraction with a suitable solvent, preferably after the cell structure has been destroyed by drying. Following this, Chapter 6 familiarises students with production of enzymes. Chapter 7 concentrates on biological fuel generation such as sources of biomass etc. Chapter 8 is devoted to single cell protein and recombinant proteins of high value. The impact of pharmaceuticals on human health care is an area where biotechnological innovations are likely to play very important role. Closely related to this, Chapter 9 focuses on biotechnology and medicine. Chapter 10 contains information on biotechnology in agricultural and forestry industries whereas Chapter 11 is devoted to biotechnology of animals.

Food biotechnology is concerned with the integration of both modern biological knowledge and techniques and current bioengineering principles in food processing and preservation. In this context, Chapter 12 focuses on food and beverage biotechnology. Chapter 13 is devoted to biotransformation, dealing with the use of biocatalyst to convert a substrate into a product in a limited number of enzymes. Chapter 14 focuses on quality assurance and quality control and Chapter 15 deals with safety and ethics in biotechnology. Chapter 16 is concerned with patenting in biotechnology. Various issues of patenting in biotechnology like fundamental requirements of patent, patenting of living things, common categories of patents, patenting of genes and DNA are discussed in detail.

Environmental biotechnology has a great potential in transformation and enhancement of resource-base, reduction of energy consumption, pollution abatement, enhancement of biomass production and conservation of non-renewable resources, generation of useful organic chemicals from bioconversion of biomass, etc. Some of these applications have indirect beneficial effects on wider problems such as global warming, acid rain and climatic changes. Keeping these in view, Chapter 17 is devoted to environmental biotechnology.

Glossary and index have been povided at the end for quick reference. Diagrams, figures and tables supplement the text. All the topics have been covered in a cogent and lucid style to help the reader grasp the information quickly and easily.

It may not be wrong to hold that the present book *Textbook of Biotechnology* is a complete treatise on biotechnology. This textbook is essential reading for all students and teachers of biotechnology, applied micro-biology and researchers in the many biotechnology industries.

The textbook also caters to the requirement of the syllabus prescribed by various Indian universities for undergraduate students persuing Biotechnology. It has been prepared with meticulous care, aiming at making the book error-free. Constructive suggestions are always welcome from users of this book.

S.C. BHATIA

Contents

1

An Introduction to Biotechnology

INTRODUCTION

Biotechnology is highly multidisciplinary; it has its foundations in many fields including biology, microbiology, biochemistry, molecular biology, genetics, chemistry and chemical and process engineering.

It has been considered as a series of enabling technologies involving the practical application of organisms or their cellular components to manufacturing and service industries and environmental management. Historically, biotechnology was an art rather than a science, exemplified in the manufacture of wines, beers, cheeses, etc., where the techniques of manufacture were well worked out and reproducible but the molecular mechanisms were not understood. With the major advances in our understanding of microbiology and biochemistry, these processes have become better understood and improved. Modern biotechnological processes now encompass a wide range of new products including antibiotics, vaccines and monoclonal antibodies, the production of which has been optimised by improved fermentation procedures. Biotechnology has been further diversified by a host of new molecular innovations, allowing unprecedented changes to be made to living systems. Transgenic plants and animals are heralding a new age in agriculture, and gene therapy in humans may eradicate many previously incapacitating diseases. In the environment, biotechnology is allowing major improvements in water and land management and also remediating the pollution generated by over industrialisation.

Thus biotechnology is a range of biological, chemical and engineering disciplines with varying degrees of application to agricultural, medical, industrial and environmental situations. Biotechnology should be viewed as a spectrum of enabling technologies that find a variety of applications in many sectors. The assessment of the viability of biotechnology in any country must be made in the context of its biological sciences and their relationships with the productive sectors. Modern biotechnology springs from universities and other research centres that generate then basic knowledge needed to solve practical problems posed by society.

Historically, the applied use of biological organisms (in particular microorganisms) has developed in a somewhat empirical manner over many years. In many ways the control of the processes was seen as more an art than a science. In more recent times most of these ancient biotechnological processes have been subjected to rigorous scientific scrutiny and analysis, which is, in many cases, influencing and replacing traditional empiricism; of particular significance has been the growth of the science of genetics.

In many long-established biotechnological processes such as brewing and cheese production, organisms that improve the process have been deliberately selected by people. Better understanding of genetics has led to more effective application of the genomic potential of the organisms used in these industries. However, in the past 10 years, major new developments in the ability to select and manipulate genetic material (recombinant DNA technologies, cell fusions, etc.) have led to unprecedented interest in the creative uses of living organisms. It is confidently anticipated that these new genetic techniques, coupled with advances in fermentation technology and downstream processing, will have major economic impact on biologically based industries, agriculture and forestry throughout the world. However, although much has been prophesied, only a very small amount of industrial success has been realised to date. Perhaps there has been too much popular emphasis on the highly futuristic aspects of biotechnology; they will almost all be realised eventually, but on a considerably longer time-scale than previously anticipated. In practice, the majority of products to be derived from biotechnology will be recognisable extensions of existing ones or stem from improvements in production. Minor improvements in processes by many of these new technologies may not be highly newsworthy but will help to give the producer companies a competitive edge. The history of biotechnology shows how intimate must be the interplay between industry and academia to maintain the competitive edge to generate new products and to expand its scope, profitability and social impact.

Over the next two decades biotechnology will have a major impact on health, pharmaceuticals, agriculture, food and the environments. Clearly the development of new drugs, human and animal vaccines, diagnostics, transgenic crops and a range of new processes to clean up and manage the environment must undoubtedly bring substantial improvements in the living standards of most people. Undoubtedly also, life expectancy will be extended.

Although there is now a vast reservoir of relevant biological and engineering knowledge and expertise waiting to be put into productive biotechnological use, the eventual rate of application will be determined not primarily by science and technology, but rather by many other equally important factors such as industrial investment policies, the establishment of market needs, economics of new processes, safety factors and government regulations, the marketing skills needed to introduce new products into commercial use and above all how the public perceive this new range of technologies.

New aspects of biotechnology such as transgenic crop plants and farm animals and gene therapy in humans will bring huge benefits to humankind but not without generating fears in some sections of the population. The ethical and moral issues raised by some aspects of biotechnology must be addressed by open discussion and informed communication. Most people are scientifically untrained and find the complexity of new biotechnology such as transgenic animals and gene therapy bewildering and threatening.

Notwithstanding the obvious difficulties, there is no doubt that biotechnology in its many forms will have an increasingly important role to play in our society. However, biotechnology will require highly trained and skilled professionals because its raw materials are knowledge and intelligence.

This newly acquired biological knowledge has already made vastly important contributions to the health and welfare of humankind. Yet few people fully recognise that the life sciences affect over 30 per cent of global economic turnover by way of healthcare, food and energy, agriculture and forestry, and that this economic impact will grow as biotechnology provides new ways of influencing raw material processing. Biotechnology will increasingly affect the efficiency of all fields involving the life sciences and it is now realistically accepted that by the early twenty-first century, it will be contributing many trillions of dollars to world markets.

In the following chapters, biotechnology will be shown to cover a multitude of different applications ranging from the very simple and traditional, such as the production of beers, wines and cheeses, to highly complex molecular processes, such as the use of recombinant DNA technologies to yield new drugs or to introduce new traits into commercial crops and animals. The association of traditional industries such as brewing with modern genetic engineering is gaining momentum and it is not for nothing that industrial giants such as Guinness, Carlsberg and Bass are heavily involved in biotechnology research. Biotechnology is developing at a phenomenal pace and will increasingly be seen as a necessary part of the advance of modern life and not simply a way to make money.

While biotechnology has been defined in many forms (Table 1.1), in essence it implies the use of microbial, animal or plant cells or enzymes to synthesise, breakdown or transform materials.

Table 1.1. Some selected definitions of biotechnology.

The application of biological organisms, systems or processes to manufacturing and service industries
The integrated use of biochemistry, microbiology and engineering sciences in order to achieve technological (industrial) application capabilities of microorganisms, cultured tissue cells and parts thereof
A technology using biological phenomena for copying and manufacturing various kinds of useful substance
The application of scientific and engineering principles to the processing of materials by biological agents to provide goods and services
The science of the production processes based on the action of micro-organisms and their active components and of processes involving the uses of cells and tissues from higher organisms. Medical technology, agriculture and traditional crop breeding are not generally regarded as biotechnology
Biotechnology is really no more than a name given to a set of techniques and processes
Biotechnology is the use of living organisms and their components in agriculture, food and other industrial processes
Biotechnology - the deciphering and use of biological knowledge

The European Federation of Biotechnology (EFB) considers biotechnology as 'the integration of natural sciences and organisms, cells, parts thereof, and molecular analogues for products and services'. The aims of this Federation are:

1. To advance biotechnology for the public benefit.
2. To promote awareness, communication and collaboration in all fields of biotechnology.
3. To provide governmental and supranational bodies with information and informed opinions on biotechnology.
4. To promote public understanding of biotechnology.

The EFB definition is applicable to both 'traditional or old' and 'new or modern' biotechnology. Traditional biotechnology refers to the conventional techniques that have been used for many centuries to produce beer, wine, cheese and many other foods, while 'new' biotechnology embraces all methods of genetic modification by recombinant DNA and cell fusion techniques, together with the modern developments of 'traditional' biotechnology processes. The difficulties of defining biotechnology and the resulting misunderstandings have led some people to suggest the abandonment of the term 'biotechnology' as too general and to replace it by the precise term of whatever specific technology or application was being used.

Unlike a single scientific discipline, biotechnology can draw upon a wide array of relevant fields such as microbiology, biochemistry, molecular biology, cell biology, immunology, protein engineering, enzymology, classified breeding techniques, and the full range of bioprocess technologies Fig. 1.1. Biotechnology is not itself a product or range of products like micro-electronics: rather it should be regarded as a range of enabling technologies that will find significant application in many industrial sectors. As will be seen in later sections, it is a technology in search of new applications and the main benefits lie in the future. New biotechnological processes will, in many instances, function at low temperature, will consume little energy and will rely mainly on inexpensive substrates for biosynthesis.

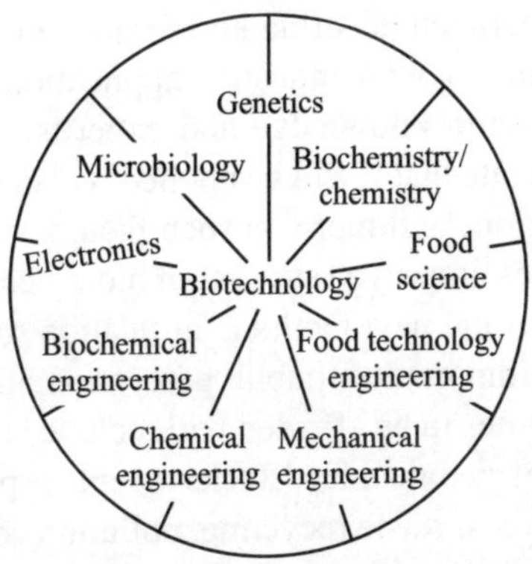

Fig. 1.1. The interdisciplinary nature of biotechnology.

However, it should be recognised that biotechnology is not something new but represents a developing and expanding series of technologies dating back (in many cases) thousands of years, to when humans first began unwittingly to use microbes to produce foods and beverages such as bread and beer and to modify plants and animals through progressive selection for desired traits. Biotechnology encompasses many traditional processes such as brewing, baking, wine-making, cheese production, the production of oriental foods such as soya sauce and tempeh, and sewage treatment where the use of microorganisms has been developed somewhat empirically over countless years. It is only relatively recently that these processes have been some time for modern scientifically based practices fully to replace traditional empiricism.

The new biotechnology revolution began in the 1970s and early 1980s, when scientists learned to alter precisely the genetic constitution of living organisms by processes out with traditional breeding practices. This 'genetic engineering' has had a profound impact on almost all areas of traditional biotechnology and further permitted breakthroughs in medicine and agriculture, in particular those that would be impossible by traditional breeding approaches. Some of the most exciting advances will be in new pharmaceutical drugs and gene therapies to treat previously incurable diseases, to produce healthier foods, safer pesticides, innovative environmental technologies and new energy sources.

There is also a considerable danger that biotechnology will be viewed as a coherent, unified body of scientific and engineering knowledge and thinking to be applied in a coherent and logical manner. This is not so: the range of biological, chemical and engineering disciplines that is involved is having varying degrees of application to the industrial scene.

Traditional biotechnology has established a huge and expanding world market and, in monetary terms, represents a major part of all biotechnology financial profits. 'New' aspects of biotechnology founded in recent advances in molecular biology, genetic engineering and fermentation process technology are now increasingly finding wide industrial application. A breadth of relevant biological and engineering knowledge and expertise is ready to be put to productive use; but the rate that it will be applied will depend less on scientific or technical considerations and more on such factors as adequate investment by the relevant industries, improved systems of biological patenting, marketing skills, the economics of the new methods in relation to current technologies and, possibly of most importance, public perception and acceptance.

The industrial activities to be affected will include human and animal food production, provision of chemical feedstocks to replace petrochemical sources, alternative energy sources, waste recycling, pollution control, agriculture and forestry. The new techniques will also revolutionise many aspects of medicine, veterinary sciences, and pharmaceutics.

Biotechnological industries will be base largely on renewable and recyclable materials and so can be adapted to the needs of a society in which energy is ever increasingly expensive and scarce. In many ways, biotechnology is a series of embryonic technologies and will require much skilful control of its development, but the potentials are vast and diverse, and undoubtedly will play an increasingly important part in many future industrial processes.

BIOTECHNOLOGY—AN INTERDISCIPLINARY PURSUIT

Biotechnology is a *priori* an interdisciplinary pursuit. In recent decades a characteristic feature of the development of science and technology has been the increasing resort to multidisciplinary strategies for the solution of various problems. This has led to the emergence of new interdisciplinary areas of study, with the eventual crystallisation of new disciplines with identifiable characteristic concepts and methodologies. Chemical engineering and biochemistry are two well-recognised examples of disciplines that have done much to clarify our understanding of chemical processes and the biochemical bases of biological systems.

The term 'multidisciplinary' describes a quantitative extension of approaches to problems that commonly occur within a given area. It involves the marshalling of concepts and methodologies from a number of separated disciplines and applying them to a specific problem in another area. In contrast, interdisciplinary application occurs when the blending of ideas that occur during multidisciplinary co-operation leads to the crystallisation of a new disciplinary area with its own concepts and methodologies. In practice multidisciplinary enterprises are almost invariably mission orientated. However, when true interdisciplinary synthesis occurs the new area will open up a novel spectrum of investigation. Many aspects of biotechnology have arised through the interaction between various parts of biology and engineering.

A biotechnologist can utilise techniques derived from chemistry, microbiology, biochemistry, chemical engineering and computer science (Fig. 1.1). The main objectives will be the innovation, development and optimal operation of processes in which biochemical catalysis has a fundamental and irreplaceable role. Biotechnologists must also aim to achieve a close working co-operation with experts from other related fields such as medicine, nutrition the pharmaceutical and chemical industries, environmental protection and waste process technology. The application of biotechnology will increasingly rely upon each of the contributing disciplines for an understanding of the technical language of the others and, above all, an understanding of the potential as well as the limitations of the other areas.

Biotechnology is a demanding industry that requires a skilled workforce and a supportive public to ensure continued growth. Economies that encourage public understanding and provide a competent labour force should achieve long-term benefits from biotechnology. The main types of company involved with biotechnology can be placed in seven categories (Table 1.2). A key factor in the distinction between biology and biotechnology is their scale of operation. The biologists, usually works in the range nanograms to milligrams. Biotechnologists working on the production of vaccines may be satisfied with milligram yields, but many other projects aim at kilograms or tonnes. Thus, one of the main aspects of biotechnology consists of scaling up biological processes.

Table 1.2. Company categories involved in biotechnology.

Therapeutics
Pharmaceutical products for the cure or control of human diseases including antibiotics, vaccines, gene therapy
Diagnostics
Clinical testing and diagnosis, food, environment, agriculture
Agriculture/forestry/horticulture
Novel crops or animal varieties, pesticides
Food
Wild range of food products, fertilisers, beverages, ingredients
Environment
Waste treatment, bioremediation, energy production
Chemical intermediates
Reagents including enzymes, DNA/RNA, speciality chemicals
Equipment
Hardware, bioreactors, software and consumables supporting biotechnology

Many present-day biotechnological processes have their origins in ancient and traditional fermentations such as the brewing of beer and the manufacture of bread, cheese, yoghurt, wine and vinegar. However, it was the discovery of antibiotics in 1929 and their subsequent large-scale production in the 1940s that created the greatest advances in fermentation technology. Since then we have witnessed a phenomenal development in this technology, not only in the production of antibiotics but in many other useful, simple or complex biochemical products, e.g., organic acids, polysaccharides, enzymes, vaccines and hormones. Inherent in the development of fermentation processes is the growing close relationship between the biochemist, the microbiologist and the chemical engineer. Thus, biotechnology is not a sudden discovery but rather a coming of age of a technology that was initiated several decades ago. Looking to the future, the *Economist*, when reporting on this new technology, stated that it may launch 'an industry as characteristic of the twenty-first century as those based on physics and chemistry have been of the twentieth century'.

If it is accepted that biotechnology has its roots in distant history and have large successful industrial outlets, why then has there been such public awareness of this subject in recent years. Undoubtedly the main dominating reason must derive from the rapid advances in molecular biology, in particular recombinant DNA technology, which is giving human beings dominance over nature. By these new techniques it is possible to manipulate directly the heritable material (DNA) of cells between different types of organism, creating new combinations of characters and abilities previously known to be present on this planet. The potential of this series of techniques first developed in academic laboratories is now being rapidly exploited in industry. The industrial benefits are immense but the inherent dangers of tampering with nature must always be appreciated and respected.

While in theory the technology is available to transfer a particular gene from any organism into any other organism, microorganism, plant or animal, in actual practice there are numerous constraining factors such as which genes are to be cloned, and how they can be selected. The single most limiting factor in the application of genetic engineering is the dearth of basic scientific knowledge of gene structure and function. For plants it is salutary to note that only about 150 plant genes, out of a conservative total of about 10,000, have so far been characterised at the DNA level.

The developments of biotechnology are proceeding at a speed similar to that of micro-electronics in the mid-1970s. Although the analogy is tempting, any expectations that biotechnology will develop commercially at the same spectacular rate should be tempered with considerable caution. While the potential realisation of 'new' biotechnology cannot be doubted, a meaningful commercial realisation is now only slowly occurring and will accelerate as we approach the end of the century. New biotechnology will have a considerable impact across all industrial uses of the life sciences. In each case the relative merits of competing means of production will influence the economics of a biotechnological route. Biotechnology will undoubtedly have great benefits in the long-term in all sectors.

The growth in awareness of modern biotechnology parallels the serious world-wide changes in the economic climate arising from the escalation of oil prices since 1973. There is a growing realisation that fossil fuels and other non-renewable resources will one day be in limited supply. This will result in the requirement of cheaper and more secure energy sources and chemical feedstocks, which biotechnology could perhaps fulfil. Countries with climatic conditions suitable for rapid biomass production could well have major economic advantages over less climatically suitable parts of the world. In particular, the tropics must hold high future potential in this respect.

Another contributory factor to the growing interest in biotechnology has been the current recession in the western world, in particular the depression of the chemical and engineering sections, in part due to the increased energy costs. Biotechnology has been considered as one important means of restimulating the economy, whether on a local, regional, national or even global basis, using new biotechnological methods and new raw materials. In part, the industrial boom of the 1950s and 1960s was due to cheap oil, whereas information technology advances in the 1970s and 1980s resulted from developments in micro-electronics. It is quite feasible that the 2010 will increasingly be seen as the era of biotechnology. There is undoubtedly a world-wide increase in molecular biological research, the formation of new biotechnological companies, large investments by nations, companies and individuals and the rapid expansion of data bases, information sources and, above all, extensive media coverage.

It is perhaps unfortunate that there has been an over-concentration on the new implications of biotechnology and less identification of the very large traditional biotechnological industrial bases that already function throughout the world and contribute considerably to most 'nations' gross national products. Indeed, many of the innovations in biotechnology will not appear *a prior* as new products but rather as improvements to organisms and processes in long-established biotechnological industries, e.g., brewing and antibiotics production.

New applications are likely to be seen earliest in the area of healthcare and medicine, followed by that of agriculture and food technology. Exciting new medical treatments and drugs based on biotechnology are appearing with ever-increasing regularity. Prior to 1982 insulin for human diabetics was derived from cow and pig pancreases. The gene for human insulin was the isolated and cloned into a micro-organism, which was then mass-produced by fermentation. This genetically engineered human insulin, identical to the natural human hormone, was the first commercial pharmaceutical product of recombinant DNA technology and now supplies millions of insulin user world wide with a safe, reliable and unlimited source of this vital hormone. Biotechnology has also made it easier to detect and diagnose human, animal and plant diseases. In clinical diagnosis there are now hundreds of specialised kits available for simple home use or for complex laboratory procedures such as blood screening. Biotechnology methods can now improve the nutrition taste and appearance of plants and various food products, enhance resistant to specific viruses and insect pests and produce safer herbicides. For food safety, new probes can rapidly detect and accurately identify specific microbia pathogens in food, e.g., the bacteria *Salmonella* and *Listeria* and fungal toxins such as aflatoxin.

Application, in chemical production, fuel and energy production, pollution control and resource recovery will possibly take longer to develop and will depend on changes in the relative economics of currently employed technologies.

Biotechnology-based industries will not be labour-intensive and although they will create valuable new employment, the need will be more for brains than muscle. Much of modern biotechnology has been developed and utilised by large companies and corporations. However, many small- and medium- sized companies are realising that biotechnology is not a science of the future but provides real benefits to their industry today. In many industries traditional technology can produce compounds causing environmental damage; whereas biotechnology methods can offer a 'green' alternative promoting a positive public image and also avoiding new environmental penalties. Knowledge of biotechnology innovations must be translated through to all sectors of industry.

Many new, high technology biotechnology companies have arisen from entrepreneurs from academia who are often dominant, charismatic individuals whose primary aim has been to develop a new technology. New biotechnology companies have certain features not often seen in others (Table 1.3). The position of new biotechnology at the interface between academia and industry creates a unique need for abstracting information from a wide rage of sources and companies spend large sums on information management.

Table 1.3. Some unique features of biotechnology companies.

Technology driven and multidisciplinary: product development can involve molecular biologists, clinical researchers, product sales force
Must manage: regulatory authorities, public perception; issues of health and safety; risk assessment
Business climate characterised by repid change and considerable risk– one biotechnology innovation may quickly supercede another
Biotechnology business growth highly dependent on venture capital–usually needs exceptionally high level of funding before profit sales return

Biotechnology is high technology *par excellence*. The most exciting and potentially profitable facets of new biotechnology in the next decade will involve R&D at the very frontiers of current knowledge and techniques. In the late 1970s molecular biologists were putting forward vague promises about the wonders of this scientific discipline while the technologies needed to realise it were still requiring immense levels of research and product development funding. Biotechnologists now make predictions with more confidence, since many of the apparently insurmountable problems have been more easily overcome than had been predicted and many transitions from laboratory experiments to large-scale industrial processes have been achieved. Truly, new biotechnology has come of age.

For biotechnology to be commercially successful and exploited there is a need both to recruit a specialist workforce and also for the technology to be understood and applied by practitioners in a wide range of other areas including law, patents, medicine, agriculture, engineering, etc. Higher education will supply the range of specialist disciplines encompassing biotechnology, while some courses will endeavour to produce 'biotechnology' graduates who have covered many of the specialist areas at a less rigorous level than the pure degree specialisation. Also many already employed in biotechnology-based industries must regularly have means of updating or even retraining. The European-based BIOTOL (Biotechnology by open learning) has now produced a wide range of learning programmes. Such programmes are designed not only for the needs of students but also for company training activities and are written in a user-friendly style of good open learning materials. The currency of biotechnology throughout the work will be an educated, skilled workforce and ready access to the ever-widening knowledge and resource base. Science has defined the world in which we live and biotechnology, in particular, will become an essential and accepted activity of our culture.

BIOTECHNOLOGY—A THREE-COMPONENT CENTRAL CORE

Many biotechnology component centre core, in which one part is concerned with obtaining the best processes may be considered as having a three biological catalyst for a specific function or process, the second part creates (by construction and technical operation), the best possible environment for the catalyst to perform, and the third part (downstream processing) is concerned with the separation and purification of an essential product or products from a fermentation process.

In the majority of examples so far developed, the most effective, stable and convenient form for the catalyst for a biotechnological process is a whole organism and it is for this reason that so much of biotechnology revolves around microbial processes. This does not exclude the use of higher organisms; in particular, plant and animal cell culture will play an increasingly important role in biotechnology.

Micro-organisms can be viewed both as primary fixers of photosynthetic energy and as systems for bringing about chemical changes in almost all types of natural and synthetic organic molecules. Collectively, they have an immense gene pool that offers almost unlimited synthetic and degradative potential. Furthermore, micro-organisms can possess extremely rapid growth rates far in excess of any of the higher organisms such as plants and animals. Thus, immense quantities can be produced under the right environmental conditions in short time periods.

The methodologies that are in general use enable the selection of improved micro-organisms from the natural environmental pool, the modification of micro-organisms by mutation and, more recently, the mobilisation of a spectacular array of new techniques, deriving from molecular biology, which may eventually permit the construction of micro-organisms, plants and animals with totally novel biochemical potentials. These new techniques have arisen largely from fundamental achievements in molecular biology, over the last two decades.

These manipulated and improved organisms must be maintained in substantially unchanged form and this involves another spectrum of techniques for the supervention of organisms, for retaining essential features during industrial processes and, above all, retaining long-term vigour and viability. In many examples the catalyst is used in a separated and purified form, namely as enzymes, and a huge amount of information has been built up on the large-scale production, isolation and purification of individual enzymes and on their stabilisation by artificial means.

The second part of the core of biotechnology encompasses all aspects of the containment system or bioreactor within which the catalysts must function. Here the combined specialist knowledge of the bioscientist and bioprocess engineer will interact, providing the design and instrumentation for the maintenance and control of the physico-chemical environment such as temperature, aeration, pH, etc., thus allowing the optimum expression of the biological properties of the catalyst. Having achieved the required end-point of the biotechnological process within the bioreactor, e.g., biomass or biochemical product, in most cases it will be necessary to separate the organic products from the predominantly aqueous environment. This third aspect of biotechnology, namely downstream processing, can be a technically difficult and expensive procedure, and is the least understood area of biotechnology.

Downstream processing is primarily concerned with initial separation of the bioreactor broth or medium into a liquid phase and a solids phase, and subsequent concentration and purification of the product. Processing will usually involve more than one stage. Downstream processing costs (as approximate proportions of selling process) of fermentation products, very considerably, e.g., with yeast biomass, penicillin G and certain enzymes processing costs as percentages of selling are 20 per cent, 20–30 per cent and 60–70 per cent respectively.

Successful involvement in a biotechnological process must draw heavily upon more than one of the input disciplines. The main areas of application of biotechnology are shown in Table 1.4 while Fig. 1.2 attempts to show the input of many disciplines into the biotechnological processes together with the differing enabling technologies.

Biotechnology will continue to create exciting new opportunities for commercial development and profits in a wide range of industrial sectors including, healthcare and medicine, agriculture and forestry, fine and bulk chemicals production, food technology, fuel and energy production, pollution control and resource recovery. Biotechnologies offers a great deal of hope for solving many of the problems our world faces!

Table 1.4. Main areas of application of biotechnology.

Bioprocess technology
Historically, the most important area of biotechnology, namely brewing, antibiotics, mammalian cell culture, etc.; extensive development in progress with new products envisaged, namely polysaccharides, medically important drugs, solvents, protein enhanced foods. Novel fermenter designs to optimise productivity
Enzyme technology
Used for the catalysis of extremely specific chemical reactions; immobillisation of enzymes; to create specific molecular converters (bioreactors). Products formed include L-amino acids, high fructose syrup, semi-synthetic penicillins, starch and cellulose hydrolysis, etc. Enzyme probes for bioassays
Waste technology
Long historical importance but more emphasis now being made to couple these processes with the conservation and recycling of resources; foods and fertilisers, biological fuels
Environmental technology
Great scope exists for the application of biotechnological concepts for solving many environmental problems—pollution control, removing toxic wastes; recovery of metals from mining wastes and low-grade ores
Renewable resource technology
The use of renewable energy sources, in particular, lignocellulose to generate new sources of chemical raw materials and energy ethanol, methane and hydrogen. Total utilisation of plant and animal material
Plant and animal agriculture
Genetically engineered plants to improve nutrition, disease resistance, keeping quality, improved yields and stress tolerance will become increasingly commercially available. Improved productivity, etc., for animal farming. Improved food quality, flavour, taste and microbial safety
Healthcare
New drugs and better treatment for delivering medicines to diseased parts. Improved disease diagnosis, understanding of the human genome

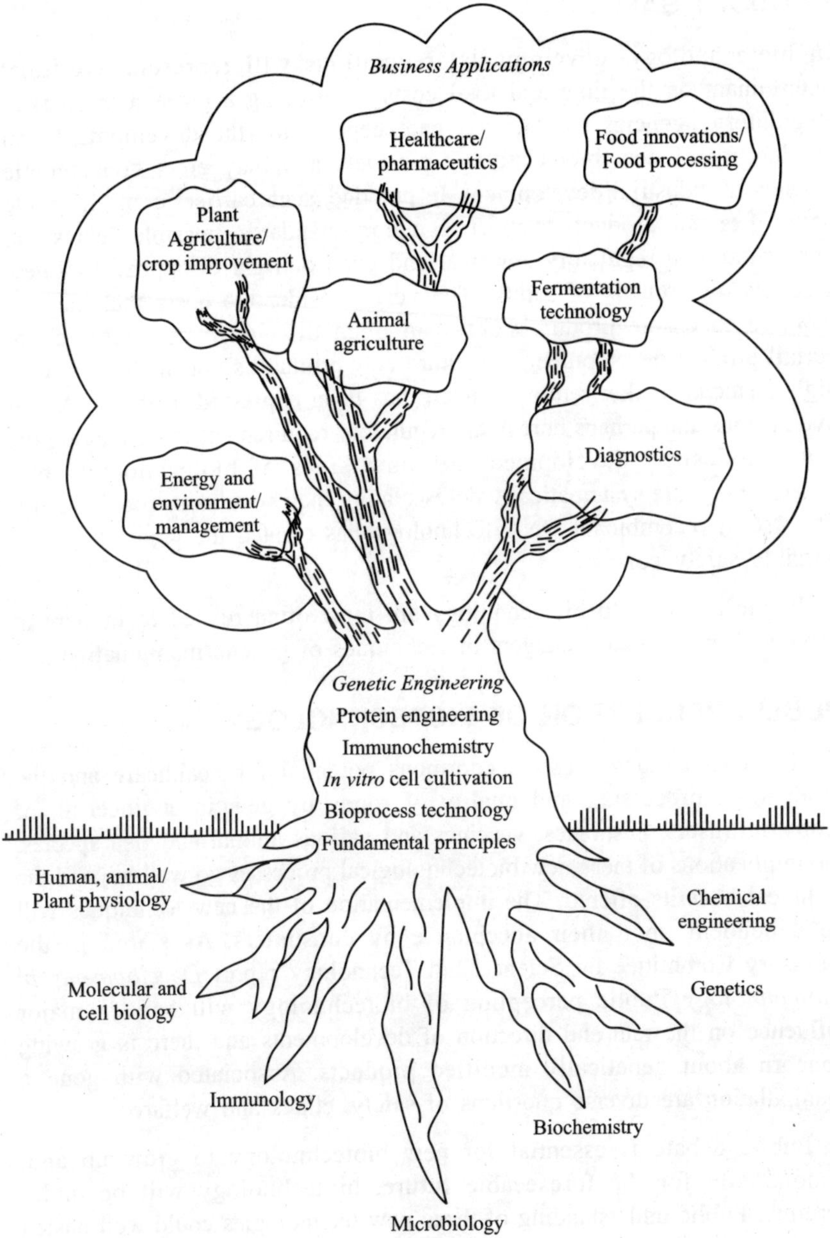

Fig. 1.2. The biotechnology tree.

PRODUCT SAFETY

In biotechnology, governmental regulations will represent a critical determinant on the time and total costs in bringing a product to market. Regulatory agencies can act as 'gatekeepers' for the development and availability of new biotechnology products and may erect considerable barriers to industrial development. In practice, such barriers come from the costs of testing products to meet regulatory standards, possible delays and uncertainties in regulatory approval and even outright disapproval of new products on grounds of safety. The very considerable costs that may be required to ensure product safety can often discourage new research or curtail product development if a future new product is not likely to have a high financial market return. Concern has been expressed in the USA that overzealous and perhaps unrealistic regulatory requirement are damaging the future industrial development of some areas of biotechnology and, consequently, are systematically de-escalating their regulatory requirements. The use of recombinant DNA technology has created the greatest area of possible safety concern.

Public attitude to biotechnology are most often related to matters of perceived or imaginary dangers in techniques of genetic manipulation.

PUBLIC PERCEPTION OF BIOTECHNOLOGY

While biotechnology presents enormous potential for healthcare and the production, processing and quality of foods by genetic engineering of crops, fertilisers, pesticides, vaccines and various animal and fish species, the implications of these new biotechnological processes go well beyond the technical benefits offered. The implementation of the new techniques will be dependent upon their acceptance by consumers. As stated in the Advisory Committee on Science and Technology report *Developments in Biotechnology* Public perception of biotechnology will have a major influence on the rate and direction of developments and there is growing concern about genetically modified products. Associated with genetic manipulation are diverse questions of safety, ethics and welfare.

Public debate is essential for new biotechnology to grow up and, undoubtedly for the foreseeable future, biotechnology will be under scrutiny. Public understanding of these new technologies could well hasten public acceptance. However, the low level of scientific literacy (e.g., in the USA, where only 7 per cent of the population are scientifically literate) does mean that most people will not be able to draw informed conclusions about this important biotechnology issue. Consequently, it is conceivable (and indeed occurring) that a small number of activists can argue the case against genetic engineering in such emotive and ill-reasoned ways that the public and the politicians are misled. The biotechnology community needs to sit up and take note of and work with the public. People influence decision-making by governments through the ballot box or through the presence of public opinion. Ultimately the benefits of biotechnology will speak for themselves as will be seen in the following chapters.

BIOTECHNOLOGY AND THE DEVELOPING WORLD

Successful agriculture holds the answer to the poverty gap between the rich and poor nations. In the developed world, agricultural sciences are well developed, producing abundance of high quality products. Agricultural biotechnology will further improve quality, variety and yield. Will these few plant species, improved by genetic engineering, find their way to the developing countries, ensuring higher productivity, greater resistance to disease and be more marketable?. It is not yet clear what will happen other than that the affluent nations will become increasingly well endowed with an abundance of food. World-wide there will be enough food for all but will it always continue to be disproportionately distributed. Biotechnology developments need high inputs of finance and skilled workforces – both of which are in short supply in most developing nations.

While many developing nations have successfully collaborated in the past with Western biotechnology companies. The ability of developing nations to avail themselves of the many promises of new biotechnology will to a large extent depend on their capacity to intergrate modern developments of biotechnology within their own research and innovation systems, in accordance with their own needs and priorities.

However, it must be appreciated that biotechnological development not only will depend on scientific and technological advances, but will also be subject to considerable political, economic and, above all, public acceptance. Finally, it has been said that most scientific disciplines pass through golden ages when new approaches open the door to rapid and fundamental expansion. Biotechnology is just now entering this golden period. A spectacular future lies ahead.

2

Substrates for Biotechnology

INTRODUCTION

It has been estimated that the annual net yield of plant biomass arising from photosynthesis is at least 120 billion tonnes of dry matter on land and around 50 billion tonnes from the world's oceans. Of the land-produced biomass, approximately 50 per cent occurs in the complex form of lignocellulose.

The highest proportion of land-based biomass (44 per cent) is produced as forest. It is surprising to note that, while agricultural crops account for only 6 per cent of the primary photosynthetic productivity, from this amount is derived a major portion of food for humans and animals as well as many essential structural materials, textiles and paper products. Many traditional agricultural products may well be further exploited with the increasing awareness of biotechnology. In particular, new technological approaches will undoubtedly be able to utilise the large volume of waste material from conventional food processing that presently finds little use.

Biomass agriculture, aquaculture and forestry may hold great economic potential for many national economies, particularly in tropical and subtropical regions. Indeed, the development of biotechnological processes in developing areas where plant growth excels could well bring about a change in the balance of economic power.

It should be noted that the non-renewable energy and petrochemical feed-stocks on which modern society is so dependent, namely oil, gas and coal, were derived from ancient types of biomass. Modern industrialised nations have come to rely heavily on fossil reserves both for energy and as feedstocks for a wide range of production processes. In little over a century the industrialised world has drawn heavily on fossil fuels that took millions of years to form beneath the beds of the oceans or in the depths of the earth. Furthermore, it is a very unequal pattern of usage. At present, the USA with 6 per cent and western Europe with 8 per cent of the world's population use 31 per cent and 20 per cent respectively, of the world's oil and gas productions.

While coal stocks may last for many hundreds of years this is not true for oil and gas and, at current usage levels, the world's known available sources of oil and gas will have been fully exploited sometime into the next century. The answer to these problems must be the use of photosynthetically derived biomass for energy and industrial feedstocks. Currently more than ten times more energy is generated annually by photosynthesis than is consumed by humankind. At present, large-scale exploitation of biomass for fuel and chemical feedstocks is restricted by the low cost of fossil alternatives, the heterogeneous nature of biomass sources and the diffuse distribution.

The use of biomass directly as a source of energy has long been practised in the less industrialised regions such as Latin America, China, India and Africa. In developed nations, biomass derived from agriculture and forestry has been directed largely towards industrial and food uses. At present biomass is used to derive many products of industrial and commercial importance (Table 2.1).

Table 2.1. Important products derived form biomass.

Fuels
Methane (biogas) especially in the developing world
Pyrolysis products (gas, charcoal)
Ethanol (via cane juice and cellulose fermentation)
Oils (from hydrogenation)
Direct combustion of waste biomass
Feedstocks
Ethanol (potential feedstock for industry)
Synthesis gas (from chemical gasification)
Fertilisers
Compost
Sludge
Feeds
Direct feed supplements
Single cell proteins

NATURAL RAW MATERIALS

Natural raw materials originate mostly from agriculture, food industry and forestry. These are mainly carbohydrates of varying chemical complexity and include sugar, starch, cellulose, hemicellulose and lignin. The wide range of by-products obtained from raw materials and of use in biotechnological processes is shown in Table 2.2.

Table 2.2. Range of by-products that could be used as substrates in biotechnology.

Agriculture	*Forestry*	*Industry*
Straw	Wood waste hydrolysate	Molasses
Bagasse	Sulphite pulp liquor	Distillery wastes
Maize cobs	Bark, sawdust branches	Whey
Coffee, cocoa and coconut hulls	Paper and cellulose	Industrial waste water from food industries (olive, palm-oil, potato, date, citrus, cassava)
Fruits peels and leaves	Fibres	
Tea wastes		
Oilseed cakes		Wash waters (dairy, canning, confectionery, bakery, soft drinks, sizing, malting, corn steep)
Cotton wastes		
Bran		
Pulp (tomato, coffee, banana, pineapple, citrus, olive)		Fishery effluent and wastes
		Meat by-products
Animal wastes		Municipal garbage
		Sewage
		Abattoir wastes

Sugar-bearing raw materials such as sugar beet, sugar cane and sugar millet are the most suitable and available to serve as feedstocks for biotechnological processing. As traditional uses of sugar are replaced by more efficient alternatives, the sugar surplus on the commodity market will give further incentive to develop new uses. Many tropical economies would collapse if the market for sugar were removed. Already cane sugar serves as the substrate for the Brazilian gasohol programme and many other nations are rapidly seeing the immense potential of these new technologies.

Starch-bearing agricultural products include the various types of grain such as maize, rice and wheat, together with potatoes and other root crops such as sweet potato and cassava. A slight disadvantage of starch is that it must usually be degraded to monosaccharides or oligosaccharides by digestion or hydrolysis before fermentation. However, many biotechnological processes using starch are being developed, including fuel production.

There can be little doubt that cellulose, both from agriculture and forestry sources, must contribute a major source of feedstock for biotechnological processes such as fuels and chemicals. However, cellulose is a very complex chemical and invariably occurs in nature in close association with lignin. The ability of lignocellulose complexes to withstand the biodegradative forces of nature is witnessed by the longevity of trees, which are composed mainly of lignocellulose.

Lingnocellulose is the most abundant and renewable natural resource available to humanity throughout the world. However, massive technological difficulties must be overcome before economic use may be made of this plentiful compound. At present, expensive energy-demanding pretreatment processes are required to open up this complex structure to wide microbial degradation. Pure cellulose can be degraded by chemical or enzymic hydrolysis to soluble sugars, which can be fermented to form ethanol, butanol, acetone, single cell protein, methane and many other products. Exciting advances are being made in laboratories throughout the world and it is only matter of time before these difficulties are overcome. It has been realistically calculated that approximately 3.3×10^{14} kg CO_2 per year are fixed on the surface of the earth and that approximately 6 per cent of this, i.e. 22 billion tonnes per year will be cellulose. On a world-wide basis land plants produce 24 tonnes cellulose per person per year. Time will surely show that lignocellulose will be the most useful carbon source for biotechnological developments.

AVAILABILITY OF BY-PRODUCTS

While biotechnological processes will use many agricultural products such as sugars, starches, oils, etc., as substrates, the vast array of waste products derived from agriculture, and currently not creatively used, will undoubtedly be subjected to detailed examination and future utilisation. Agricultural and forestry wastes come in many diverse types: cereal straws, corn husks and cobs, soya wastes, coconut shells, rice husks, coffee bean husks, wheat bran, sugar cane bagasse and forestry wastes including trimmings, sawdust, bark, etc. (Table 2.2). Only a modest fraction of these wastes are utilised on a large scale primarily because of economic and logistical factors.

A primary objective of biotechnology is to improve the management and utilisation of the vast volumes of agricultural, industrial and domestic waste organic materials to be found throughout the world. The biotechnological utilisation of these wastes will eliminate a source of pollution, in particular water pollution, and convert some of these wastes into useful by-products.

Not all processes will involve biosystems. In particular, the processes of reverse osmosis and ultrafiltration are finding increasing uses. Reverse osmosis is a method of concentrating liquid solutions: a porous membrane allows water to pass through but not the salts dissolved in it. Ultrafiltration is a method of separating the high and low molecular weight compounds in a liquid by allowing the liquid and low molecular weight compounds to pass through while holding back the high molecular weight compounds and suspended solids. Some current applications of these technologies include concentration of dilute factory effluents, concentration of dilute food products, sterilisation of water, purification of brackish water, and separation of edible solids from dilute effluents.

Waste materials are frequently important for economic and environmental reasons. For example, many by-products of the food industry are of low economic value and are often discharged into waterways, creating serious environmental pollution problems. An attractive feature of carbohydrate waste as a raw materials is that, if its low cost can be coupled with suitable low handling costs, an economic process may be obtained. Furthermore, the world-wide trend towards stricter effluent control measures, or parallel increase in effluent disposal charges, can lead to the concept of waste as a 'negative cost' raw material. However, the composition or dilution of the waste may be so dispersed that transport to a production centre may be prohibitive. On these occasions biotechnology may serve only to reduce a pollution hazard. Each waste material must be assessed for its suitability for biotechnological processing. Only when a waste is available in large quantities and preferably over a prolonged period of time can a suitable method of utilisation be considered (Table 2.3).

Table 2.3. Biotechnological strategies for utilisation of suitable organic waste materials.

Upgrading the food waste quality to make it suitable for human consumption
Feeding the food waste directly or after processing to poultry, pigs, fish or other singly stomach animals that can utilise it directly
Feeding the food waste to cattle or other ruminants if unsuitable for single stomach animals because of high fibre content, toxins or other reasons
Production of biogas (methane) and other fermentation products if unsuitable for feeding without expensive pretreatments
Selective other purposes such as direct use as fuel, building materials, chemical extraction, etc.

Two widely occurring wastes that already find considerable fermentation uses are molasses and whey. Molasses is a by-product of the sugar industry and has a sugar content of approximately 50 per cent. Molasses is widely used as a fermentation feedstock for the production of antibiotics, organic acids and commercial yeasts for baking, and is directly used in animal feeding. Whey, obtained during the production of cheese, could also become a major fermentation feedstock.

More complex wastes such as straw and bagasse are widely available and will be increasingly used as improved processes for lignocellulose breakdown become available (Table 2.4). Wood wastes will include low grade wood, bark and sawdust, as well as waste liquors such as sulphite waste from pulp production, which already finds considerable biotechnological processing in Europe and communist countries.

Table 2.4. Pretreatments required before substrates are suitable for fermentation.

Substrate	*Pretreatment*
Sugary materials	
Sugar cane, beet, molasses, fruit juices, whey	Minimal requirements dilution and sterilisation
Starchy materials	
Cereals, rice, vegetables, process liquid wastes	Some measure of hydrolysis by acid or enzymes. Initial separation on non-starch components may be required
Lignocellulosic materials	
Corn cobs, oat hulls, straw, bagasse, wood wastes, sulphite liquor, paper wastes	Normally requires complex pretreatment involving reduction in particle size followed by various chemical or enzymic hydrolyses. Energy intensive and costly

The largest proportion of total volume of waste matter is from animal and rearing (faeces, urine), then agricultural wastes, wastes from food industries and finally domestic wastes. The disposal of many waste materials, particularly animal wastes, is no problem in traditional agriculture and is particularly well exemplified in China, where recycling by composting has long been practised. However, where intensive animal rearing is undertaken, serious pollution problems do arise.

CHEMICAL AND PETROCHEMICAL FEEDSTOCKS

With the development of commercial processes for the production of singly cell protein (SCP) and other organic products, a number of chemical and petrochemical feedstocks have become particularly important for fermentation processes, since these materials have the advantage of being available in large quantities and in the same quality in most parts of the world. Thus, natural gas or methane and gas oil have been preferred as raw material because of their easy processing and universal availability. Main commercial interest has been concerned with *n*-paraffins, methanol and ethanol. Their involvement in various aspects of biotechnology, but particularly in SCP production, will be considered later in the chapter.

Future biotechnological processes will increasingly make use of organic materials that are renewable in nature or occur as low value wastes that may presently cause environmental pollution. Table 2.5 summarises the many technical considerations that must be made when approaching the utilisation of waste materials. Some processes may also more economically utilise specific fractions of fossil fuels as feedstocks for biotechnological processes.

Table 2.5. Technical considerations for the utilisation of waste materials.

Biological availability	Low (cellulosics)
	Moderate (starch, lactose)
	High (molasses, pulping sugars)
Concentration	Solid (milling residues, garbage)
	Concentrated (molasses)
	Weak (lactose, pulping sugars)
	Very dilute (process and plant wash liquors)
Quality	Clean (molasses, lactose)
	Moderate (straw)
	Dirty (garbage, feedlot waste)
Location	Collected (large installation, small centres)
	Collected specialised (olive, palm oil, date, rubber, fruit, vegetable)
	Dispersed (straw, forestry)
Seasonally	Prolonged (palm oil, lactose)
	Very short (vegetable cannery waste)
Alternative uses	Some (straw)
	None (garbage)
	Negative (costly effluents)
Local technology potential	High (USA)
	Middle (Brazil)
	Low (Malaysia)

RAW MATERIALS AND THE FUTURE OF BIOTECHNOLOGY

It is now clear that the future development of large-scale biotechnological processes is inseparable from the supply and cost of raw materials. During the early and middle parts of this century the availability of cheap oil led to an explosive development of the petrochemical industry and many products formerly derived from the fermentation ability of micro-organisms were instead derived from cheaper and more efficient chemical methods. However, escalating oil prices in the 1970s created profound reappraisals of these processes and as the price of crude oil approached that of some major cereal products there was a reawakening of interest in many fermentation processes for the production of ethanol and related products. However, the decrease in oil prices in 1986 again widened the gap and left uncertainty in the minds of industrial planners.

The most important criteria that will determine the selection of a raw material for a biotechnological process will include price, availability, composition and form, and oxidation state of the carbon source.

There is little doubt that cereal crops, particularly maize, rice and wheat, will be the main short- and medium- term raw materials for biotechnological processes. Furthermore, it is believed that this can be achieved without seriously disturbing food supplies for humans and animals. Throughout the world there is an uneven distribution of cereal production capacity and demand. Overproduction of cereals occurs mostly where extensive biotechnological processes are in practice. Areas of poor cereal production will undoubtedly benefit from the developments in agricultural biotechnology now at an early but highly optimistic stage.

Although much attention has been given to the uses of wastes in biotechnology there are many major obstacles to be overcome. For instance, availability of agricultural wastes is seasonal and geographic availability problematic; they are also often dilute and may contain toxic wastes. However, their buildup in the environment can present serious pollution problems and therefore their utilisation in biotechnological processes, albeit at little economic gain, can have overall community value.

Although in the long term biotechnology must seek to utilise the components of cellulose and lignocelluloses as fuels or feedstocks, the technological difficulties are still considerable. The chemical complexities of these molecules is legendary and it is proving more difficult than had been expected to break them down economically to usable primary molecules. Biotechnology will have profound effects on agriculture and forestry by enabling production costs to be decreased, quality and consistency of products to be increased and novel products generated.

Wood is extensively harvested to provide fuel, materials for construction and to supply pulp for paper manufacture. Supplies throughout the world are rapidly decreasing, resulting in extensive examples of deforestation: deforestation can often then be followed by desertification and soil erosion. Selective breeding programmes are now producing 'elite' fast growing hardwood trees, such as eucalyptus and acacia, derived from genetic manipulation programmes coupled with new methods of macro- and micro-propagation. Such fast growing trees could even be used for electricity generation, giving less net overall CO_2 production than the use of fossil fuels and so minimising and potential 'greenhouse' effect.

There may also be an increased non-food use of many agriculturally derived substances such as sugars, starches, oils, and fats. Biotechnology will significantly aid their overall production by improved plant growth, influence biosynthetic pathways to change production, change chemical structure of molecules, and improve processing by enzyme technology. The ability of biotechnology to improve disease resistance, quality composition and in some cases to grow crops in marginal lands could surely lead to higher yields of products, and in developing economies this could have profound beneficial effects. Supplies in excess of food needs could allow new industries to develop and therefore reduce poverty.

Cotton is a major agricultural crop in many developing nations and is a major natural textile fibre, with world production exceeding that of synthetic textile fibres. It is not well recognised that cotton accounts for about 10 per cent of world's annual use of agrochemicals, mainly insecticides. Devolvement of disease-resistant cotton plants by new molecular methods could have major economical and environmental impact. Major yield increases would have considerable benefits to the producing nations.

How successful will biomass be as a crucial raw material for biotechnology? An OECD report has set out the factors that will determine the competitiveness of natural (biomass) and synthetic (fossil biomass) derived products, namely:

1. The relative price of the basic raw materials.
2. Quality, variability, regularity of supply and safety of the raw materials.
3. Relative costs of chemical base material conversion compared to conversion of agricultural products.
4. Premium accorded by the market to 'natural' as compared to synthetic products and increasing requirement for products to be biodegradable.

3

Genetics and Biotechnology

INTRODUCTION

In essence, all properties of organisms depend on the sum of their gene potential. There are two broad categories of genes: structural and regulatory. Structural genes encode for amino acid sequences of proteins, which, as enzymes, determine the biochemical capabilities of the organisms by catalysing particular synthetic or catabolic reactions or alternatively, play more static roles as components of cellular structures. In contrast, the regulatory genes control the expression of the structural genes by determining the rate of production of their protein-products in response to intra- or extra-cellular signals. The derivation of these principles has been achieved using well-known genetic techniques.

James Watson and *Francis Crick* led to the construction of the double helix model depicting the molecular structure of DNA and subsequent hypotheses on its implications for the understanding of gene replication. Since then there has been an spectacular unravelling of the complex interactions required to express the coded chemical information of the DNA molecule into cellular and organism components. Changes in the DNA molecule making up the genetic complement of an organisms is the means by which organisms evolve and adapt themselves to new environments. In natural, changes in the DNA of an organism can occur in two ways:

1. By *mutation*, which is a chemical deletion or addition of one or more of the chemical parts of the DNA molecule.
2. By the interchange of genetic information or DNA between like organism normally by *sexual reproduction*. This is achieved by a process of conjugation in which there is a donor, called male and a recipient, called fcmale.

Classical genetics was, until recently, the only way in which heredity could be studied and manipulated. However, in recent years, new techniques have permitted unprecedented alterations in the genetic make-up of organisms, even allowing exchange of DNA between unlike organisms.

MANIPULATION OF THE GENETIC MATERIAL

The manipulation of the genetic material in organisms can now be achieved in three clearly definable ways: organismal, cellular and molecular.

Organismal

Genetic manipulation of whole organisms has been happening naturally by sexual reproduction since the beginning of time. The evolutionary progress of almost all living creatures has involved active interaction between the genomes and the environment. Active control of sexual reproduction has been practised in agriculture for decades–even centuries. In more recent times it has been used with several industrial micro-organisms, e.g., yeasts. It involves selection, mutation, sexual crosses, hybridisation, etc. However, it is a very random process and can take a long time to achieve the desired results–if at all in some cases. In agriculture, the benefits have been immense, with much improved species of plants and animals, while in the biotechnological industries there have been greatly improved productivities, e.g., of antibiotics and enzymes.

Cellular

Cellular manipulations of DNA have been used for over two decades and involve either cell fusion or the culture of cells and the regeneration of whole plants from these cells. This is a semi-random or directed process, in contrast to organismal manipulations and the changes can be more readily identified. Successful biotechnological examples or these methods include monoclonal antibodies and the cloning of many important plant species.

Molecular

Molecular manipulations of DNA and RNA first occurred nearly two decades ago and heralded a new era of genetic manipulations enabling–for the first time in biological history–a directed control of the changes. This is the much publicised area of *genetic engineering* or *recombination DNA* technology, which is now bringing dramatic changes to biotechnology. In these techniques the experimenter is able to know much more about the genetic changes being made. It is now possible to add or delete parts of the

DNA molecule with a high degree of precision and the product can be easily identified. Current industrial ventures are concerned with the production of new types of organism and numerous compounds ranging from pharmaceuticals to commodity chemicals.

INDUSTRIAL GENETICS

Biotechnology has so far been considered as an interplay between two components, one of which is the selection of the best biocatalyst for a particular process, while the other is the construction and operation of the best environment for the catalyst to achieve optimum operation.

The most effective, stable and convenient form for the biocatalyst is a whole organism; in most cases it is some type of microbe, e.g., a bacterium, yeast or mould, although mammalian cell cultures and (to a lesser extent) plant cell cultures are finding ever-increasing uses in biotechnology.

Most micro-organisms used in current biotechnological processes were originally isolated from the natural environment and have subsequently been modified by the industrial geneticist into specialised organisms for specific productivity. The success of strain selection and improvement programmes practised by all biologically based industries (e.g., brewing, antibiotics, etc.) is a direct result of the close cooperation between the technologist and the geneticist.

In future, this relationship will be even more necessary in formulating the specific physiological and biochemical characteristics that are sought in new organisms in order to give the fullest range of biological activities to biotechnology.

In biotechnological processes, the aim is primarily to optimise the particular characteristics sought in an organism, e.g., specific enzyme production or by-product formation. Genetic modification to improve productivity has been widely practised. The task of improving yields of some primary metabolites and macromolecules (e.g., enzymes) is simpler than trying to improve the yields of complex products such as antibiotics.

Advances have been achieved in this area by using *screening* and *selection* techniques to obtain better organisms. In a selection system all rare or novel strains grow while the rest do not; in a screening system all strains grow, but certain strains or cultures are chosen because they show the desired qualities required by the industry in question.

In most industrial genetics the basis for changing the organism's genome has been by mutation using X-rays and mutagenic chemicals. However, such methods normally lead only to the loss of undesired characters or increased production due to loss of control functions. It has rarely led to the appearance of a new function or property. Thus, an organisms with a desired feature will be selected from the natural environment, propagated and subjected to a mutational programme, then screened to select the best progeny.

Unfortunately, many of the micro-organisms that have gained industrial importance do not have a clearly defined sexual cycle. In particular, this has been the case in antibiotic-producing micro-organisms; this has meant that the only way to change the genome with a view to enhancing productivity has been to indulge in massive mutational programmes, followed by screening and selection to detect the new variants that might arise.

Once a high producing strain has been found, great care is required in maintaining that strain. Undesired spontaneous mutations can sometimes occur at a high rate, giving rise to degeneration of the strain's industrial importance. Strain instability is a constant problem in industrial utilisation of micro-organisms. Industry has always placed great emphasis on strain viability and productivity potential of the preserved biological material. Most industrially important micro-organisms can be stored for long periods, e.g., in liquid nitrogen, by lyophilisation (freez-drying) or under oil and still retain their desired biological properties.

However, despite elaborate preservation and propagation methods, a strain has generally to be grown in a large production bioreactor in which the chances of genetic changes through spontaneous mutation and selection are very high. The chance of a high rate of spontaneous mutation is probably greater when the industrial strains in use have resulted from many years of mutagen treatment. Great secrecy surrounds the use of industrial micro-organisms and immense care is taken to ensure that they do not unwittingly pass to outside agencies.

There is now a growing movement away from the extreme empiricism that characterised the early days of the fermentation industries. Fundamental studies of the genetics of micro-organisms now provide a background of knowledge for the experimental solution of industrial problems and increasingly contribute to progress in industrial strain selection.

In recent years, industrial genetics has come to depend increasingly on two new ways of manipulating DNA–protoplast and cell fusion and recombinant DNA technology. These are now important additions to the technical repertoire of the geneticists involved with biotechnological industries. A brief examination of these techniques will attempt to show their increasingly indispensable relevance to modern biotechnology.

PROTOPLAST AND CELL FUSION TECHNOLOGIES

Plants and most microbial cells are characterised by a distinct outer wall or exoskeleton that gives the shape characteristic to the cell or organism. Immediately within the cell wall is the living membrane, or plasma membrane retaining all the cellular components such as nuclei, mitochondria, vesicles, etc. For some years now it has been possible, using special techniques (in particular, hydrolytic enzymes), to remove the cell wall, releasing spherical membrane-bound structures known as protoplasts. These protoplasts are extremely fragile but can be maintained in isolation for variable periods of time. Isolated protoplasts cannot propagate themselves as such, requiring first to regenerate a cell wall before regaining reproductive capacity.

In practice, it is the cell wall that largely hinders the sexual conjugation of unlike organisms. Only with completely sexually compatible strains does the wall degenerate, allowing protoplasmic interchange. Thus, natural sexual mating barriers in micro-organisms, may, in part, be due to cell wall limitations and by removing this cell wall, the likelihood of cellular fusions may increase.

Protoplasts can be obtained routinely from many plant species, bacteria, yeasts and filamentous fungi. Protoplasts from different strains can sometimes be persuaded to fuse and so overcome the natural sexual mating barriers. However, the range of protoplast fusions is severely limited by the need for DNA compatibility between the strains concerned. Fusion of protoplasts can be enhanced by treatment with the chemical polyethylene glycol, which, under optimum conditions, can lead to extremely high frequencies of recombinant formation. This can be increased still further by ultraviolet irradiation of the parental protoplast preparations. Protoplast fusion can also occur with human or animal cell types.

Protoplast fusion has obvious empirical applications in yield improvement of antibiotics by combining yield-enhancing mutations from different strains or even species. Protoplasts will also be an important part of genetic engineering, in facilitating recombinant DNA transfer. Fusion may provide a method of reassorting whole groups of genes between different strains of macro- and micro-organisms.

One of the most exciting and commercially rewarding areas of biotechnology involves a form of mammalian cell fusion leading to the formation of monoclonal antibodies. It has long been recognised that certain cells (B-lymphocytes) within the body of vertebrates have the ability to secrete antibodies that can inactivate contaminating or foreign molecules (the antigen) within the animal system. The antibody has a Y-shaped molecular structure and uses one part of this structure to bind the invading antigen and the other parts to trigger the body's response to eliminate the antigen–antibody complex. It has been calculated that a mammalian species can generate up to 100 million different antibodies, thereby ensuring that most invading foreign antigens will be bound by some antibody. Antibodies have high binding affinities and specificity against the chosen antigen. For the mammalian system it is the major defence against disease-causing organisms and other toxic molecules.

Attempts to cultivate the antibody-producing cells in artificial media have generally proved unsuccessful, the cells either dying or ceasing to produce the antibodies. It is now known that individual B-lymphocyte cells produce single antibody types. However, in 1975 *George Kohler* and *Cesar Milstein* successfully demonstrated the production of pure or *monoclonal antibodies* from the fusion product (*hybridoma*) of B-lymphocytes (antibody-producing cells) and *myeloma* tumour cells).

The monoclonal antibody techniques changes antibody-secreting cells (with limited life span) into cells capable or continuous growth, while maintaining their specific antibody-secreting potential. This 'immortalisation' is achieved by a fusion technique, whereby B-lymphocyte cells are fused to 'immortal' cancer or myeloma cells in one-to-one ratio, forming hybrids or hybridomas capable of continuous growth and antibody secretion in culture. Single hybrid cells can then be selected and grown as clones or pure cultures of the hybridomas. Such cells continue to secrete antibody and the antibody is of one particular specificity as opposed to the mixture of antibodies that occurs in an animal's bloodstream after conventional methods of immunisation.

Monoclonal antibody formation is achieved by injecting a mouse or rabbit with an antigen, later removing the spleen and then allowing fusion of individual spleen cells with individual myeloma cells. Approximately 1 per cent of the spleen cells are antibody-secreting cells and 10 per cent of the final hybridomas consists of antibody-secreting cells (Fig. 3.1). Techniques are available to identify the right antibody-secreting hybridoma cell and then cloning or propagating that cell into large populations, with subsequent formation of large quantities of the desired antibody. These cells may be frozen and later reused.

Monoclonal antibodies have now gained wide application in many diagnostic techniques, which require a high degree of specificity. Specific monoclonal antibodies have been combined into test kits for diagnostic purposes, in healthcare, plant and animal agriculture and in food manufacture. Monoclonal antibodies may also be used in the future to carry cytotoxic drugs to the site of cancer cells, while in the fermentation industry they are already widely used as separating agents to bind and purify expensive products.

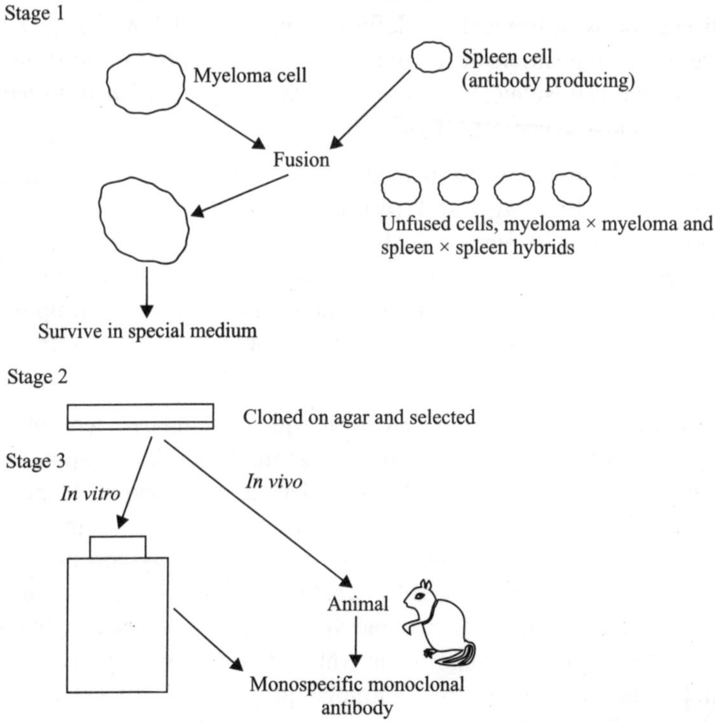

Fig. 3.1. Diagram illustrating the formation of antibody-producing hybridomas by fusion techniques. Stage 1: myeloma cells and antibody-producing cells (derived from immunised animal or human) are incubated in a special medium containing polyethylene glycol, which enhances fusion. Stage 2: the myeloma spleen hybridoma cells are selected out and cultured in closed agar dishes. State 3: the specific antibody-producing hybridoma is selected and propagated in culture vessels (*in vitro*) or in an animal (*in vivo*) and monoclonal antibodies harvested.

In the 20 years since the development of the first monoclonal antibody the methodology has developed from a purely scientific tool into one of the fastest expanding fields of biotechnology and has revolutionised, expanded and diversified the diagnostic industry. The monoclonal antibody market is expected to continue to grow at a very high rate and in healthcare alone the anticipated annual world market could be several billion dollars by the year

2007. It is undoubtedly one of the most commercially successful and useful areas of modern biotechnology.

GENETIC ENGINEERING

Genes comprise the fundamental basis of all life, determine the properties of all living forms of life and are defined segments of DNA. Because DNA structure and composition of all living forms is essentially the same, any technology that can isolate, change or reproduce a gene is likely to have an impact on almost every aspect of society.

Genetic recombination, as occurs during normal sexual reproduction, consists of the breakage and rejoining of DNA molecules of the chromosomes and is of fundamental importance to living organisms for the reassortment of genetic material. Genetic manipulation has been performed for centuries by selective breeding of plants and animals superimposed on natural variation. The potential for genetic variation has, thus, been limited to close taxonomic relatives.

In contrast, recombinant DNA techniques, popularly termed 'gene cloning' or 'genetic engineering', offer potentially unlimited opportunities for creating new combinations of genes that at the moment do not exist under natural conditions. Genetic engineering can be defined as the formation of new combinations of heritable material by the insertion of nucleic acid molecules, produced by whatever means outside the cell, into any virus, bacterial plasmid or other vector system so as to allow their incorporation into a host organisms in which they do not naturally occur but in which they are capable of continued propagation. In essence, gene technology is the modification of the genetic properties of an organism by the use of recombinant DNA technology. Genes may be viewed as the biological software and are the programmes that drive the growth, development and functioning of an organism. By changing the software in a precise and controlled manner, it becomes possible to produce desired changes in the characteristics of the organism.

These techniques allow the splicing of DNA molecules of quite diverse origin and when combined with techniques of genetic transformation, etc., facilitate the introduction of foreign DNA into other organisms. The foreign DNA or gene construct is introduced into the genome of the recipient organism hote in such a way that the total genome of the hote if unchanged except for the single manipulated gene.

Thus, DNA can be isolated from cells of plants, animals or micro-organisms (the donors) and can be fragmented into groups of one or more genes. Such fragments can then be coupled to another piece of DNA (the *vector*) and then passed into the host or recipient cell, becoming part of the genetic complement of the new host. The host cell can then be propagated in mass to form novel genetic properties and chemical abilities that were unattainable by conventional ways of selective breeding or mutation. While traditional plant and animal genetic breeding techniques also change the genetic code, this change is achieved in a less direct and controlled manner. Genetic engineering will now enable the breeder to select the particular gene required for a desired characteristic and modify only that gene.

Although much work to date has involved bacteria, the techniques are evolving at an astonishing rate and ways have been developed for introducing DNA into other organisms such as yeasts and plant and animal cell cultures. Provided that the genetic material transferred in this manner can replicate and be expressed in the new cell type there are virtually no limits to the range of organisms with new properties that could be produced by genetic engineering. Life forms containing 'foreign' DNA are termed 'transgenic'.

These methods potentially allow totally new functions to be added to the capabilities of organisms and open up vistas for the genetic engineering of industrial micro-organisms and agricultural plants and animals that are quite breathtaking in their scope. This is undoubtedly the most significant new technology in modern biotechnology. In industrial microbiology it will permit the production, in micro-organisms, of a wide range of hitherto unachievable products such as human and animal proteins and enzymes such as insulin and chymosin (rennet); in medicine there will be better vaccines, hormones and improved therapy of diseases; in agriculture improved plants and animals will achieve greater productivity, quality of products, disease resistance etc.; in food production we can expect improved quality, flavour, taste and safety; and in environmental aspects there will be a wide range of benefits such as pollution control. It should be noted that genetic engineering is a way of doing things rather than an end in itself. Genetic engineering will add to, rather then displace, traditional ways of developing products. However, there are many people who view genetic engineering as a transgression of normal life processes that goes well beyond normal evolution.

Genetic engineering holds the potential to extend the range and power of every aspect of biotechnology. In the first instance these techniques will be used widely to improve existing microbial processes by improving stability of existing cultures and eliminating unwanted side-products. It is confidently anticipated that within this decade recombinant DNA techniques will form the basis of new micro-organisms with new and unusual metabolic properties. In this way fermentations based on these technical advances could become competitive with petrochemicals for producing a whole range of chemical compounds, e.g., ethylene glycol (used in the plastics industry). In the food industry, improved strains of bacteria and fungi are now influencing such traditional processes as baking and cheese-making and bringing greater control and reproducibility of flavour and texture.

Understanding fully the working concepts of recombinant DNA technology requires a good knowledge of molecular biology. The basic molecular requirements for the *in vitro* transfer and expression of foreign DNA in a host cell (gene transfer technology) are as follows:

Vector or Carrier System

Two broad categories of vector molecules have been developed as vehicles for gene transfer, namely *plasmids* (small units of DNA distinct from chromosomes) and *bacteriophages* (or bacterial viruses). Vector molecules normally exist within a cell in an independent or extrachromosomal form and do not become part of the chromosomal system of the organism. Vector molecules should be capable of entering the host cell and replicating within it. Ideally, the vector should be small, easily prepared and must contain as least one site where integration of foreign DNA will not destroy an essential function. Plasmids will undoubtedly offer the greatest potential in biotechnology and have been found in an increasingly wide range of organisms, e.g., bacteria, yeast and mould fungi; they have been studied mostly in Gram-negative bacteria.

Splicing Genes

The most significant advances towards the construction of hybrid DNA molecules *in vitro* have come from the discovery that site-specific *restriction endonuclease enzymes* produce specific DNA fragments that can be joined to any similarly treated DNA molecule using another enzyme, *DNA ligase*. Restriction enzymes are present in a wide range of bacteria and can distinguish between DNA from their own cells and foreign DNA by recognising certain sequences of nucleotides. There are techniques available for breaking open a length of DNA into shorter fragments that contain a number of genes determined by the enzyme used. Such DNA fragments can then be separated from each other on the basis of differing molecular weights and can subsequently be joined together in number of ways, provided that the ends are complementary. The sources of DNA can be quite different, giving an opportunity to replicate the DNA biologically by inserting it into other cells. The composite molecules into which DNA has been inserted have also been termed 'DNA chimaeras' because of the analogy with the Chimaera of Greek mythology, a creature with the head of a lion, the body of a goat and the tail of a serpent.

Introduction of Vector DNA Recombinants

The new recombinant DNA can now be introduced into the host cell by *transformation* (the direct uptake of DNA by a cell from its environment) or *transduction* (DNA transferred from one organism to another by way of a carrier or vector system) and if acceptable the new DNA will be cloned with the propagation of the host cell. Novel methods of ensuring DNA uptake into cells include *electroporation* and *mechanical particle delivery* or *biolistics*. Electroporation is a process of creating transient pores in the cell membrane by application of a pulsed electric field. Creation of such pores in a membrane allows introduction of foreign molecules such as DNA, RNA, antibodies, drugs, etc., into the cell cytoplasm. Development of this technology has arisen from synergy of biophysics, bioengineering and cell and molecular biology. While the techniques is now used widely to create transgenic micro-organisms, plants and animals, it is also being used increasingly for application of therapeutics and gene therapy. The mechanical particle delivery or 'gene gun' methods deliver DNA on microscopic particles into target tissue or cells. This process is increasingly used to introduce new genes into a range of bacterial, fungal, plant and mammalian species and has become a main method of choice for genetic engineering of many plant species including rice, corn, wheat, cotton and soyabean.

Assaying Foreign Gene Products

Many methods are now available for the detection of hybrid DNA. The strategies involved in genetic engineering are outlined in Table 3.1 and Fig. 3.2.

Table 3.1. Strategies involved in genetic engineering.

Formation of DNA fragments
Extracted DNA can be cut into small sequences by specific enzymes, restriction endonucleases found in many species of bacteria
Splicing of DNA into vectors
The small sequences of DNA can be joined or spliced into the vector DNA molecules by an enzyme, DNA ligase, creating an artificial DNA molecule

(Contd...)

Introduction of vectors into host cells
The vectors are either viruses or plasmids, are replicons and can exist in an extrachromosomal state; transfer is normally by transduction or transformation
Selection of newly acquired DNA
Selection and ultimate characterisation of the recombinant clone

Although the theory underlying both the exchange of genetic information between unrelated organisms and their propagation is becoming better understood, difficulties still persist at the level of some applications. Further research is required before such exchanges become commonplace and the host organisms propagated in large quantities. Early studies on genetic engineering were carried out mainly with the bacterium *Escherichia coli*, but increasingly other bacteria, yeast and filamentous fungi have been used. Mammalian systems have been increasingly developed using the simian virus (SV40) and oncogenes (genes that cause cancer), while several successful methods are available for plant cells, in particular the *Agrobacterium* system.

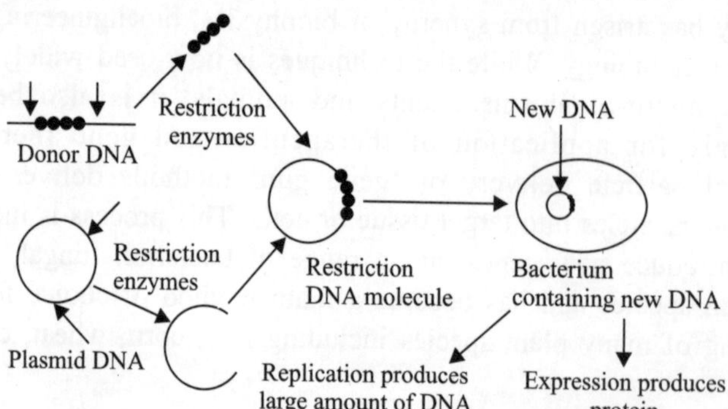

Fig. 3.2. Recombinant DNA: the techniques of recombining genes from one species with those from another.

The ability to clone DNA and to study its expression in controlled environments is a development of quite unparalleled importance in modern bioscience. The further expression of genetic engineering in the context of biotechnology, where newly created gene pools in organisms can be expressed in large quantities, will offer unbounded opportunities for the well-being of humankind.

POTENTIAL LABORATORY BIOHAZARDS OF GENETIC ENGINEERING

The early studies on gene manipulation provoked wide discussion and considerable concern at the possible risks that could arise with certain types of experiment. Thus, it was believed by some people that the construction of recombinant DNA molecules and their insertion into micro-organisms could create novel organisms that might inadvertently be released from the laboratory and become a biohazard to human beings or the environment. In contrast, others considered that newly designed organisms, with their additional genetic material, would not be able to compete with the normal strains present in nature. The present views of gene manipulation studies are becoming more moderate as experiments have shown that this work can proceed within a strict safety code, when necessary involving physical and biological containment of the organism.

The standards of containment enforced in the early years of recombinant DNA studies were unnecessarily restrictive and there has been a steady relaxation of the regulations governing much of the routine genetic engineering activities. However, for many types of study, particularly with parthogenic micro-organisms, the standards will remain stringent. Thus, for strict physical containment laboratories involved in this type of study must have highly skilled personnel and correct physical containment equipment, e.g., negative pressure laboratories, autoclaves and safety cabinets.

Biological containment can be achieved or enhanced by selecting non-pathogenic organisms as the cloning agents of foreign DNA or by the deliberate genetic manipulation of a micro-organism to reduce the probability of survival and propagation in the environment. *Escherichia coli*, a bacterium extremely prevalent in the intestinal tracts of warm-blooded and cold-blooded animals as well as in humans, is the most widely used cloning agent. To offset the risk of this cloning agent becoming a danger in the environment a special strain of *E. coli* has been constructed by genetic manipulation that incorporates many fail-safe features. This strain can grow only under special laboratory conditions and there is no possibility that it can constitute a biohazard if it escapes out of the laboratory.

The government-controlled Health and Safety Executive regulates and monitors recombinant DNA work within the UK. This committee seeks advice from the Genetic Manipulation Advisory Group (GMAG), who formulate realistic procedural guidelines, which in general, have proved widely acceptable to the experimenting scientific community. Most other advanced scientific nations involved in recombinant DNA studies have set up similar advisory committees.

POLYMERASE CHAIN REACTION

Undoubtedly the most far reaching development in modern biology during the last 10 years has been the invention of the *polymerase chain reaction*

(PCR). This is basically a technique that allows the selective amplification of any fragment of DNA, provided the DNA sequences flanking the fragment are known. It is described as a technique that finds a needle in a haystack and then produces a haystack of needles by specific amplification.

The PCR process relies on the sequence of 'base-pairs' along the length of the two strands that make the complete DNA molecule. In DNA there are four deoxynucleotides derived from the four bases adenosine (A) thymidine (T) guanine (G) and Cytidine (C). The strands or polymers that comprise the DNA molecule are held to each other by hydrogen bonds between the base-pairs. In this arrangement A binds only to T while G binds only to C and this unique system folds the entire molecule into the now well-recognised double helix structure.

PCR involves three processing steps: *denaturation*, *annealing* and then *extension* by DNA polymerase (Figs. 3.3a and 3.3b). In step 1, the double-standard DNA is heated (95–98°C) and separates into two complementary single strands. In step 2 (60°C) the synthetic oligonucleotide primers (chemically synthesised short-chain nucleotides), short sequences of nucleotides (usually about 20 base-pairs long), are added and bind to the single strands in places where the strand's DNA complements their own. In step 3 (37°C) the primers are extended by DNA polymerase in the presence of all four deoxynucleoside triphosphates resulting in the synthesis of new DNA strands complementary to the template strands. The completion of the three steps comprises a cycle and the real power of PCR is that with 25–30 cycles this experimental synthesis leads to massive amplification of DNA, which can then be used for analytical purposes. A major recent advance has been the development of automated thermal cyclers (PCR machines), which allow the entire PCR to be performed automatically in several hours.

The applications of PCR increase almost daily and include molecular biology/genetic engineering, infections and parasitic disease diagnosis, human genetic disease diagnosis, forensic validation, plant and animal breeding and environmental monitoring. PCR has been used extensively in the well-known procedure of genetic (or DNA) fingerprinting, the fallibility of which is now being challenged in courts of law (Fig. 3.4).

While PCR is finding considerable and unique use in archaeology, it is doubtful whether we will ever be able to resurrect woolly mammoths and dinosaurs from ancient animal remains, as recently epitomised in Michael Crighton's *Jurassic Park*.

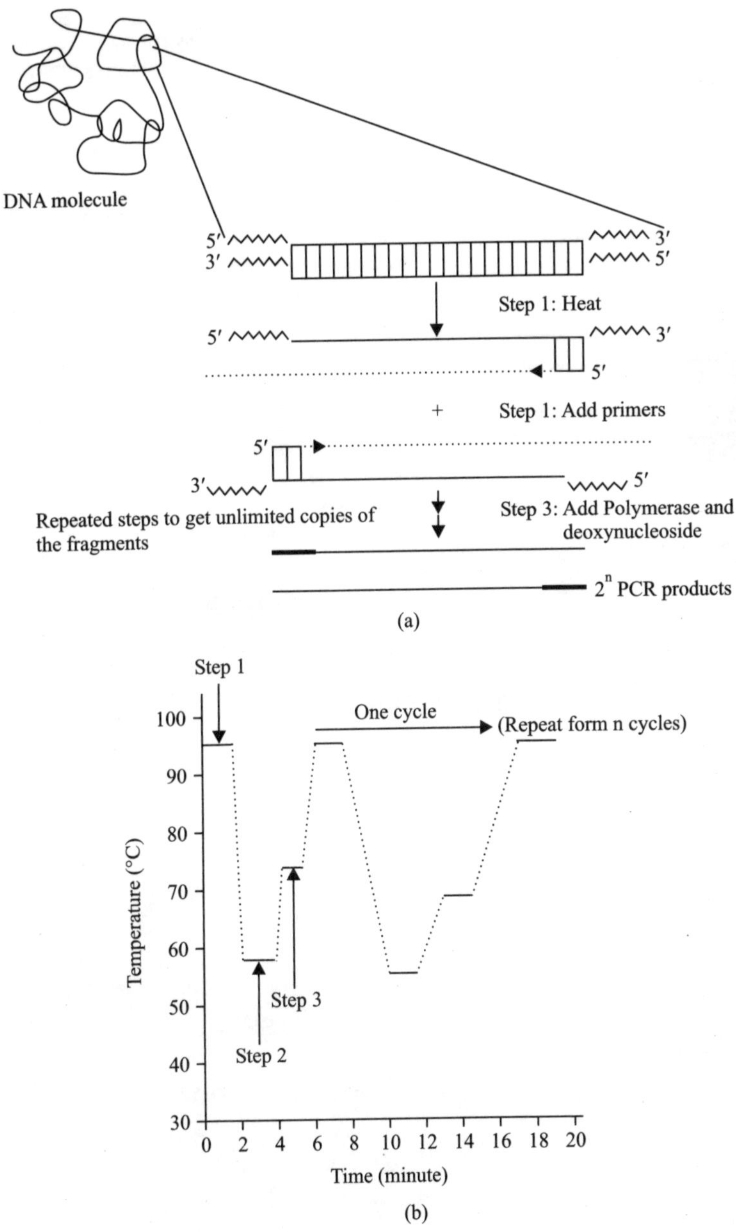

Fig. 3.3. (a) The polymerase chain reaction. The double-standard DNA is heated and separates into two single strands. The synthetic oligonucleotide primers then bind to their complementary sequence and are extended in the direction of the arrows, giving a new strand of DNA identical to the template's original partner. (b) PCR temperature cycling profile.

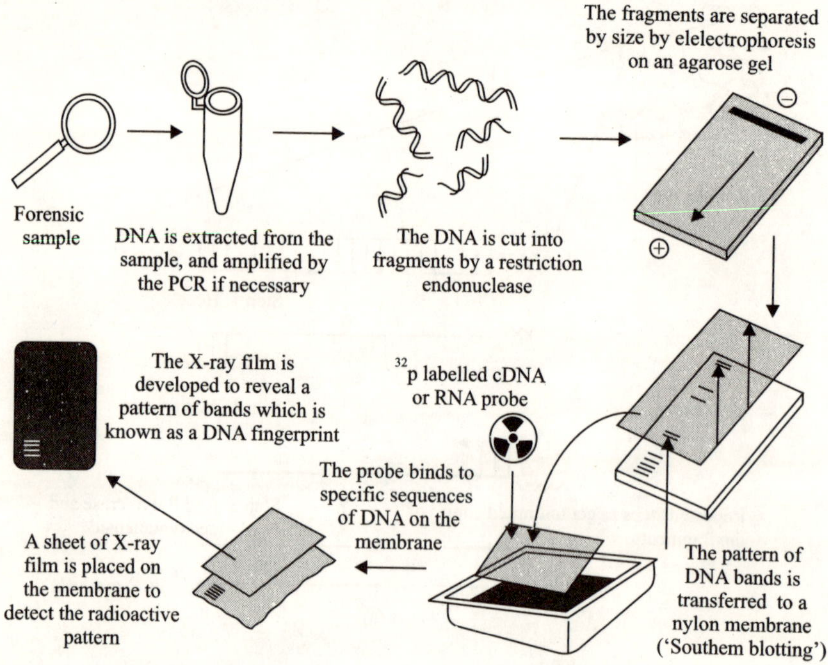

Fig. 3.4. DNA fingerprinting.

4

Fermentation Technology and Industrial Microbiology

INTRODUCTION

The very beginnings of fermentation technology or, as it is now better recognised, bioprocess technology were derived in part from the use of micro-organisms for the production of foods such as cheeses yoghurts, sauerkraut, fermented pickles and sausages, soya sauce, and other oriental products, and beverages such as beers, wines and derived spirits (Table 4.1). In many cases, the present day production processes for such products are still remarkably similar. These forms of bioprocessing were long viewed as arts or crafts but are now increasingly subjected to the full array of modern science and technology. Paralleling these useful product formations was the identification of the roles micro-organisms could play in removing obnoxious and unhealthful wastes, which has resulted in the world-wide service industries involved in water purification, effluent treatment and solid waste management.

Bioprocessing in its many forms involves a multitude of complex enzymecatalysed reactions within specific micro-organisms and these reactions are critically dependent on the physical and chemical conditions that exist in their immediate environment. Successful bioprocessing will occur only when all the essential factors are brought together. Although the traditional forms of bioprocess technology related to foods and beverages still represent the major commercial bioproducts, new products are increasingly being derived from microbial fermentations, namely:

1. To overproduce essential primary metabolites such as acetic and lactic acids, glycerol, acetone, butyl alcohol, organic acids, amino acids, vitamins and polysaccharides.

2. To produce secondary metabolites (metabolites that do not appear to have an obvious role in the metabolism of the producer organism) such as penicillin, streptomycin, cephalosporin, giberellins, etc.
3. To produce many forms of industrially useful enzymes, e.g., exocellular enzymes such as amylases, pectinases and proteases and intracellular enzymes such as invertase, asparaginase, restriction endonucleases, etc.

All of these type's of microbial product now command large industrial markets and are essential to modern society (Table 4.1).

Table 4.1. Fermentation products according to industrial sectors.

Sector	*Activities*
Chemicals	
Organic (bulk)	Ethanol, acetone, butanol
	Organic acids (citric, itaconic)
Organic (fine)	Enzymes
	Perfumeries
	Polymers (mainly polysaccharides)
Inorganic	Metal beneficiation, bioaccumulation and leaching (Cu, U)
Pharmaceuticals	Antibiotics
	Diagnostic agents (enzymes, monoclonal antibodies)
	Enzyme inhibitors
	Steroids
	Vaccines
Energy	Ethanol (gasohol)
	Methane (biogas)
	Biomass
Food	Dairy products (cheeses, yogurts, fish and meat products)
	Beverages (alcoholic, tea and coffee)
	Baker's yeast
	Food additives (antioxidants, colours, flavours, stabilisers)
	Novel foods (soya sauce, tempeh, miso)
	Mushroom products
	Amino acids, vitamins
	Starch products
	Glucose and high fructose syrups
	Functional modifications of proteins, pectins

(Contd...)

Sector	*Activities*
Agriculture	Animal feedstuffs (SCP)
	Veterinary vaccines
	Ensilage and composting processes
	Microbial pesticides
	Rhizobium and other N-fixing bacterial inoculants
	Mycorrhizal inoculants
	Plant cell and tissue culture (vegetative propagation, embryo production, genetic improvement)

More recently, bioprocess technology is increasingly using cells derived from higher plants and animals to produce many important products. Plant cell culture is aimed largely at secondary product formations such as flavours, perfumes and drugs whereas mammalian cell culture has been concerned with vaccine and antibody formation and the production of protein molecules such as interferon, interleukins, etc.

The future market growth of these bioproducts is largely assured because, with limited exceptions, most cannot be produced economically by other chemical processes. It will also be possible to make further economies in production by genetically engineering organisms to higher or unique productivities and utilising new technological advances in processing. The advantages of producing organic products by biological as opposed to purely chemical methods are listed in Table 4.2.

Table 4.2. Advantages and disadvantages of producing organic compounds by biological rather than chemical means.

Advantages	*Disadvantages*
Complex molecules such as proteins and antibodies cannot be produced by chemical means	Can be easily contaminated with foreign unwanted micro-organisms, etc.
Bioconversions give higher yields	The desired product will usually be present in a complex product mixture requiring separation
Biological systems operate at lower temperatures, near neutral pH etc.	Need to provide, handle and dispose of large volumes of water
Much greater specificity of catalytic reaction	Bioprocesses are usually extremely slow when compared with conventional chemical processes
Can achieve exclusive production of an isomeric compound	

The product formation stages in bioprocess technology are essentially very similar no matter what organism is selected, what medium is used and what product formed. In all examples, large numbers of cells are grown under defined controlled conditions. The organisms must be cultivated and motivated to form the desired products by means of a physical/technical containment system (the *bioreactor*), and the correct medium composition and environmental growth regulating parameters such as temperature and aeration. Optimisation of the bioprocess spans both the biosystems and the technical systems. The proper exploitation of an organism's potential to form distinct products of defined quality and in large amounts will need a detailed knowledge of the biochemical mechanisms of product formation.

The same apparatus, with modifications, can be used to produce an enzyme, an antibiotic, an amino acid or single cell protein. In its simplest form, the bioprocess can be seen as just mixing micro-organisms with a nutrient broth and allowing the components to react, e.g., yeast cells with a sugar solution to give alcohol.

More advanced and sophisticated processes operating at large scale need to control the entire system so that the bioprocess can proceed efficiently and be readily and exactly repeated with the same amounts of raw materials and inoculum to produce precisely the same amount of product. All biotechnological processes are essentially performed within containment systems or bioreactors.

Large numbers of cells are invariably involved in these processes and the bioreactor ensures their close involvement with the correct medium and conditions for growth and product formation. It also should restrict the release of the cells into the environment. A main function of a bioreactor is o minimise the cost of producing a product or service. Examples of the iverse product categories produced industrially in bioreactors are given in ıble 4.3.

ɔle 4.3. Examples of products in different categories in biotechnological ıstries.

ory	*Example*
ıass[a]	Baker's yeast, single cell protein
ɔmponents[b]	Intracellular proteins
hetic products[b]	Antibiotics, vitamins, amino and organic acids
c products[a]	Ethanol, methane, lactic acid
rsion[a]	High fructose corn syrup, 6–aminopenicillanic acid
ıtment	Activated sludge, anaerobic digestion

conversion of feedstock cost-intensive processes.
recovery cost-intensive processes.

PRINCIPLES OF MICROBIAL GROWTH

The growth of organisms may be seen as the increase of cell material expressed in terms of mass or cell number and results from a highly complicated and coordinated series of enzyme-catalysed biological steps. Growth will be dependent on the availability and transport of necessary nutrients to the cell and subsequent uptake and on environmental parameters such as temperature, pH and aeration being optimally maintained.

The quantity of biomass or specific cellular component in a bioreactor can be determined gravimetrically (by dry weight, wet weight, DNA or protein) or numerically for unicellular systems (by number of cells). Doubling time refers to the period of time required for the doubling in the weight of biomass while generation time relates to the period necessary for the doubling of cell numbers. Average doubling times increase with increasing cell size and complexity, e.g., bacteria 0.25–1 hrs., yeast 1–2 hrs. mould fungi 2–6.5 hrs., plant cells 20–70 hrs., and animal cells 15–48 hrs.

In normal practice an organism will seldom have totally ideal conditions for unlimited growth; rather, growth will be dependent on a limiting factor, e.g., an essential nutrient. As the concentration of this factor drops, so also will the growth potential of the organism decrease.

In biotechnological processes there are three main ways of growing micro-organisms in the bioreactor, namely batch, semi-continuous or continuous. Within the bioreactor reactions can occur with static or agitated cultures, in the presence or absence of oxygen, and in liquid or low moisture conditions (e.g., on solid substrates). The micro-organisms can be free or can be attached to surfaces by immobilisation or by natural adherence.

In a *batch culture* the macro-organisms are inoculated into a fixed volume of medium and as growth takes place nutrients are consumed and products of growth (biomass, metabolites) accumulate. The nutrient environment within the bioreactor is continuously changing and thus, in turn, enforcing changes to cell metabolism. Eventually, cell multiplication ceases because of exhaustion or limitation of nutrient (s) and accumulation of toxic excreted waste products.

The complex nature of batch growth of micro-organisms is shown in Fig. 4.1. The initial *lag* phase is a time of no apparent growth but actual biochemical analyses show metabolic turnover indicating that the cells are in the process of adapting to the environmental conditions and that new growth will eventually begin. There is then a *transient acceleration* phase as the inoculum begins to grow, to be followed quickly by the *exponential phase*. In the exponential phase microbial growth proceeds at the maximum possible rate for that organism, with nutrients in excess, ideal environmental parameters and growth inhibitors absent.

However, in batch cultivations exponential growth is of limited duration and, as nutrient conditions change, growth rate decreases, entering the *deceleration phase* to be followed by the *stationary phase* when overall growth can no longer be obtained because of nutrient exhaustion. The final phase of the cycle is the *death phase* when growth rate has ceased. Most biotechnological batch processes are stopped before this stage because of decreasing metabolism and cell lysis.

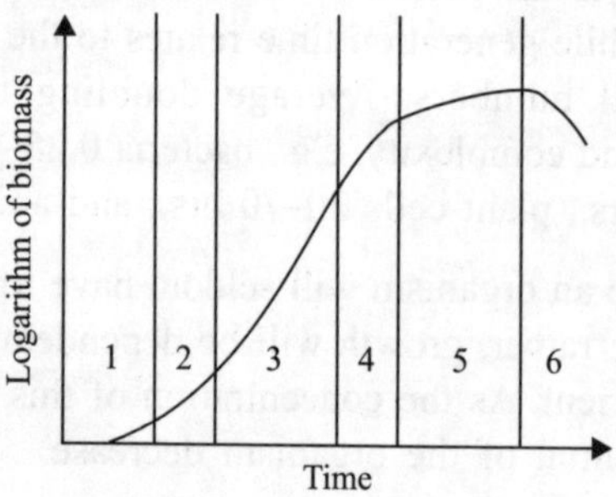

Fig. 4.1. Growth characteristics in a batch culture of a micro-organism. 1, lag phase; 2, transient acceleration; 3, exponential phase; 4, deceleration phase; 5, stationary phrase; 6 death phase.

In industrial usage, batch cultivation has been operated to optimise organism or biomass production and then to allow the organism to perform specific biochemical transformation such as end-product formation (e.g., amino acids, enzymes) or decomposition of substances (sewage treatment, bioremediation). Many important products such as antibiotics are optimally formed during the stationary phase of the growth cycle in batch cultivation.

However, there are means of prolonging the life of a batch culture and thus increasing the yield by various substrate feed methods, namely:

1. Gradual addition of concentrated components of the nutrient, e.g., carbohydrates, so increasing the volume of the culture *(fed batch)*– used for industrial production of baker's yeast.
2. Addition of medium to the culture (perfusion) and withdrawal of an equal volume of used cell-free medium–used in animal cell cultivations.

In contrast to batch conditions the practice of *continuous cultivation* gives near-balanced growth, with little fluctuation of nutrients, metabolites, cell numbers or biomass. This practice depends on fresh medium entering a batch system at the exponential phase of growth with a corresponding withdrawal of medium *plus* cells. Continuous methods of cultivation will permit organisms to grow under steady state (unchanging) conditions in which growth occurs at a constant rate and in a constant environment. In a completely mixed continuous culture system sterile medium is passed into the bioreactor (Fig. 4.2) at a steady flow rate, and culture broth (medium, waste products and organisms) emerges from it at the same rate, keeping the volume of the total culture in the bioreactor constant.

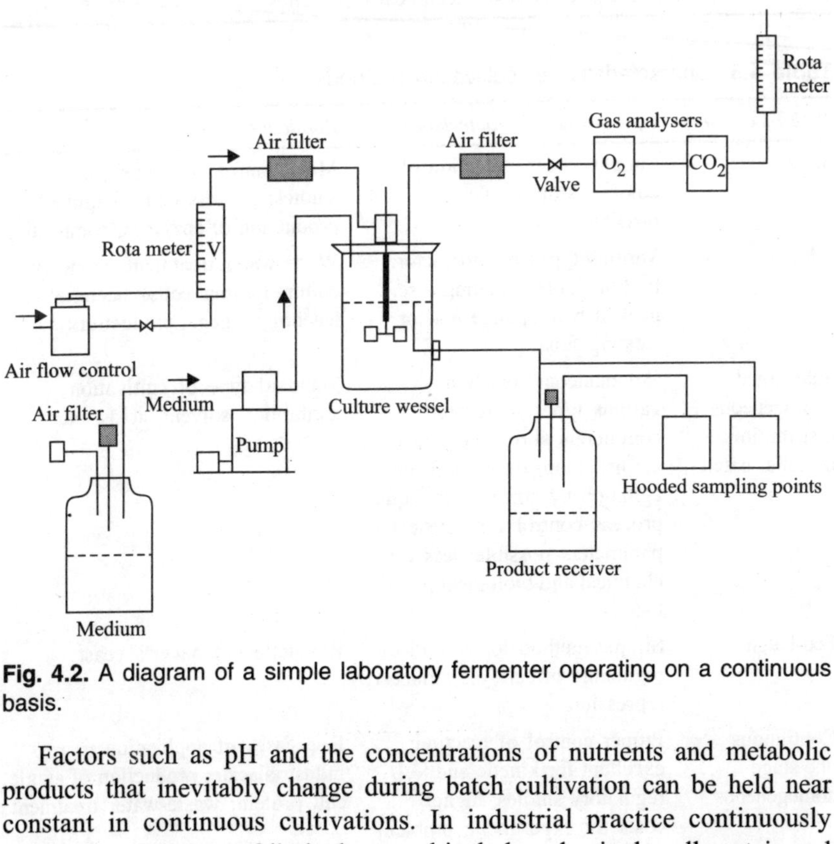

Fig. 4.2. A diagram of a simple laboratory fermenter operating on a continuous basis.

Factors such as pH and the concentrations of nutrients and metabolic products that inevitably change during batch cultivation can be held near constant in continuous cultivations. In industrial practice continuously operated systems are of limited use and include only single cell protein and ethanol productions and some forms of waste-water treatment processes. However, for many reasons (Table 4.4) batch cultivation systems represent the dominant form of industrial usage.

The full range of cultivation methods for micro-organisms in shown in Table 4.5.

Table 4.4. Advantages of batch and fed-batch culture techniques in industry.

Products may be required only in relatively small quantities at any given time
Market needs may be intermittent
Shelf-life of certain products is short
High product concentration is required in broth to optimise downstream processing operations
Some metabolic products are produced only during the stationary phase of the growth cycle
Instability of some production strains require their regular renewal
Continuous processes can offer many technical difficulties

Table 4.5. Characteristics of cultivation methods.

Type of culture	*Operational characteristics*	*Application*
Solid	Simple, cheap, selection of colonies from single cell possible	Maintenance of strains, genetic studies; process control limited production of enzymes; composting
Film	Various types of bioreactor: trickling filter, rotating disc, packed bed, sponge reactor, rotating tube	Waste-water treatment, monolayer culture (animal cells); bacterial leaching; vinegar production
Submerged homogeneous distribution of cells; batch	'Spontaneous' reaction, various types of reactor: continuous stirred tank reactor, air lift, loop, deep shaft, etc.; agitation by stirrers, air, liquid process control for physical parameters possible; less for chemical and biological parameters	Standard type of cultivation antibiotics, solvent, acids, etc.
Fed-batch	Simple method for control of regulatory effects, e.g., glucose repression	Production of baker's yeast
Continuous one stage homogeneous	Proper control of reaction; excellent for kinetic and regulatory studies; higher costs for experiment; problem of aseptic operation; the need for highly trained operators	Few cases of application in industrialscale; production of single cell protein; waste-water treatment

Bioreactor/Fermenter

For each biotechnological process the most suitable containment system must be designed to give the correct environment for optimisation of organism growth and metabolic activity. Bioreactors range from simple stirred or nonstirred open containers to complex, aseptic, integrated systems involving varying levels of advanced computer inputs (Fig. 4.3). For many advanced fermentation processes there is a need to avoid contamination of the process by unwanted micro-organisms and at the same time to avoid undue release of the culture organism into the environment. Indeed, when operating with genetically engineered microbes it is a legal necessity to retain the organisms within the bioreactor system.

Bioreactors are of two distinct types (Fig. 4.3). In the first instance they are primarily non-aseptic systems where it is not absolutely essential to operate with entirely pure cultures, e.g., brewing, effluent disposal systems; in the second type, aseptic conditions are a prerequisite for successful product formation, e.g., antibiotics, vitamins, polysaccharides. This type of process involves considerable challenges on the part of engineering construction and operation.

The physical form of many of the most widely used bioreactors has not altered much over the past 40 years; however, in recent years, novel forms of bioreactor have been developed to suit the needs of specific bioprocesses and such innovations are finding increasing specialised roles in bioprocess technology (Fig. 4.3). Fig. 4.4 shows a gigantic specialised bioreactor used for single cell protein production.

In all forms of fermentation the ultimate aim is to ensure that all parts of the system are subject to the same conditions. Within the bioreactor the micro-organisms are suspended in the aqueous nutrient medium containing the necessary substrates for growth of the organism and required product formation. All nutrients, including oxygen, must be provided to diffuse into each cell, and waste products such as heat, CO_2 and waste metabolites removed.

The concentration of the nutrients in the vicinity of the organism must be held within a definite range, since low values will limit the rate of organism metabolism whereas excessive concentrations can be toxic. Biological reactions run most efficiently within optimum ranges of environmental parameters and in biotechnological processes these conditions must be provided on a microscale so that each cell is equally provided for. When the large scale of many bioreactor systems is considered it will be realised how difficult it is to achieve these conditions in a whole population. It is here that the skills of the process or chemical engineer and the microbiologist must come together.

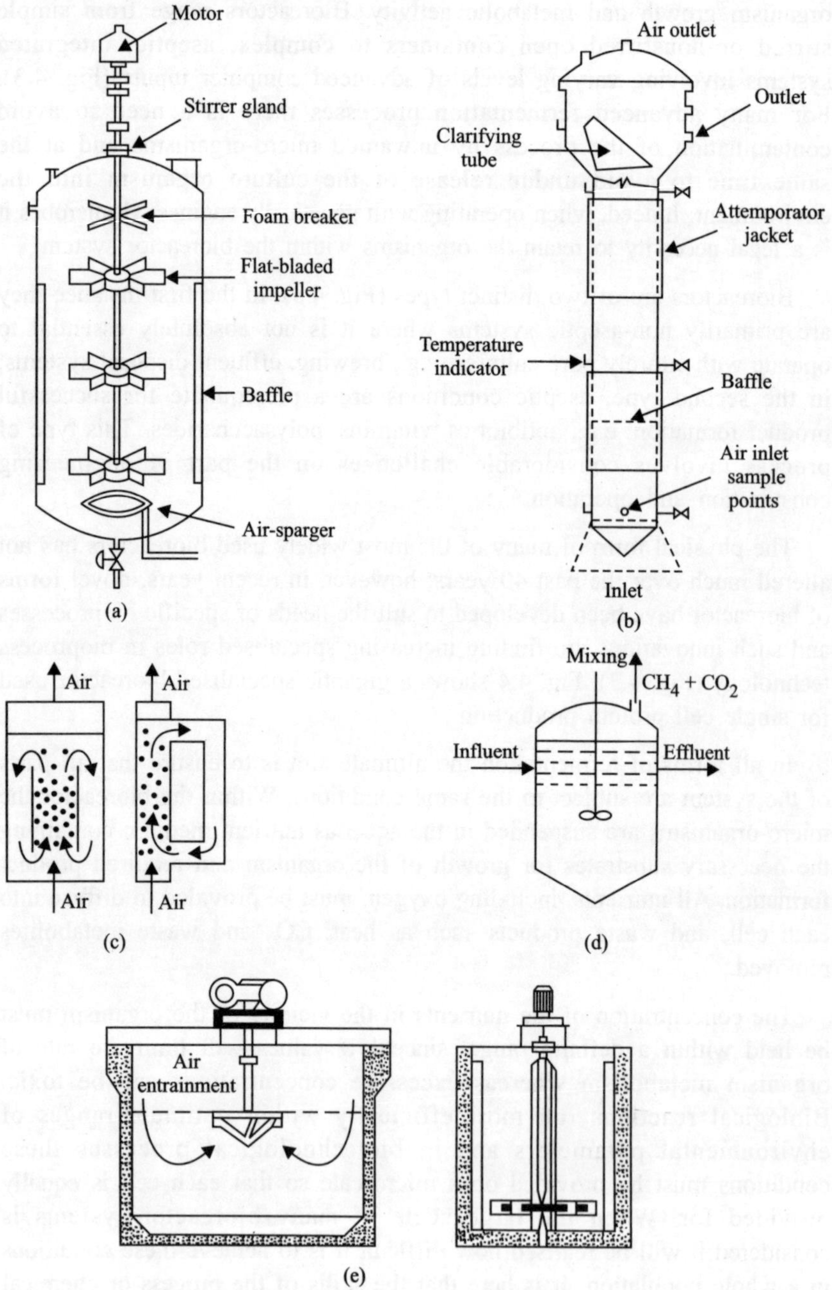

Fig. 4.3. Various forms of bioreactor: (a) Continuous stirred tank bioreactor; (b) Tower bioreactor; (c) Loop (recycle) bioreactor; (d) Anaerobic digester or bioreactor; and (e) Activated sludge bioreactor.

Fermentation reactions are multiphase, involving a gas phase (containing N_2, O_2 and CO_2), one or more liquid phases (aqueous medium and liquid substrate) and solid microphase (the micro-organisms and possibly solid substrates). All phases must be kept in close contact to achieve rapid mass and heat transfer. In a perfectly mixed bioreactor all reactants entering the system must be immediately mixed and uniformly distributed to ensure homogeneity inside the bioreactor.

To achieve optimisation of the bioreactor system, the following operating guidelines must be closely adhered to:

1. The bioreactor should be designed to exclude entrance of contaminating organisms as well as containing the desired organisms.
2. The culture volume should remain constant, i.e., no leakage or evaporation.
3. The dissolved oxygen level must be maintained above critical levels of aeration and culture agitation for aerobic organisms.
4. Environmental parameters such as temperature, pH, etc., must be controlled; and the culture volume must be well mixed.

The standard of materials used in the construction of sophisticated fermenters is also important (Table 4.6).

Table 4.6. Standards of materials used in sophisticated fermenter design.

All materials coming into contact with the solutions entering the bioreactor or the actual organism culture must be corrosion resistant to prevent trace metal contamination of the process
The materials must be non-toxic so that slight dissolution of the material or components does not inhibit culture growth
The materials of the bioreactor must withstand repeated sterilisation with high pressure steam
The bioreactor stirrer system, entry ports and end plates must be easily machinable and sufficiently rigid not to be deformed or broken under mechanical stress
Visual inspection of the medium and culture is advantageous, transparent materials should be used wherever possible

Fermentation technologists seek to achieve a maximisation of culture potential by accurate control of the bioreactor environment. But still there is a great lack of true understanding of just what environmental conditions will produce an optimal yield of organism or product.

To understand and control a fermentation process it is necessary to know the state of the process over a small time increment and, further, to know how the organism responds to a set of measurable environmental conditions. Process optimisation requires accurate and rapid feedback control. In the future, the computer will be an integral part of most bioreactor systems. However, there is a lack of good sensor probes that will allow on-line analysis to be made on the chemical components of the fermentation process.

A large world-wide market exists for the development of new rapid methods of monitoring the many reactions within a bioreactor. In particular, the greatest need is for innovatory microelectronic designs.

Bioreactor configurations have changed considerably over the last few decades. The original fermentation system was a shallow tank agitated or stirred manually. From this has developed the basic aeration tower system that now dominates industrial usage. As fermentation systems were further developed, two design solutions to the problems of aeration and agitation have been implemented. The first approach uses mechanical aeration and agitation devices, with relatively high power requirements; the standard example is the continuously stirred tank reactor (CSTR), widely used throughout conventional laboratory and industrial fermentations. Such bioreactors ensure good gas/liquid mass transfer, have reasonable heat transfer, and ensure good mixing of the bioreactor contents.

The second main approach to aerobic bioreactor design uses air distribution (with low power consumption) to create forced and controlled liquid flow in a recycle or loop bioreactor. In this way the contents are subjected to a controlled recycle flow, either within the bioreactor or involving an external recycle loop. Thus, stirring has been replaced by pumping, which may be mechanical or pneumatic, as in the case of the airlift bioreactor.

The stirred tank bioreactor system is still the most widely used but most new designs are dominated by the recycle principles and several new designs have found ready acceptance because of their high efficiency in new areas of fermentation technology, for e.g., single cell protein production from *n*-alkanes, methanol, and also waste-water treatment.

Large amounts of organic waste waters from domestic and industrial sources are routinely treated in aerobic and anaerobic systems. Activated sludge processes are widely used for the oxidative treatment of sewage and other liquid waste (Fig. 4.3a). Such processes use batch or continuously agitated bioreactor systems to increase the entrainment of air to optimise oxidative breakdown of the organic material. These bioreactors are large and for optimum functioning will have several or many agitator units to facilitate mixing and oxygen uptake. They are widely used in most municipal sewage treatment plants.

Anaerobic bioreactors or digesters have long been used to treat sewage matter. In the absence of free oxygen certain microbial consortia are able to convert biodegradable organic material to methane, CO_2 and new microbial biomass. Most common anaerobic digesters work in a continuous or semicontinuous manner.

An outstanding example of methane generation is the Chinese biogas programme, where millions of family size anaerobic bioreactors are in operation. Such bioreactors are used for treatment of manure, human excreta, etc., producing biogas for cooking and lighting and the sanitisation of the waste, which then becomes an excellent fertiliser.

In almost all fermentation processes performed in a bioreactor there is generally a need to measure specific growth-related and environmental parameters, record them and then use the information to improve and optimise the process. Bioreactor control measurements are made in either an on-line or an off-line manner. With an on-line measurement the sensor is placed directly with the process stream whereas for off-line measurement a sample is removed aseptically from the process stream and analysed. Bioreactor processing is still severely limited by a shortage of reliable instruments capable of on-line measurement of important variables such as DNA, RNA, enzymes and biomass. Off-line analysis is still essential for these compounds and, since the results of these analyses are usually not available until several hrs., after sampling, they cannot be used for immediate control purposes. However, on-line measurement is readily available for temperature, pH, dissolved oxygen and CO_2 analyses.

Scale-up

Fermentation processes are normally developed in three stages or scales. In the initial stage basic screening procedures are carried out using relatively simple microbiological techniques, such as Petri dishes, Erlenmeyer flasks, etc. This is followed by a pilot plant investigation to determine the optimal operating conditions in a volume capacity of 5 to 200 litres. The final stage is the transfer of the study to plant or production scale and final economic realisation.

Throughout these stages of development the biotechnologist aims to maintain the optimal environmental conditions for the process at all levels of development. These environmental conditions involve both chemical factors (substrate concentrations, etc.) and physical factors (mass transfer ability, mixing ability, power dissipation, etc.). In particular, the physical factors create the greatest problems when the process is moving from one scale to another. It is in this area that the skills of the chemical or process engineer find greatest expression.

Design for Fermentation Processes

Water is at the centre of all biotechnological processes and in most cases will be the dominant component of the media in which micro-organisms grow. After liquid fermentation processes have achieved optimum production, the removal of water is a major factor in the cost of bioproduct recovery and downstream processing. The quality of water is highly relevant because it affects microbial growth and the production of specific bioproducts. In the past, traditional brewing centres were established in localities where natural sources provided water of high quality without the need to resort to extensive pretreatment.

In media production there is usually quality control of the raw materials. It is increasingly being realised that, in respect of volume, water is one of the most important raw materials in many biotechnological processes and that its supply and use must be carefully monitored and controlled.

The basic nutritional requirements of micro-organisms are an energy or carbon source, an available nitrogen source, inorganic elements and, for some cell types, specific growth factors. In most biotechnological processes carbon and nitrogen sources are more often derived from relatively complex mixtures of cheap natural products or by-products (Table 4.7.).

Table 4.7. Sources of carbohydrate and nitrogen for industrial media.

Sources of carbohydrate	*Form*	*Sources of nitrogen (% nitrogen by weight)*
Glucose	Pure glucose monohydrate, hydrolysed starch	Barley (1.5–2.0) Beet molasses (1.5–2.0)
Lactose	Pure lactose, whey powder	Corn-steep liquor (4.5)
Starch	Barley, groundnut meal, oat flour, rye flour, soya bean meal	Groundnut meal (8.0) Oat flour (1.5–2.0) Pharmamedia (8.0)
Sucrose	Beet molasses, cane molasses, crude brown sugar, pure white sugar	Rye flour (1.5–2.0) Soya bean meal (8.0) Whey powder (4.0)

Availability and type of nutrient can exert strong physiological control over fermentation reactors and product formation. Raw material input to a fermentation will be dependent largely on the cost of the material at a particular time, since commodity market prices do alter with seasonal and other variables.

Sterilisation practices for biotechnological media must achieve maximum kill of contaminating micro-organisms, with minimum temperature damage to medium components. Batchwise sterilisation in the bioreactor is still the most widely used method, although continuous methods are gaining increased acceptability.

Media preparation may seem to be a relatively uninteresting part of the overall bioprocess but it is in fact the cornerstone of the whole operation. Poor media design will lead to low efficiency of growth and concomitant poor product formation.

Solid Substrate Fermentation

There are many biotechnological processes that involve the growth of micro-organisms on solid substrates in the absence or near absence of free water (Table 4.8). The most regularly used solid substrates are cereal grains, legume seeds, wheat bran, lignocellulose materials such as straws, sawdust or wood shavings, and a wide range of plant and animal materials. Most of these compounds are invariably polymeric molecules, insoluble or sparingly soluble in water, but are mostly cheap, easily obtainable and represent a concentrated source of nutrients for microbial growth.

Table 4.8. Some examples of solid substrate fermentations.

Example	*Substrate*	*Micro-organism(s) involved*
Mushroom production (European and oriental)	Straw, manure	*Agaricus bisporus* *Lentinula edodes* *Volvariella volvaceae*
Sauerkraut	Cabbage	*Lactic acid bacteria*
Soya sauce	Soya beans and wheat	*Aspergillus oryzae*
Tempeh	Soya beans	*Rhizopus oligosporus*
Ontjom	Peanut press cake	*Neurospora sitophila*
Cheeses	Milk curd	*Penicillium roquefortii*
Leaching of metals	Low grade ores	*Thiobacillus sp.*
Organic acids	Cane sugar, molasses	*Aspergillus niger*
Enzymes	Wheat bran, etc.	*Aspergillus niger*
Compositing	Mixed organic material	Fungi, bacteria, actinomycetes
Sewage treatment	Components of sewage	Bacteria, fungi and protozoa

Many of these fermentations have great antiquity and in many instances there are records dating back hundreds of years. In the Orient there is a wide array of food fermentations including soya sauce and tempeh as well as many large industrial enzyme processes. In the West, the fermentation processes have centred on the production of silage, mushroom cultivation, cheese and sauerkraut production, and the composting of plant and animal waste. Solid substrate fermentations using recyclable raw materials such as straw, wood and other waste materials could well be industries of the future, producing ethanol, methane and edible biomass.

The microbiological components of solid substrate fermentations can occur as single pure cultures, mixed identifiable cultures or totally mixed indigenous micro-organisms. In many solid substrate fermentations there is a need to pretreat the substrate raw materials to enhance the availability of the bound nutrients and also to reduce the size of the components, e.g., pulverised straw, shredded vegetable materials to optimise the physical aspects of the process. However, cost aspects of pretreatment must be balanced with eventual product value.

Bioreactor designs for solid substrate fermentations are inherently more simple than for liquid cultivations. They are classified into fermentations (i) without agitation; (ii) with occasional agitation; and (iii) with continuous agitation. The relative advantages and disadvantages of solid substrate fermentations when compared with liquid fermentations are represented in Table 4.9.

Table 4.9. Comparison of solid substrate fermentation with liquid fermentations.

Advantages	*Disadvantages*
Simple media with cheaper natural rather than costly fossil-derived components	Processes limited mainly to moulds that tolerate low moisture levels
Low moisture content of materials gives economy of bioreactor space, low liquid effluent treatment, less microbial contamination, often no need to sterilise, easier downstream processing	Metabolic heat production in large-scale operation creates problems Process monitoring, e.g., moisture levels, biomass, O_2 and CO_2 levels, is difficult to achieve accurately
Aeration requirements can be met by simple gas diffusion or by aerating intermittently, rather than continuously	Bioreactor design not well developed. Product limitation. Slower growth rate of micro-organisms
Yields of products can be high	
Low energy expenditure compared with stirred tank bioreactors	

SMALL-SCALE LIQUID FERMENTATIONS

Small-scale liquid fermentations are undoubtedly the most frequent fermentations, because they are operationally simple and inexpensive to conduct, allowing many to be run simultaneously. While small-scale fermentations can meet the requirements for many fermentation objectives, they have limitations that prevent them from solving all problems (Table 4.10). However, they can often save much expense by pointing the way for more effective use of innately more expensive larger equipment. The most frequent reason for disappointment in the results from small fermentations is in misalignment of expectations with this tool. The goal of this section is to describe the standard use of small-scale liquid fermentation, including reasonable expectations for this tool, and to put the use of this tool in the broader context of all fermentations especially relative to alternative scales, vessels and approaches. This section is intended to provide background on the basics of small-scale liquid fermentation as commonly practiced in industry today. Ultimately, however, fermentation science is an empirical and intelligent process, in which an intelligent and educated human works from general principles and specific empirical data to solve each unique problem.

Table 4.10. Strengths and weaknesses of small-scale liquid fermentation.

Strengths

- Potential for large numbers of parallel fermentations
  - Low-cost vessels
  - Low operating costs
  - Vessels take little space and require little infrastructure
- Easy to scale by increasing the number of identical vessels

Weaknesses

- Effectively limited to batch fermentation processes (or punctuated batch)
- Limited maximum oxygen transfer rate
- Discontinuous monitoring only
- Non adaptive control of pH or substrate concentration

'Fermentation' is used in the common way, meaning simply the cultivation of micro-organisms. This includes aerobic and anaerobic cultivation of bacteria and fungi but excludes cultivation of cells that would naturally be part of a plant or animal (i.e., cell culture). Discussion will be heavily weighted toward aerobic culture, because most of the higher value products are from aerobic industrial processes and because the scaling of aerobic processes is innately more complicated than the scaling of anaerobic process. 'Small scale' has been interpreted as individual fermentations ranging from microliter volumes, such as in microwell plates, up to the liter range in flasks or carboys. These scales are unified as much by a practical limitation to batch operations as by volume. In batch fermentations the sterilised media components are supplied at the beginning of the fermentation with no additional 'feeds' after inoculation. This batch operation is often the greatest limitation of small-scale liquid fermentation. However, the cost of each fermentation vessel is very low compared with typical stirred and fed vessels.

There are many reasons to do small-scale liquid fermentations and they can be conveniently divided into three major categories based on the 'product'. The three classes of products are (i) products associated with the growth of the cells, including whole cells (biomass) and subcellular components; (ii) microbial products produced after the growth phase (secondary metabolites or idiolites); and (iii) data.

Although batch fermentations are simple to set up and run, they re dynamic processes that are essentially never in a steady state; thus, they may be far more difficult to understand than more 'controlled' processes such as those in stirred, monitored, and fed tank processes. The concept of control in fermentation has its roots in the operation of a chemostat, in which a continuous fermentation process is designed so that one nutrient will be 'limiting' or 'controlling.' That is, the fermentation growth rate is proportional to the dilution rate of the medium. A corollary to the controlled growth rate achieved in a chemostat in true steady state is that the physiology is also controlled. Although small-batch fermentations cannot be in steady state, it is often productive to think of them as a succession of states in which some nutrient is limiting the process. If the goal is a reproducible process, then the immediate objective is to achieve the same sequence and duration of states, and to understand the nature of changes to this succession when experimental variables are introduced. Highly variable fermentation processes may constructively be viewed as the result of a lack of awareness of some parameter critical to the particular process, that is, a failure to search for, find, and control some parameter critical to that fermentation. Often in small liquid batch processes, that critical parameter is gas exchange or balance between respiration rate (oxygen demand) and oxygen transfer.

A reasonably full mechanistic understanding of any fermentation process is costly. This is especially true when complex medium components are used. However, after appropriate background work it is possible to empirically define many aspects of the specific fermentation system. This requires beginning with a basic system that meets the basic needs of the micro-organism and provides a solid, reproducible foundation for future research (Table 4.11). If improvements are sought, the scientist must learn, through experimental definition, what limits the cells of a fermentation at each stage of that process, e.g., what limits growth first in the process, and product yield later, if the product is not growth-associated. An empirical understanding of process may be obtained by systematically exploring the effect of various levels of each nutrient. Selection of the base fermentation system, including the vessel, closure, medium, associated equipment, and operating parameters, necessarily goes hand-in-hand with the objective of the fermentation and with the biological requirements of the micro-organism to be cultured.

Table 4.11. Components of a small fermentation system.

Parameter selected	*Parameters affected*
Vessel	Culture volume, gas-liquid mass transfer, mixing
Colour	Gas transfer out-to-in, in-to-out; evaporation
Medium composition	Kinetics and extent of growth and metabolism
Temperature	Kinetics of growth and metabolism, evaporation
Agitation rate (cycles per minute and throw)	Gas-liquid mass transfer, mixing time

Shaken Erlenmeyer flasks with cotton plug closures are the classical small-scale liquid fermentation system (Fig. 4.4a), but variations on all aspects of this basic system have been developed. The basic shake flask system has been modified to increase mixing and mass transfer with indentations (Fig. 4.4b) and 'corners.' Alternative closures have been selected because they allow more gas transport in and out of the vessel, and simple sparged and lightly stirred carboys Fig. 4.4c are used for very large volumes when shaking is impractical or deleterious to the microbe.

Fermentation Vessels

The standard Erlenmeyer flask has the virtue of being readily available in a variety of sizes at low cost. However, since it was not designed for fermentation, many other vessels have been used for small-scale batch fermentations to meet special objectives. Alternatives to the basic shaken flask have included vessels with a square base, a tetrahedron shape, and a more barrel-like shape. Most often the vessel seems to be selected for oxygen transport, liquid mixing, convenience, and familiarity. However, it is important to periodically evaluate whether your standard fermentation vessel is appropriate to your current objectives.

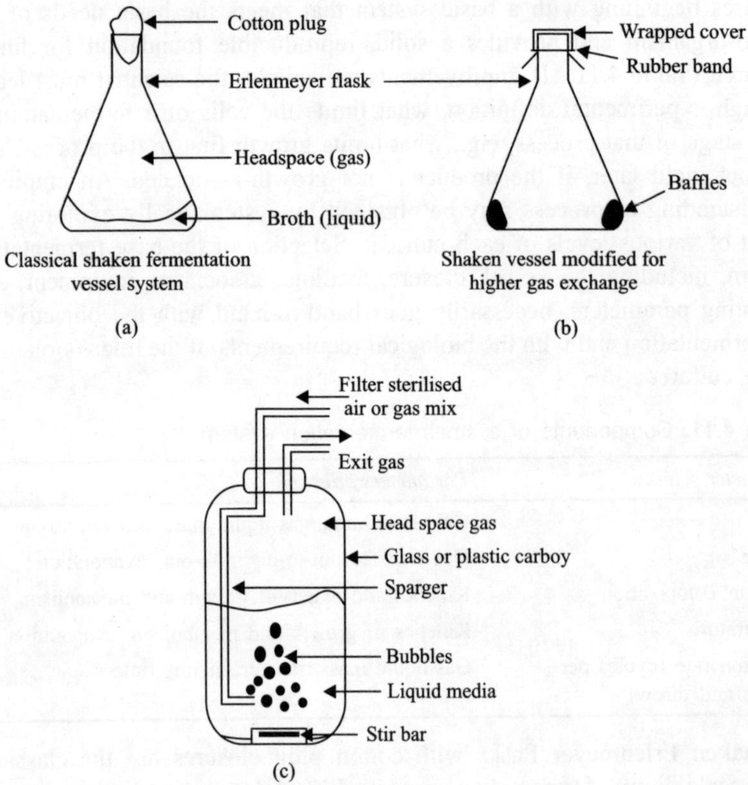

Fig. 4.4. Schematic representations of common small-scale liquid fermentation systems.

There are no absolute limits on the range of volumes that could be considered small-scale liquid fermentations. The lower limit has historically been determined by requirements for ease of reproducible and aseptic operations and for reliable sampling and analysis methods. As liquid handling and analytical methods have been miniaturised, some fermentations have been similarly miniaturised. An entire industry has developed around automated liquid handling and biochemical analysis in microwell plates especially the 96-well microwell plates. This microwell format-based industry has increased the precision, accuracy, and ease of handling small volumes and measuring outcomes from reactions in these microwell plates. Those same plates and liquid handling technologies are now used for cultivation of microbes is some situations. They are especially useful when large numbers of strains need to be evaluated simultaneously, and growth (e.g., turbidity or pH change) is the measured end point. For example, enumeration of viable cells in a solution is sometimes more convenient by most probable number determination in microwell plates then by plating on agar. Furthermore, BioLog has developed and markets a culture identification system based entirely on growth of unknown microbes in 96-well plates containing a fixed array of carbon sources. In addition to clinical culture identification, variations on the standard BioLog system have been used to survey microbial communities and to screen for micro-organisms from nature that are capable of degrading toxic chemicals. Generic microwell plates have been used to evaluate the physiology of many strains in parallel and to select improved strains with a variety of traits. Evaporation remains a significant limitation on miniaturisation of lengthy fermentations. However, this barrier can be overcome by incubation in a high humidity chamber or ignored if the incubation time is short enough.

Test tubes of many sorts have been used effectively for volumes up to ca. 10 ml. They are cheap and easy to handle with common lab equipment. Traditionally, culture tubes were glass but there are now many disposable plastic tubes, with caps, that are extremely convenient for use in the culture of microbes. Even the ubiquitous 1.5-ml microcentrifuge tubes have been used as shaken fermentation vessels to evaluate many strains in parallel fermentations. Gas transport was through an open hole with no barrier to the entrance of foreign microbes, but strain evaluation was successful because it was brief.

The upper limit on the volume of small-batch fermentations is typically set by oxygen uptake requirements of the culture to be fermented. Generally, the gas transfer rate at the gas-liquid interface is less as the volume gets larger. Shaken vessels with up to about 1 litre of working volume are fairly common and can be scaled for larger harvests by simultaneous parallel batches. In addition, larger volume carboys with a bubble sparger are excellent fermentation vessels for microbes with low demand for gas exchange. These would include anaerobes, facultative organisms with anaerobic processes, aerobic organisms at low densities or low metabolic rates, and some photosynthetic processes. Carboys are not shaken, but rather are mixed with a magnetically coupled stir bar or simply by the lift of the bubble column (air lift mode). Maximum oxygen transfer rates would be much less than those in typical shaken flasks and stirred fermentors. Small-batch fermentations between these two extremes are traditional and convenient for most purposes. If the goal is biomass and the organism is aerobic, several well-balanced, shaken Erlenmeyer flask fermentations are likely to produce more cells than a harvest volume 10-fold greater from a bubbled carboy due to the substantially greater oxygen transfer rates possible in the shaken vessel.

Vessel Closures

The purpose of the closure is to serve as a selectively permeable barrier. Foremost, the closure must provide a sterile barrier to maintain an axenic culture and to prevent aerosols from the culture from escaping. The second set of requirements is to allow exchange of gases. A high rate of exchange of oxygen and carbon dioxide is most frequently critical, but the ideal closure would simultaneously limit evaporation. Limitation of gas exchange past the closure has two major consequences. First, it may limit oxygen availability to aerobic cells as their density and oxygen demand increase. Second, the gas mixture within the headspace of the flask will differ in CO_2 and N_2, as well as O_2, from the ambient gas mixture as the oxygen is consumed and converted to carbon dioxide. Elevated CO_2 has the potential to alter the medium pH or have other specific biological effects. Few materials are ideal as closures, but many are in common use.

Cotton fibre plugs are cheap and simple and therefore remain in common use. They are often wrapped in cotton cheesecloth to improve handling and to prevent fibres from sticking to the inside of the neck of the vessel. They are an excellent choice if high and reproducible rates of gas transport are not critical to your fermentation system. Cotton plugs are known to severely limit gas transport in and out of baffled fermentations vessels, especially if wet from condensation following autoclaving or splash from vigorous mixing. Variability in the density of the hand-made cotton fibre plugs combines with variability in dryness to vary the maximum gas transport rate between ostensibly equivalent flasks. Foam plugs suffer from the same potential for wetting but may be more consistent with respect to density since they are commercially manufactured. They have the clear advantage over cotton plugs that they come ready to insert, with no preparation.

Solid caps for aerobic fermentation are typically friction-fit for ease of handling. They allow free air flow through a path between the cap and the neck of the vessel. Gas transport is limited by the annular area of that path and is often small enough to limit the maximum gas transport rate in the fermentation (i.e., gas transfer in and out of the vessel may be lower than that between the gas and liquid phases within the vessel). Since there is an open path, with no sterile barrier, there is a risk of airborne microbes or spores finding their way in or out of the vessel. Solid caps are most often used for test tube vessels in short fermentations because of their ease in removal and reclosure. Screw-on solid caps and septum-type caps are commonly used for anaerobic fermentations.

Shakers

Machines for holding small liquid fermentation vessels and moving them to induce mixing are generically called shakers. There are many sizes, designs, and options available. Before purchasing shakers, it is vital to specify the functional requirements so that you can carefully compare features, options, upgradability, price, and maintenance costs.

Environmental control, minimally temperature and often humidity, is generally necessary for fermentation process control. Shakers can be purchased with or without environmental controls; if purchased without controls, the shaker would normally be placed in an environmentally controlled room or temperature-controlled box. Space, cost, and flexibility would determine which of these two alternatives is better. Maintenance of temperatures below 25 to 28°C generally requires chilling. This can be accomplished by built-in refrigeration or with a coil-type heat exchanger and externally supplied chilled water. Humidity control becomes important with long fermentation times, when much of the liquid volume will be lost to evaporation without elevated ambient humidity. However, high humidity may result in the unintended growth of mould on the walls of the environmental chamber due to condensation. Some humidified environmental chambers have heaters in their walls to prevent condensation.

Fermentation broth temperature can be controlled by bathing the vessels in constant-temperature water or air. Water bath shakers have the advantages of efficient heat transfer and the potential for sampling without removing the vessel from the bath; this may result in less effect on fermentation. But the shaker speed of water bath incubators is limited by splash outside of vessels and substantial maintenance is required to keep the bath water clean. Culture vessels obviously come out wet, so they need to be dried prior to aseptic operations. Air incubators are more commonly used for shakers because of their overall greater ease of handling and because of their higher potential shaking speeds. With either type of temperature control, but especially with air incubators, there is a very real potential for unevenness in temperature within the chamber.

Shaker motion is in two common patterns: orbital (also called rotary or gyrotory) and reciprocal (also called linear). Shaking vigor (which determines gas-liquid mass transfer and shear) is a function of the diameter of the orbit or stroke of a reciprocal shaker (commonly called 'throw' and measured in inches or centimetres) and the cycle time (or speed, measured in rpm or cpm).

Shakers come in a variety of physical size and configurations. Vessel capacity requirements, space availability, and cost are the major considerations to balance in selection of the basic shaker. The highest capacity per shaker and per unit of floor space comes from multi-tier shakers.

Liquid Media Composition

At first look, selection of medium for small-scale liquid fermentations is the same as that for larger vessels (including animals). The ideal starting point is to supply a 'diet' balanced for carbon, nitrogen phosphorus, minerals and essential vitamins. However, for many processes the selection of a fermentation medium is inextricably linked to selection of the other fermentation parameters. Therefore, if fermentation performance (efficiency, titer, etc.) is a goal, optimisation will require empirical determination.

Medium composition can be divided into two extreme categories: (i) nutrient-balanced media typically used for biomass and growth-associated products (primary metabolism) and (ii) unbalanced media that support only low growth rates but ongoing metabolism. Hybrid media compositions between these two extremes are most common. In this instance, balanced nutrients support rapid growth to an initial cell mass (a vegetative stage); the remaining nutrients are unbalanced to trigger a secondary metabolic phase or sequence of secondary phases.

Nutrient-balanced media for growth in small-scale batch processes are typically no different from those that would be used for stirred and fed fermentors, except that it is important to adjust the 'strength' to be commensurate with the oxygen transfer capability of the fermentation system. If not adjusted, the fermentation process cannot be expected to be similar between scales, despite the use of the same medium composition. The whole medium can be adjusted by dilution or concentration of all ingredients, or the concentration of a key ingredient (e.g., glucose or ammonium) that is intended to limit rapid growth at all scales can be adjusted.

In larger, stirred tank fermentations, a secondary phase is often sustained and controlled by delivery of an exogenous feed, supplying limiting quantities of an essential nutrient especially carbon, nitrogen, or phosphorus. With the restriction to batch operations of small liquid fermentations, a similar nutrient limitation may be achievable by introduction of a kinetic limit using a slowly metabolised nutrient source. The limitation may be directly on the rate of metabolism or it may be related to transport into the cells or availability for transport. Introduction of a limit to availability is commonly accomplished with ingredients such as starch, flour, oils, and proteins that must be degraded by extracellular enzymes before their subunits can be transported and further metabolised. Ingredients engineered into sustained-release formulations constitute a possible alternative approach to this kinetic limitation.

Sterilisation of media can be a cause of variability. Although it is common to sterilize media by autoclaving, too often the operation is done with inadequate control, resulting in failure to sterilise (especially *Bacillus* spores, if present in high numbers) or in inconsistency due to differences in the chemical reactions driven by the heat of autoclaving. Common mistakes include (i) failure to presteam long enough to bring the liquid volume to 100°C prior to pressurising the chamber so the liquid never reaches 121°C; (ii) overcrowding the autoclave chamber or using closed-basin trays so steam flow is restricted, effectively insulating the material to be sterilised from the steam; and (iii) failure to deliver an equivalent amount of heat per volume to media in different sizes of vessels. Effective sterilisation can be proved by including commercial spore test ampoules in a 'dummy' load, equivalent to the largest liquid load in the cycle. Sterilisation time can be varied experimentally to test whether your particular fermentation system is significantly affected by the heat of sterilisation.

Gas Exchange in Small Liquid Fermentations

The mass transfer rate between phases (e.g., oxygen transfer from gas to liquid) is very often one of the most critical operational parameters for any aerobic fermentation. Too often it is neglected as small fermentation systems are set up and is therefore undoubtedly a frequent cause of irreproducibility. Gas-liquid mass transfer, or an estimate of the maximum potential mass transfer (expressed as $K_L a$, is a standard parameter defining operating limits in sparged fermentors but has equal potential to limit small batch cultures. The importance of oxygen mass transfer in aerobic fermentations should be obvious since the maximum oxygen transfer rate will determine the cell mass that can be supported with a maximally aerobic physiology. The mass transfer that ultimately matters is that transferred to and from the cells per unit time. It is best to take a whole-system view toward this. For example, whether oxygen reaches the cells depends on, sequentially (Fig. 4.5); (i) diffusion of oxygen from the ambient air outside the vessel through the closure to the interior compartment (vessel headspace), (ii) oxygen transfer from the gas phase to the liquid phase within the vessel and (iii) soluble oxygen transfer from the liquid to the cells. Any one of these steps can be limiting in an aerobic shaken fermentation and different ones can limit at different times. The direction of diffusion of the product carbon dioxide will be opposite that of oxygen (i.e., from cells to liquid to gas and out), again depending on mass action as the driving force in this reverse pathway.

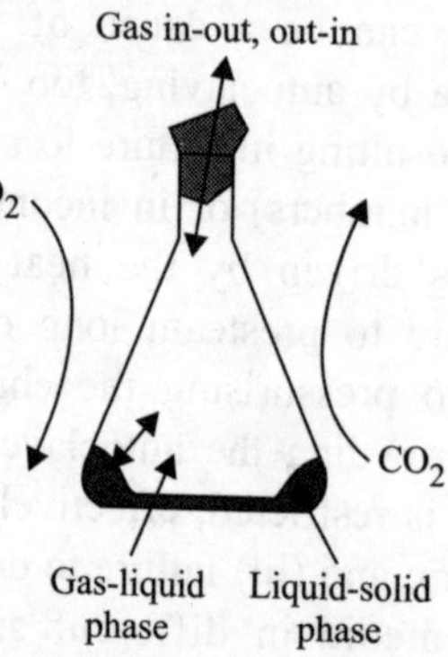

Fig. 4.5. Schematic representation of gas exchange in a respiring shaken flask.

The rate of diffusion of gases through the vessel closure of a shaken fermentation depends on the diameter of the vessel opening, the depth of the closure, the nature of the closure, and the difference in concentrations of the specific gas in the inside and outside compartments. Diffusion through the closure has been reported to limit oxygen availability in some fermentation systems based on inference from oxygen absorption rate experiments and on cell yield. Furthermore, even when the oxygen diffusion rate through the closure has not become rate limiting, that closure may cause very large increases in the concentration of CO_2 (to 15 per cent) in the vessel headspace. This situation is fundamentally different from that for sparged vessels, in which CO_2 is constantly stripped by the excess of N_2 in the sparge air and thus may be a barrier to scale-up. Generally, it is best to have greater mass transfer at the vessel closure than across the gas-liquid interface, although this does not eliminate this fundamental difference between shaken and sparged vessels.

Liquid-solid mass transfer in fermentations generally will limit only when mixing is poor. This may occur with poor agitation, in highly viscous broths, or in cultures growing in large colonies (clumps, balls, or pellets). Mixing times in nonviscous solutions ranged from 52 second to less than 2 seconds in several standard fermentations configuration and were affected by liquid volume, vessel type (e.g., smooth versus baffled flask), agitation rate, and agitation pattern.

Foam Control

Many fermentation media or fermentation broths will develop a foam head when mixed vigorously. A stable foam constitutes another potential barrier to gas exchange between the vessel's headspace and the liquid phase. Antifoams generally decrease gas-liquid mass transfer rates under simple circumstances, but may either increase or decrease overall fermentation mass transfer rates depending on where the limitation exists. Sparged and stirred fermentors do not depend primarily on gas diffusion from the headspace to the liquid, hence are more likely to lose mass transfer potential with antifoam addition. Although stirred fermentors have the option of either chemical or mechanical foam control, shaken vessels are limited to chemical control with antifoams. Common criteria for selection of an antifoam are (i) effectiveness in controlling foam; (ii) absence of toxicity to the specific fermentation; (iii) compatibility with scale-up or harvest procedures; (vi) a minimal negative effect or a positive effect on gas-liquid oxygen transport; and (v) cost of use. Silicon-based emulsions seem to be most effective on the basis of final concentration of the active antifoam component, but natural oils are often used to control foam while serving as a carbon source to the microbe so the final concentration can be allowed to decline to very low levels to minimise impact on harvest.

Fermentation Sampling and Analyses

The importance of analysis of fermentations cannot be overstated; clearly, without valid measurements of outcomes it is impossible to understand, improve, or scale a fermentation and it may not be possible to reproduce it. There are two critical components to effective fermentation analysis. First is sample preparation. A representative sample of the culture must be taken and prepared appropriately for the specific analysis. Second, the specific analysis must be proved valid in the presence of the fermentation broth matrix; media ingredients often interfere with chemical analyses. Since analysis of small-scale fermentations is not fundamentally different from analysis of any other fermentation, the following is a limited discussion focusing on small-scale batch process. Matrix effects depend primarily on the medium composition and are otherwise common to all scales. Sampling methods and sample preparation are often linked to the scale of fermentation.

It is best if small fermentations are sampled only once, at harvest. Time points can be obtained by harvesting replicate flasks over time. Miniaturised on-line analysis is possible but may not be sensible since it invariably adds complexity to a system whose strength is simplicity. 'Grab' samples, if small relative to the total fermentation volume, can be taken throughout the fermentation but carry several risks; (i) sampling normally requires that the vessel be opened, exposing the fermentation to additional risk of contamination; (ii) each sample removed will lower the volume of the fermenation, thus changing the mass transfer characteristics of the fermentation; (iii) interrupting shaking while sampling may induce a stress response in the microbes (e.g., due to oxygen limitation or temperature change during handling) that is not reproducible.

Whole fermentation broth samples often need to be further processed prior to analysis. Most frequently, the cells (along with other particulates) and a clear liquid fraction are separated from one another by centrifugation or filtration for different analyses. Centrifugation is effective for essentially all fermentation processes. If only small volumes are needed, the method is very rapid and convenient with the use of disposable labware and a microcentrifuge or other small centrifuge. Filtration can be convenient with some micro-organisms, but many fermentations will clog the filter quickly, making the method inconvenient. If your fermentation can be easily filtered, the method and the advantage that many broths can be filtered in parallel with the use of a manifold filter holder and the cells can be washed easily.

Fermentations invariably have one primary end point to be measured that is aligned with the objective of the process. This may be biomass, a cellular product, or data, as outlined above. This is the obvious first analytical objective, and if there is no intention to scale or improve the process, there may be no need for any further analyses. Conversely, if there is an intention to scale or transfer the process to a different fermentation system, it will likely be necessary to do additional analyses to characterise the process and the kinetics of events in the fermentation (Table 4.12).

Estimates of biomass are very useful when viewed temporally because they allow inferences about growth phase (when plotted versus fermentation time), cell yield on substrates (e.g., g of cells per g of glucose consumed) and specific product yields (gram of product per gram of cells). The best estimate of biomass would come from direct measurement of the dry weight after harvesting cells by centrifugation or filtration from a known volume of soluble broth, washing those cells to remove residual soluble materials, and weighing the heat-dried cell pellet. For many unicellular micro-organisms a rapid turbidimetric determination can substitute effectively for dry weight determinations after establishing a standard curve (e.g., relating optical density around 600 nm to dry cell weight). The biomass of filamentous microbes grown on soluble media can sometimes be estimated from simple packed cell volume measurements (fraction of broth volume pelleted by centrifugation), again following establishment of a standard curve relating packed cell volume to dry cell weight. Biomass estimates of small fermentations may be inaccurate or impossible by any means if the broth contains large amounts of insoluble ingredients (e.g., flours, grits, or fish meal), as are common in many antibiotic fermentations. In practice, industrial scientists often forgo estimates of biomass until processes with insoluble media are scaled to stirred tanks and then use exit-gas measurements (especially CO_2 and O_2) for data to infer the growth stage; yield calculations that depend on biomass determination may still not be straightforward.

Table 4.12. Common generic measurements of fermenations.

Measurement (units)	*Method*
Dry cell weight (g/litre)	Gravimetric (packed, washed cells from a known harvest volume are dried fully and weighed) or indirect (optical density is related to dry weight in an empirical standard curve and measured as a surrogate)
Packed cell volume (%)	Centrifiuge in calibrated tube, read directly
pH	pH dip strips or meter
Ammonium (μm or ppm)	Ion chromatography or blood analyser[a]
Phosphate (μm or ppm)	Ion chromatography or blood analyser[a]
Glucose (mM or g/litre)	Test strips (e.g., blood glucose) for semiquantitative estimate; HPLC or glucose oxidase-based analysers for quantitative estimate
Amino and organic acids (mM or g/litre)	Ion chromatography or specialised assays

[a] Clinical blood analyzers and Kodak biolyzer (Eastman Kodak Co.) can do enzyme based determination of many chemicals (e.g., glucose, ammonia, phosphate) and enzymes (e.g., amyiase, lipase) of potential interest to fermentations.

Determination of residual soluble components of the fermentation, such as glucose, organic acids, ammonium, or phosphate (present in a supernatant or filtrate fraction), constitutes the other broad class of common analyses. These may be for general characterisation of the process (for example, to help determine whether carbon or nitrogen is limiting first) or for quantification of a specific component thought to be critical to the process.

5

Downstream Processing in Biotechnology

INTRODUCTION

The isolation and purification of a biotechnological product to a form suitable for its intended use is popularly termed 'downstream processing' (DSP). In most cases this means recovery of a product from a dilute aqueous solution. The complexity of downstream processing is determined by the required purity of the product, in turn determined by its application. The products of biotechnology include whole cells, organic acids, amino acids, solvents, antibiotics, industrial enzymes, therapeutic proteins, vaccines, gums, etc. As the products vary greatly in size and nature, different separation principles are required for their recovery and purification. The liability or sensitivity of many of the bioproducts, particularly the proteins, to the environmental conditions, places further demands on the characteristics of the separation processes used for their production. It is also imperative to minimise the number of steps and to maintain high yields in the different stages, since the ultimate result will be poor if several low-yielding steps are combined.

DOWNSTREAM PROCESSING: A MULTISTAGE OPERATION

Many of the steps in DSP are traditional unit processes used extensively in chemical industry and have been described in chemical engineering literature. This chapter is devoted to DSP processes especially adapted for biotechnology. The downstream processing scheme normally employed for isolation and purification of biomolecules can be divided into the following stages: (i) solid-liquid separation or clarification; (ii) concentration; (iii) purification; and (iv) formulation.

Sometimes, as in the case of industrial enzymes, the concentration stage may provide purification sufficient for the final product quality, obviating the need for a separate purification stage. A wide range of unit operations are available for each stage (Fig. 5.1).

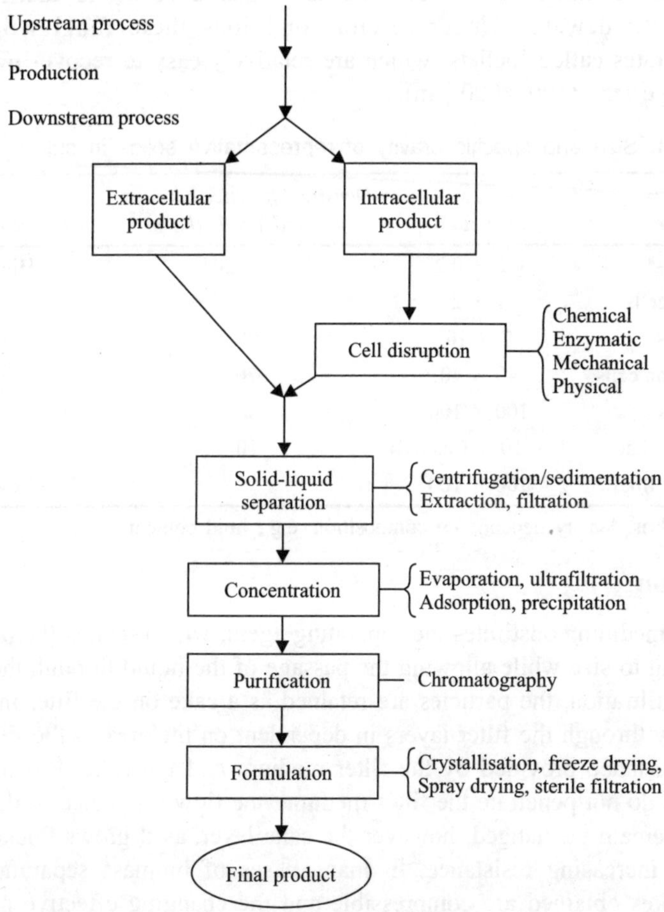

Fig. 5.1. Downstream processing. Different stages in the isolation and purification of the products of biotechnology.

SOLID-LIQUID SEPARATION

Solid-liquid separation is a primary recovery operation for the separation of whole cells from the culture broth, removal of cell debris, collection of protein precipitate, collection of inclusion bodies, etc. The unit operations commonly used are centrifugation and filtration. Table 5.1 lists the characteristics of some of the culture broths used for production of biomolecules. Yeasts and bacteria are usually homogeneously suspended in the fermentation broths. Some bacteria may form slime layers depending on the strain and fermentation conditions and this will lead to separation problems. Filamentous fungi are frequently characterised by a network of intertwined filaments, producing viscous fermentation broths that may be difficult to dewater. Under certain conditions these fungi will form agglomerates called 'pellets' which are relatively easy to recover owing to their large size (100–4000 μm).

Table 5.1. Size and specific gravity of representative solids in culture broths.

Solids type	*Size (μm)*	*Density difference between solids and broth (kg m^{-3})*	*Cost of recovery*
Cell debris[a]	0.2 × 0.2	0 – 120[a]	Highest
Bacterial cells	1 × 2	70	<
Yeast cells	7 × 10	90	<
Mammalian cells	40 × 40	70	<
Plant cells	100 × 100	50	<
Fungal hyphae	1 × 10 × (matted)	10	<
Microbial flocs	100 × 100	–	Lowest

[a] Cell debris density depends on composition, e.g., lipid content.

Filtration

A filter medium constitutes the separating agent, which retains the particles according to size while allowing the passage of the liquid through the filter. In cake filtration, the particles are retained as a cake on the filter medium. The flow through the filter layers in dependent on the area of the filter and flow resistance provided by the filter medium and the cake. Provided the particles do not penetrate the filter medium, the flow resistance of the latter should remain unchanged; however the cake-layer, as it grows thicker, will provide increasing resistance. In many cases of biomass separation, the filter cakes obtained are compressible and the changing effective pressure difference will influence the flow through the filter. Examples of filter media are perforated sintered metal, cloth, synthetic fibres, cellulose, glass wool, ceramics and synthetic membranes.

Many types of filtration equipment are available. Because of simplicity of operation and low costs, vacuum filters are frequently used for clarification of fermentation broths containing 10–40 per cent solids by volume and particles with sizes of 0.5–10 μm. The best known are rotary drum vacuum filter and filter press. The former type is commonly employed for filtration of filamentous fungi and yeast cells, and is schematically presented in

Fig. 5.2. Besides simplicity and effectiveness of filtration, its attractive features are low power consumption and a contained operation. The filter press is built up of a sequence of perforated plates alternating with hollow frames mounted on suitable supports. The plates are covered with a filter medium (cloth) to create a series of cloth-walled chambers into which slurry can be forced under pressure. The solids are retained within the chambers, while the filtrate discharges into hollows on the plate surface, and hence to drain points. At the end of the filtration cycle, the hydraulic pressure is released and the cake manually removed from the cloth. Typical filter sizes available range from 120 cm^2 for laboratory scale to 14400 cm^2 (per plate) for a production scale equipment. In membrane filter presses, the cake chamber is covered with a rubber membrane that is inflated using air or water, allowing *in situ* compacting of the cake. These provide higher yield and drier cake, but involve a higher capital investment. Filtration using microporous membrane operated under pressure has developed into a viable alternative to centrifugation as a means to remove suspended particles from process fluids. The technique is described in later section of Chapter.

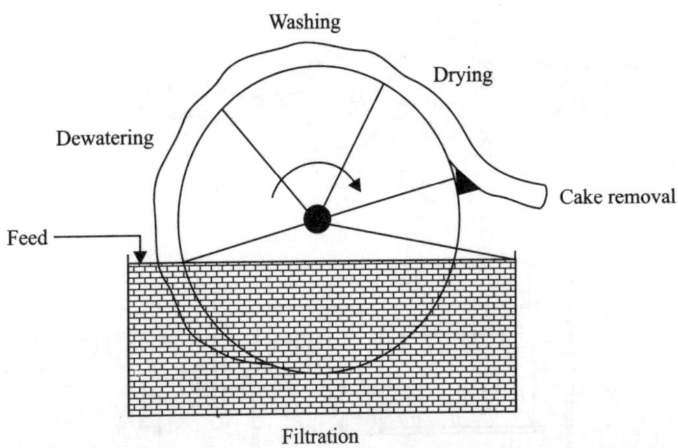

Fig. 5.2. A rotary drum vacuum filter. The filtration element comprises a rotating drum, maintained under reduced internal pressure, partially immersed in a tank of broth.

Centrifugation

In centrifugation the removal of solids relies on the density difference between the particles to be separated and the surrounding medium. Batch centrifugation is commonly applied at a laboratory scale. Although providing high centrifugal forces, low processing capacity limits the use of such centrifuges on an industrial scale, where continuous flow centrifuges are the norm. Here, feeding of the slurry and collection of the clarified solution is continuous, whilst the deposited solids can either be continuously or intermittently removed from the centrifuge. Some of the centrifuge types used in the isolation of bioproducts are shown in Fig. 5.3.

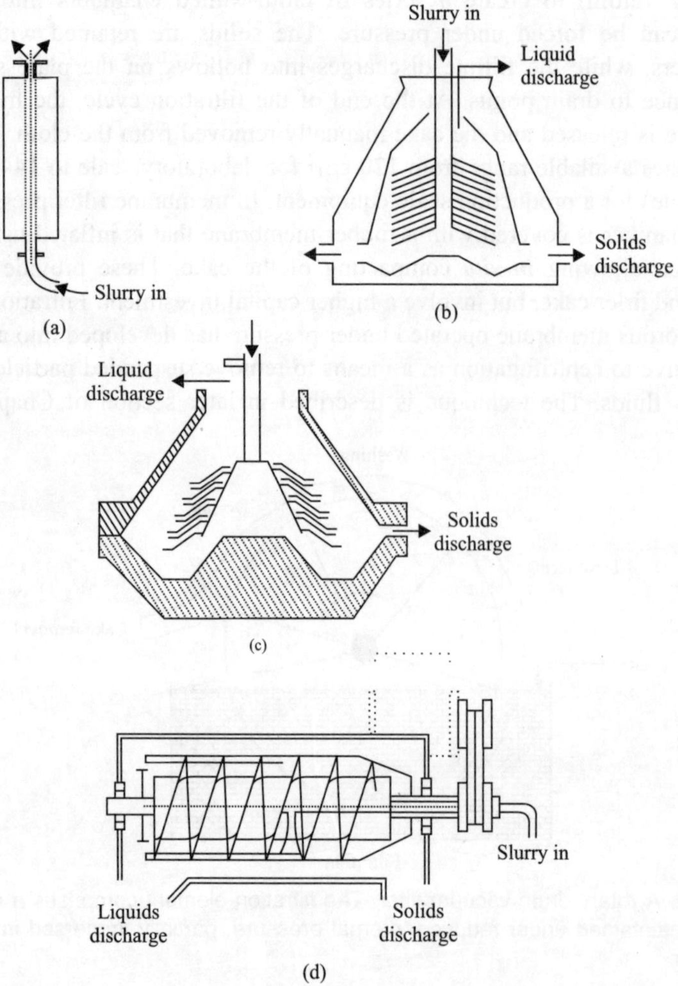

Fig. 5.3. Centrifuges. (a) The tubular bowl type is centrifuge used in pilot plants which can be operated both in batch or continuous mode, with manual discharge of the solids. High centrifugal field can be achieved owing to its slender shape and small volume. (b) the multichamber bowl type is a modification of the tubular type, containing a number of concentric tubes connected to allow zigzag flow of the particulate feed. (c) The disc stack centrifuge is used widely in biotechnology. It incorporates numerous discs (typically between 30 and 200) at an angle of 35–50° and kept 0.4–2 mm apart, dividing the bowl into separate settling zones.

Traditionally, centrifugation is used for removal of microbial cells and other discrete large particles. The supernatant obtained by centrifugation is not free of cells (10^3–10^5 cells ml^{-1}), and costs of maintenance and power consumption are both high. Separation of particulate debris from cell homogenates is quite inefficient by centrifugation.

Pretreatment of Broth to Facilitate Clarification

Pretreatment or conditioning of the broth by changing the biomass particle size, the fermentation liquor viscosity, and the interactions between the biomass particles are sometimes required to ensure efficient solid-liquid separation. This implies use of a filter aid (body-feed) in the broth and/or for precoating the filter medium. Filter aids are incompressible, discrete particles of high permeability with size ranging from 2–20 μm. Filter aids should be inert to the broth being treated; the most frequently used are Diatomite (skeletal remains of aquatic plants), Perlite (volcanic rock processed to give expanded structure) and inactive carbon.

Agglomeration of individual cells are cell-particles into large flocs, which can be easily separated at low centrifugal forces, is done by the addition of flocculating agents such as polycations, either cellulosic or based on synthetic polymers, inorganic salts or mineral hydrocolloids. Besides the flocculating agent itself, factors like the physiological state of the cells, the ionic environment, temperature, and nature of the organism influence flocculation. Interestingly, cationic filter aids also reduce the load of pyrogen, nucleic acid and acidic protein, which normally foul chromatography columns.

Flotation

In flotation, particles are adsorbed on gas bubbles, get trapped in a foam layer and can be collected. The gas may either be sparged into the particulate feed, or very fine bubbles can be generated from dissolved gases by releasing the overpressure or by electrolysis. Formation of stable foam is supported by the presence of 'collector substances' such as long chain fatty acids or amines.

RELEASE OF INTRACELLULAR COMPONENTS

The primary goal for the first step in a sequence for release of intracellular compounds is to liberate maximum amount of the product in an active state. The inactivating effects of shear, temperature and proteases need to be borne in mind at this stage. The choice of the disruption method has to be made empirically, at the same time taking into consideration the subsequent processing steps.

Disruption of Microbial Cells

The different strategies that may be used for rupture of microbial cells include breaking the cells' structure by mechanical forces, damaging preferentially the cell wall e.g., by drying or enzymatic lysis, or lysing primarily the membranes e.g., by treatment with chemicals (Table 5.2).

Table 5.2. Methods for disruption of cells and tissues.

Mechanical	*Non-mechanical*
Microbial cells	
Ultrasonication	Drying
High-pressure homogenisation	Heat shock
Agitation with abrasives	Osmotic shock
	Freeze-thaw
	Organic solvents
	Chaotropic agents
	Alkali
	Detergents
	Enzymes
Animal tissue	
Homogenisation	
Plant tissue	
Maceration	Enzymes
Freezing and grinding	

Mechanical disruption of cells is the most common means of releasing intracellular products both at the laboratory and industrial scale. Ultrasonication disrupts the cells by cavitation, and is commonly used at laboratory scale; removal of the heat generated is difficult on a larger scale. Industrial scale disruption of microbial cells is achieved by high-pressure homogenisation or by vigorous agitation with abrasives. In the former case, the cell suspension is forced at high pressure through an orifice of narrow internal diameter to emerge at atmospheric pressure. The sudden release of pressure creates a liquid shear capable of disrupting the cells. Agitation with glass in bead mills ruptures the cells by a combination of high shear and impact with the cells. The size of the beads varies from 0.2–0.5 mm for bacteria to 0.4–0.7 mm for yeasts. Recently microfluidisation has been introduced for cell disintegration. Here, the cell suspension is fed under high pressure through a chamber where it is split into two streams, to be directed at each other at high velocity before emerging at atmospheric pressure.

Non-mechanical disruption of cells is possible by physical, chemical or enzymatic means. Drying is a widely applied method of cell lysis, causing changes in the structure of the cell wall and making it possible to subsequently extract the cell contents in buffer or salt solution. Osmotic shock ruptures the cell membrane and is particularly effective for animal cells that lack a cell wall. The presence of a cell wall in the case of microbial cells may necessitate application of additional measures. Cell breakage by heat shock (thermolysis) is relatively easy and cheap but is only useful for heat-stable products. Treatment with organic solvents and detergents to render cells more permeable is often used in the laboratory scale. Toluene is used frequently, acting by dissolving membrane phospholipids and creating pores in the cell membrance. A selective liberation of enzymes from the periplasmic space is possible by treatment with water-miscible solvents such as methanol, ethanol, isopropanol or tert-butanol. The use of inflammable compounds, however, requires spark-proof equipment and special precautions with regard to fire safety.

The advantage of using enzymes for cell lysis, besides selectivity during the product release, is their use under mild conditions. The lytic enzyme used extensively, including large-scale operation, is lysozyme. This enzyme damages the peptidoglycan layer and, as a result, the internal osmotic pressure of the cells bursts the periplasmic membrane, Gram-positive bacteria are more susceptible to attack by lysozyme, while the lysis of Gram-negative bacteria requires passage of lysozyme through the other membrane, which is aided by the addition of EDTA. Glucanase and mannanase, in combination with proteases, are used for the degradation of yeast cell walls. Availability, costs and the need for removal in subsequent purification steps presently limit application of lytic enzymes. A combination of enzymatic/chemical lysis with mechanical disintegration has been suggested in enhancing the efficiencies of the respective methods, with savings in time and energy and the facilitation of subsequent processing.

Homogenisation of Animal/Plant Tissue

Animal cells are normally easy to break because of the absence of cell walls A typical procedure for homogenising animal tissue is to cut it into smal pieces, suspend in ice-cold homogenisation buffer, and grind in a blendeı Plant cells are more resistant particularly when they have tough cell walls.

The extract of a fleshy or non-fibrous plant tissue is prepared by rapid homogenisation of the material suspended in a partially frozen suitable buffer in a Waring blender pre-cooled to –20°C. The more fibrous material, which is difficult to macerate, is frozen and ground to a dry powder before adding the extraction buffer for homogenisation. Concentrated buffers with pH values around 6.5–7.2 are used in order to neutralise the acidic materials including the phenols, and also contain phenol scavengers such as polyvinylpyrrolidone (PVP) and/or Amberlite, a hydrophobic polystyrene-based adsorbent, and reducing agents such as ascorbate and thiols to prevent the accumulation of quinones and hence the inactivation of enzymes during extraction. Use of cellulases and pectinases for digestion of plant cell walls is an attractive alternative for achieving selective release of the protoplasmic material and not the vacuole contents but is limited on the large scale by the high enzyme costs.

CONCENTRATION OF BIOLOGICAL PRODUCTS

After separating the cells from the whole broth, the filtrate contains 85–98 per cent of water, with the product forming only a minor constituent. Removing large amounts of water is costly. It is done in different ways: evaporation, membrane filtration, liquid-liquid extraction, precipitation and adsorption. The choice of technique is normally dictated by the nature of the product, with minimising the loss of product activity as the main criterion.

Evaporation

Evaporation is a simple, but in most cases, an energy-consuming way of water removal. Because of its reliability and simplicity it is often applied on a large scale, normally using steam as the heat source. In biotechnology, the evaporator must often serve as a multi-purpose equipment, and should be able to handle a broad range of product viscosity (1–10000 mPa), heat sensitive products, and give a minimum of scale formation, fouling and foaming. The basic unit of evaporators comprises a heating section to which the steam is fed, a section where the concentrate and vapour are separated, a condenser for condensing the vapour and the required vacuum and product pumps, control equipment, etc. Several different types of equipment have been developed, ranging from laboratory scale (0.5–1.0 $hr.^{-1}$) water evaporation capacity) to very large industrial scale (150 m^3 $hr.^{-1}$). where the liquid flows through a number of stages (maximum seven) using the vapour of one stage as a heating source for the next.

In falling film evaporators, the liquid to be concentrated flows down long tubes and distributes uniformly over the heating surface as a thin film. The vapours flowing in the same direction increase the linear velocity of the liquid, thereby improving the heat transfer. Residence times in the evaporator are in the order of minutes. These evaporators are suited for concentrating viscous products (up to 200 mPa), and are frequently used in the fermentation industry. Plate evaporators, where the heating surface is a plate as opposed to a tube, have a relatively large evaporation area in a small volume, but the possibility of treating viscous and solids-containing fluids is limited. For higher viscosities, forced film evaporators with mechanically driven liquid films are suitable, in some cases producing a dry product. The residence time ranges from a few seconds to a few minutes. Centrifugal forced-film evaporators permit a further reduction of the residence time so that even the heat labile substances can be concentrated under gentle conditions. Evaporation takes place on a heated conical surface or plates over which the transport of the liquid takes place through the centrifugal force produced by the rotating bowl.

Liquid-Liquid Extraction

Liquid-liquid extraction is applied on a large scale in biotechnology both for concentration and for purification. It involves the transfer of solute from one liquid phase to another. The efficiency of an extraction process is governed by the distribution of substances between the two phases, defined by the partition coefficient K (concentration of substance in extract phase/ concentration of substance in raffinate phase). A partition coefficient well removed from unity is desirable. The physico-chemical properties of the product influence the demands on a liquid-liquid extraction process, as illustrated below:

Extraction of low molecular weight products

Small lipophilic target molecules are extracted using an organic solvent, whereas for hydrophilic compounds it may be difficult to design an efficient extraction process. Extraction in organic solvent can be done in one of three ways:

Physical extraction

The compound distributes itself between the two phases according to its physical preference. This applies to non-ionising compounds and the extraction is optimised by screening for the solvents that would lead to a high K value and also show a maximal difference in K for the different components present in the crude mixture.

Dissociative extraction

Difference in the dissociation constant of the ionisable components is exploited to achieve separation, and these differences are often large enough to overcome an adverse ratio of partition coefficients. Extraction of penicillin and some other antibiotics are typical for this type of extraction principle.

Reactive extraction

A carrier, such as an aliphatic amine or a phosphorous compound, is added to the organic solvent which forms selective solvation bonds or stoichiometric complexes that are also insoluble in the aqueous phase. Thus, the compound is carried from the aqueous to the organic phase. This type of extraction is advantageous for compounds that have a high solubility in aqueous medium, e.g., organic acids.

In most cases, cells and other particulates are removed prior to extraction to avoid the formation of emulsions at the interface. After extraction, the product is recovered from the solvent either by distilling off the product in case of a high-boiling solvent or distilling off the solvent when this is low boiling. If the product is heat sensitive, it is recovered by back-extraction into a new aqueous phase under conditions different from the first extraction e.g., penicillin is extracted into butyl acetate or amyl acetate from the fermentation medium at pH 2.5–3.0 and back-extracted into aqueous phosphate buffer at pH 5–7.5.

For high extraction yields, multi-step extraction in a countercurrent mode is used, which provides also savings in both solvent and time. Different kinds of extraction equipment are available including mixer-settlers, columns and centrifugal extractors. The latter are often used in extraction of antibiotics and steroids; representative examples being the Podbielniak extractor, Delaval contactor and Westfalia extraction-decanter.

Supercritical fluid (SCF) extraction

Supercritical fluid (SCF) extraction has been considered as an alternative technique to conventional extraction which has the disadvantage of toxicity and flammability of organic solvents. SCFs are materials that exist as fluids above their critical temperature and pressure, respectively. Many of the properties of SCFs are intermediate between those of gases and liquids; e.g., their diffusivity is higher than those of liquids while viscosity is lower. The attractive feature of SCFs as extractants is that their solvent properties are highly sensitive to changes in both temperature and pressure, which provides the opportunity of tailoring the solvent strength to a given application. Supercritical CO_2 is most commonly used for extractions because of its relatively low critical temperature (31.3°C) and pressure (72.9 bars). Supercritical extraction is still rather expensive though a few large-scale applications in the area of food processing have demonstrated the possibility of economically viable operations. By adding a co-solvent, such as ethanol, in small amounts, it is possible to modify the properties of the system and thereby influence the extraction behaviour. In a typical extraction process, the compressed SCF is contacted with the feedstock to be extracted in an extraction column, and the loaded SCF is transferred via an expansion valve to a separator. On lowering the pressure, the liquid is turned into gas, releasing the product as a precipitate. The gas is repressurised and recycled to the extraction column.

Extraction of proteins

Aqueous two-phase systems

Aqueous two-phase systems (ATPS), prepared by mixing two different polymers, or a polymer and a salt above certain concentrations with water as the major component (80–95 per cent), present suitable extraction systems for proteins. The two phases formed are enriched in the respective phase components, e.g., polyethylene glycol (PEG) is the main component in the top phase, while dextran or salt constitutes the bottom phase. The interfacial tension between the two phases is significantly lower than that in water-organic solvent systems. Phase separation is also slower, varying between a few minutes and 1–2 hr.; but is often speeded up by centrifugation at low g values. Partitioning of a component in aqueous two-phase systems is based on its surface characteristics, nature of the phase components and the ionic composition. A common observation is that small molecules are more or less equally distributed between the two phases; the partitioning of particulate matter is invariably one-sided and that of macromolecules covers a wide range.

Use of extraction circumvents many of the shortcomings of centrifugation and filtration for large-scale isolation of proteins from crude homogenates having high viscosity and heterogeneous distribution of particle sizes. Other attractive features of the technique are high capacity (biomass to volume ratio) and straightforward scale-up. The process is easily adapted to the extraction equipment used for water-organic solvent systems. Fig. 5.4 shows a schematic presentation of protein isolation and partial purification in ATPS. For industrial scale separations PEG/salt systems are used because of their relatively low cost. Inexpensive and biodegradable polymers may also be considered as replacement for salt in the bottom phase. The selectively of protein extraction in ATPS may be increased by incorporation of a suitable ligand molecule into the extracting phase (by being coupled to the polymer forming the phase) which selectively pulls up the target protein; the latter is subsequently released into a new bottom phase made up of either salt or fresh polymer supplemented with free ligand.

Another extraction system having potential in the downstream processing of proteins is that of reverse micelles which are thermodynamically stable aggregates of surfactant molecules and water in organic solvents. The polar surfactant head groups point toward the interior of the aggregates forming a polar core in which water can be solubilised to generate water pools. The aliphatic chains of the surfactant protrude into the surrounding organic phase. These systems have generated great interest among biotechnologists as they can be used to solubilise enzymes without loss of activity, which could then be used to catalyse many organic transformations. Reverse micellar systems have also shown potential for the separation and recovery of proteins by extraction from aqueous feed. The extraction seems to involve electrostatic interactions between the protein surface and charged surfactant, and is thus dependent on pH and ionic strength.

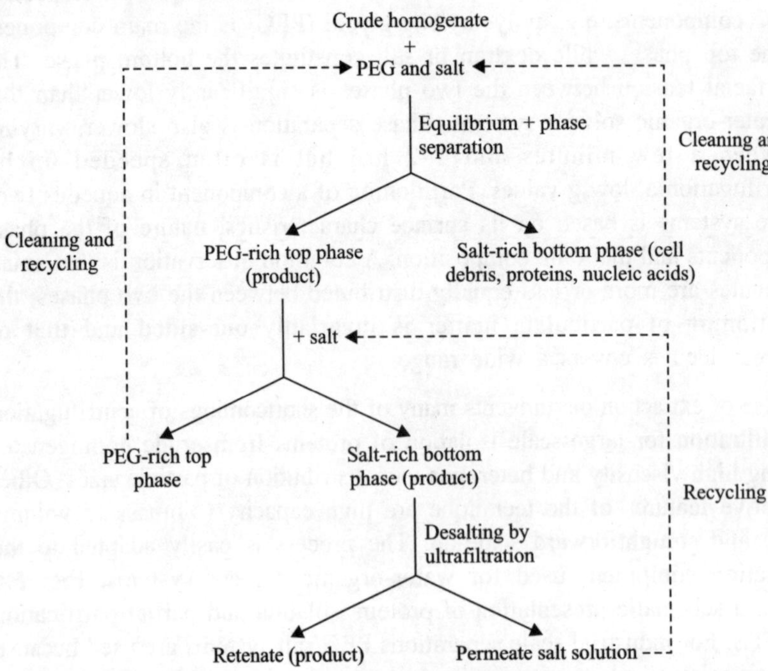

Fig. 5.4. Extraction in aqueous two-phase system. The components of the two-phase system are mixed directly with the crude homogenate.

Membrane Filtration

The use of membrane technology for separation of biomolecules and particles and concentration of process fluids has expanded dramatically in recent years. The membrane function has been made more versatile by integrating it with other separation principles.

Microfiltration and ultrafiltration

During separation, a semi-permeable membrane acts as a selective barrier retaining the molecules/particles bigger than the pore size while allowing the smaller molecules to permeate through the pores. Membrane filtration processes can be distinguished according to the type of force driving the transport through the membrane, and are named by the size of pores in the filter (Table 5.3). Microfiltration is used for separation of particles, typically 0.02–10 μm in diameter, while ultrafiltration separates polymeric solutes in the 0.001–0.02 μm range. Reverse osmosis, or hyperfiltration, separates ionic solutes typically less than 0.001 μm.

Table 5.3. Applications of membrane processes in biotechnology.

Type of membrane	*Driving force*	*Application*
Microfiltration	Hydrostatic pressure	Concentration of bacteria and viruses
	Δp 0.5–2 bar	Harvesting of cells
		Clarification of fermentation broth
Ultrafiltration	Δp 2–10 bar	Fractionation of biomolecules
		Desalting
		Production of enzymes
		Processing of whey
Hyperfiltration	Δp 20–100 bar	Concentration of pharmaceuticals
		Production of lactose
		Part desalination of solutions
Electrodialysis	ΔElectric field	Purification of charged small molecules e.g., organic acids
Pervaporation	ΔPartial vapour pressure	Selective removal of solvents (ethanol, pressure acetone-butanol) during fermentation
		Purification of solvents from azeotropic mixtures
Perstraction	Partition	Extraction of small molecules from aqueous/organic solutions

Microfiltration and ultrafiltration are widely used in the primary recovery stages of downstream processing. Selectivity of membranes, often expressed in terms of molecular weight cut off (MWCO), is mainly determined by the sieve action of the pores, but also hydrophilic/hydrophobic and charge interactions.

Separation with membranes was initially done by dead-end filtration, in which the feed flows on to the membrane. Settling of particles, deposition of colloidal species, adsorption of macromolecular solutes, precipitiation of small solutes, etc., on the surface and in the pores of the filter leads to a deposition of a cake which grows in thickness with time, reducing the flow though the filter. Cross-flow or tangential flow filtration has now completely replaced the dead end filtration for harvesting cells on a large scale. Here, a flow of the feed stream is maintained parallel to separation surface, with the aim to provide sufficient shear force close to the membrane surface, thereby preventing particulate matter from settling on, or within, the membrane structure. In practice, the membranes in cross-flow filtration are also subject to fouling, but the cake thickness is limited to a thin layer as compared to the dead end mode. Although most filtration media are relatively inert, the formation of gel layer is inevitable. The problem of permeate flux reduction can be minimised by optimising the filter selection, operating pressure, flow properties of feed, and frequent back-flushing.

Microfilters and ultrafilters are available in materials such as ceramics and steel that can be aggressively cleaned and sterilised in place. Membranes composed of polymeric materials such as polyvinyldifluoride (PVDF) and polyethersulphone (PES) are also used, but are more difficult to clean and may require chemical rather than steam sterilisation. Membrane filters are commonly plate and frame systems, employing cartridge filters within which the membrane is present in a highly folded format. This gives a large filtration surface area in a compact space with no dead spaces. Another form is the hollow-fibre system, which comprises a bundle of hollow capillaries packed in a tube. The liquid to be filtered is pumped through the central core of the hollow fibres. The permeate passing through the capillary walls can be drained as a permeate from one end of the cartridge, while the concentrated retentate emerges from the other end.

Membrane adsorbers

New micro/macroporous membrane matrices with ion exchange groups and affinity ligands, called membrane adsorbers, have been developed which bind proteins from the clarified feed pumped over them. Desorption of the protein is later performed using solutions as in chromatography. A stack of membranes provides a total surface area for adsorption equivalent to chromatography gels, giving similar high resolution separation as chromatograpic methods. In membranes, liquid transport is by convection as opposed to the diffusional flow in gels which increases the speed of separation tremendously.

Pervaporation

Pervaporation is a membrane based process having potential for recovery and concentration of volatile products, (Fig. 5.5). The major limitations of the technique for large-scale work are high energy consumption, insufficient selectivity of the membrane and difficult process design due to a temperature drop across the membrane.

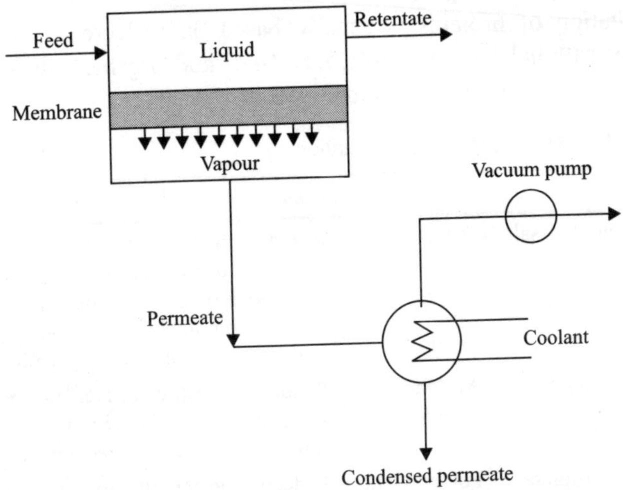

Fig. 5.5. Pervapouration. The process allows separation of volatile products by a combination of permeation through a membrane and evaporation achieved by a low pressure on the downstream side of the membrane.

Perstraction

Perstraction is another sophisticated technique for product concentration and recovery that combines membrane processes and solvent extraction. By using the membrane as a barrier between an aqueous feed and an organic solvent it is possible to transfer molecules that partition into the liquid filling the membrane pores. This technology has been used for recovery of hydrophobic substances during fermentations. The membrane protects the cells from the toxic or inhibitory effects of the extraction solvent. Depending on the direction of mass transfer, the membranes could be hydrophobic or hydrophilic. The hydrophobic membranes, which have pores filled with the organic phase, are used to extract the non-polar product from the aqueous medium. In hydrophilic membranes the pores are filled with a suitable aqueous buffer, which facilitates the removal of the product from the solvent.

Precipitation

Precipitation is an established separation technique in industry for concentration of proteins and polysaccharides. Often it constitutes the only unit operation after solid-liquid separation in the recovery of bulk proteins. Precipitation can also be used to achieve the removal of undesired by-products such as nucleic acids, pigments and other residual components from a crude extract. Precipitation is usually carried out in a batch mode in stirred tanks. Aggregates settle to the bottom of the tank the mother liquor is removed and the aggregate slurry is fed to a centrifuge or a filter.

Precipitation of proteins is usually based on a decrease in solubility induced by external factors (Table 5.4). Salts and organic solvents are the precipitating agents commonly employed in industry.

Table 5.4. Modes of protein precipitation.

Mode	*Example*	*Comments*
Addition of neutral salt	$(NH_4)_2SO_4$	Increased hydrophobic interactions between neutral protein molecules; salt is removed prior to next purification step (except for hydrophobic interaction chromatography) by dialysis, ultrafiltration or gel filtration
Addition of organic solvent	Acetone, ethanol	Reduced dielectric constant enhances electrostatic interactions between protein molecules low temperature required for operation
Addition of non-ionic polymer	Polyethylene glycol	Reduction in the effective quantity of water available for protein solvation; polymer has often a stabilising effect on proteins
Addition of charged polymer	Polyethylene-imine, poly-acrylic acid	Complex formation between oppositely charged molecules leads to charge neutralisation and precipitation
Increase in temperature		Increased hydrophobic interactions; used for precipitation of heat sensitive proteins
Change in pH		Low solubility of protein at isoelectric point; extremes of pH denature and precipitates sensitive proteins

These precipitation processes are non-specific in the sense that they exploit the ionic and hydrophobic interactions, which are common to all proteins. There are a few examples of precipitations that are semi-specific, e.g., protein-carbohydrate complexes have been precipitated by means of borax additions. Selectivity in precipitation has been introduced by use of affinity interactions. Creation of large complexes as a result of affinity interactions, as between antigen and antibody, is one mode of affinity precipitation, which is used in immuno precipitation. This concept has been generalised to some extent for selective precipitation of multimeric proteins (having more than one binding site for a ligand) by using homobifunctional ligands (synthesised by coupling two ligand molecules by a spacer). The modified ligand is able to bridge different protein molecules thereby forming aggregates. The precipitation of the affinity complex occurs only at a definite ratio of the ligand and the protein, Transition metal ions are able to precipitate proteins by bonding with the surface histidine residues, the more the residues the easier is the precipitation.

With heterobifunctional ligands, where one functionality is responsible for the affinity binding and the other is exploited for the precipitation, it becomes possible to operate affinity precipitation in a more general mode. The precipitating component of the heaterobifunctional ligand is a 'smart polymer' which responds to minor changes in an environmental parameter, e.g., pH, temperature, ionic strength, etc., by a visible change is solubility. Examples of such polymers are chitosan, alginate, galactomannan, poly (*N*-isopropylacrylamide), hydroxypropyl-methyl cellulose, etc.

Adsorption to Chromatographic Particles

Another strategy often used for concentrating a particular molecule from a crude extract is capturing on high-capacity solid adsorbent particles. Active charcoal was the first reported adsorbent material for product concentration. Ion exchange resins have since been widely used for initial capturing of low molecular weight products and proteins from crude extracts, because of their high binding capacity, their applicability to harsh cleaning-in-place (CIP) protocols and their relatively low cost. Synthetic adsorbents with hydrophobic adsorption characteristics have also been developed which can be used for extraction of organic compounds. The advantage over conventional solvent extraction is that adsorbent processes require much smaller amounts of the toxic and flammable organic solvents. Additionally, the adsorbents have a molecular sieving function based on their pore structure so that fractionation by molecular size can take place during the process. Likewise, hydrophobic adsorbents may also be used for protein concentration.

A broad range of adsorbent matrices is used for industrial processes. For capture of low molecular weight compounds like antibiotics, vitamins, and peptides, polystyrene, mathacrylate and acrylate based matrices are used. Cellulose-based adsorbents are commonly employed for protein concentration. Owing to the crude nature of the feed, the adsorption is mostly carried out in a batch mode, after which the slurry is transferred to a column for elution of the product. It may be possible to achieve significant purification by employing selective elution. Adsorption from a whole broth is also conveniently done in a fluidised or expanded bed mode (Fig. 5.6). Here, the culture broth is plumed through the resin packed column, equipped with sieve plates, in an upward direction at a velocity high enough to fluidise/expand the adsorbent particles and allowing the particulate matter to pass through. Fouling of the adsorbent by irreversible adsorption of proteins and pigments could be a problem. This necessitates intermittent cleaning of the resin under strong conditions e.g., using NaOH/ propan–2-ol, or sodium hypochlorite.

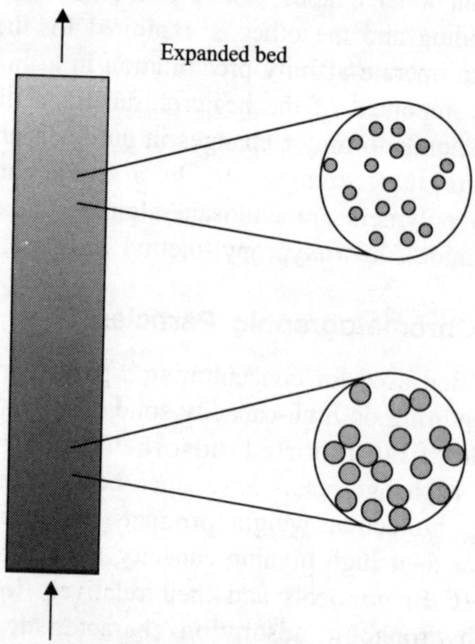

Fig. 5.6. Expanded bed adsorption chromatography. The bed of adsorbent beads in a column is expanded by an upward flow of liquid.

PURIFICATION BY CHROMATOGRAPHY

Chromatography is the technique used for high-resolution purification. The components to be the separated are distributed between a stationary and a mobile phase. The stationary phase, usually composed of uniformly sized particles, is packed in a column and equilibrated with a suitable solvent. The mixture to be separated is loaded on the column followed by the mobile phase. Elution of the components is achieved either in an isocratic mode, i.e., the same mobile phase is maintained throughout and separation of the components depends on differences in retention time in the column, or by gradient elution where the mobile phase is continuously changed to facilitate the release of the components bound to the stationary phase by non-covalent forces. The eluate from the column can be monitored continuously, e.g., by monitoring the ultraviolet absorption at 280 nm for proteins, and is collected in fractions of definite volume using a fraction collector. Various modes of chromatography are used for protein separations, (Table 5.5). These separations may be based on size (size exclusion chromatography), characteristics of surface groups, or recognition properties of protein molecules (adsorption chromatography).

Table 5.5. Various chromatographic techniques for protein separation.

Chromatography	*Separation principle*
Size-exclusion (gel filtration)	Size and shape
Ion-exchange chromatography	Net charge
Chromatofocusing	Net charge
Hydrophobic interaction chromatography	Hydrophobicity
Affinity chromatography	Molecular recognition
Immobilised metal-ion affinity chromatography	Metal ion binding
Covalent chromatography	Content of free-SH groups

Some of the matrices available for protein chromatography are listed in Table 5.6. Ideally, the chromatographic matrix should be: *inert-* to prevent any non-specific adsorption, rigid-to resist compression at high flow rates, chemically stable - to withstand harsh cleansing procedures, *bead shaped* - for good flow properties, and porous—to provide a high surface area and to allow free passage of molecules. The chromatographic materials are both organic and inorganic, synthetic and natural. During the past several years, impressive developments in chromatography media with improved selectivity and physical properties allowing high flowthroughs, have taken place. Fast protein liquid chromatography (FPLC) and high performance liquid chromatography (HPLC), utilising rigid matrices for operation with relatively high pressure, are now common tools for purification of commercial proteins. All the normal chromatographic techniques are commercially available in the HPLC-mode.

Flow-through or perfusion chromatography uses a chromatography matrix based on poly (styrene-divinylbenzene) or agarose or methyl methacrylate containing two types of pores; the big transecting through pores allowing convective flow through the particle and smaller diffusive pores lining the through pores provide a large adsorption surface area. This combination provides a rapid mass transport of sample molecules without sacrificing resolution and capacity, in contrast to conventional chromatography where the sample transport is diffusion-limited.

An innovative design of columns is used in radial flow chromatography in which the sample is applied to the outer wall and allowed to move radially by sideways flow of the eluant. The capacity of the column is thus determined by both length and radius of the column; the latter influencing also the resolution. Scale-up to larger volumes does not need to be done by increasing the bed height, and thus the problem of bed compression of soft gels obtained in conventional downward flow of eluant is overcome.

Table 5.6. Matrices used for large-scale chromatography.

Matrix	*Trade name*
Cross-linked dextran	Sephadex
Cellulose	Whatman TM, Cellufine, Sephacel, Cellex
Agarose	Sepharose, Sepharose CL, Fast flow Sepharose, Ultrogel, Superose
Cross-linked polyacrylamide	Bio gel P
Composite of polyacrylamide and dextran	Sephacryl
Composite of agarose and dextran	Superdex
Composite of agarose and polyacrylamide	Ultrogel AcA
Composite of porous kieselguhr and agarose	Macrosorb KA
Hydroxylated acrylic polymers	Trisacryl
Hydroxyethyl methancrylate polymer	Spheron
Ethyleneglycol-methacrylate co-polymer	Fractogel TSK Toyopearl
Polyacrylamide	Eupergit C
Porous silica	Spherosil, Accell
Rigid organic polymers	Monobeads, TSK-PW
Polystyrene/divinyl benzene	Poros

Size-exclusion Chromatography

This involves partitioning of proteins between the stationary liquid held by pores of the gel particles and the mobile liquid in the void volume between the particles. The gel matrices used for size exclusion chromatography have a defined pore size range to allow small molecules into the pores while the larger molecules are excluded and pass through the column with the mobile liquid. Size-exclusion chromatography is effective only with small sample volumes (equivalent to 2–5 per cent of the total bed volume), and is thus suitably used as a final polishing step in a purification protocol to separate the protein from aggregates, degradation products etc. and even for buffer exchange.

Adsorption Chromatography

In adsorption chromatography, resolution of the macromoleules is a surface-mediated process i.e., there is differential adsorption of the molecules at the surface of the matrix (Table 5.5). The matrices employed are derivatised to contain covalently attached functional groups, which adsorb and separate proteins by different mechanisms.

Ion exchange chromatography

Ion exchange chromatography is by far the most widely used technique because of its general applicability, good resolution and high capacity. Moreover, it is insensitive to sample volumes and is often used in the initial phase of downstream processing to provide both product purification and volume reduction of the process fluid. Compounds are separated according to the difference in their surface charges. Hence, pH of the medium is one of the most important parameters for binding of the target molecule, as it determines the effective charge on both the target molecule and the ion exchanger. Ionic bound molecules are eluted from the matrix, either by increasing the concentration of salt ions which compete for the same binding sites on the ion exchanger, or by changing the pH of the eluant so that the molecules lose their charges. Displacement chromatography is an alternative mode wherein a more heavily charged molecule is used to 'displace' the bound material.

Ion exchangers are grouped into anion exchangers with positively charged groups like diethylaminoethyl (DEAE) and quarternary amino (Q), and cation exchangers which have negatively charged groups like carboxymethyl (CM) and sulphonate (S).

Hydrophobic interaction chromatography

Hydrophobic interaction chromatography (HIC) is also a robust, high capacity method. It may be used early with dilute product streams, leading to concentration and purification. HIC is analogous to reverse phase chromatography (RPC) but relies on comparatively weak hydrophobic interactions between hydrophobic ligands, alkyl or aryl side chains on the gel matrix and the accessible hydrophobic amino acids on protein surface. Differences in the content of these amino acid residues can be used for separation of proteins. Binding is performed in a medium favouring hydrophobic interactions, e.g., a solution of high salt concentration. Elution of bound material is achieved by reducing the hydrophobic interactions, either by lowering the salt concentration, or the temperature, or by decreasing the polarity of the medium (by inclusion of solvents like ethylene glycol or ethanol in the buffer). Matrices with different hydrophobic groups like butyl-, octyl- and phenyl- are commercially available.

Affinity chromatography

Molecular recognition forms the basis of adsorption and separation by affinity chromatography. One of the reactants in an 'affinity pair', the ligand, is immobilised on a solid matrix and is used under suitable conditions to fish out the complementary structure (the ligate). The binding is reversible and can be broken by changing the buffer conditions. Potentially, such methods possess very high resolving power. Table 5.7 shows that a variety of ligands, with specificity for one or a group of proteins, can be used for affinity chromatography. A trend in downstream processing has been to exploit the specificity of affinity interactions earlier in the separation train so as to reduce the number of purification steps; however this puts extra demands on the ligands with respect to chemical and biological stability. This has led to increased interest in the use of pseudo-biospecific ligands like dyes and metal ions.

Table 5.7. Examples of ligands used for affinity chromatorgraphy of proteins.

Ligand type	*Protein type*
Biospecific ligands	
Mono-specific	
Receptor	Hormone
Antibody	Antigen
Hapten	Antibody
Substrate/substrate analogue, inhibitor, cofactor	Enzyme
Group-specific	
Cofactor	Enzymes
Lectins	Glycoproteins
Sugar derivatives	Lectins
Protein A/G	Immunoglobulins
Heparin	Coagulation factors, protein kinases
Pseudobiospecific ligands	
Triazine dyes	Dehydrogenases, kinases and other proteins
Metal ions	Metal ion binding proteins
Hydrophobic groups	Various proteins

The chemical coupling procedure for immobilisation of a ligand is chosen so as to provide satisfactory yields, strong linkage to minimise ligand leakage during chromatographic operation, and minimal nonspecific interactions with biomolecules. The adsorbed protein is generally eluted under conditions that minimise its interactions with the ligand, e.g., by increasing the ionic strength or changing the pH of the buffer, or by a free ligand molecule. In the latter case, a subsequent step would be to separate the protein from the free ligand.

PRODUCT FORMULATION

The commercial viability of a biotechnological product is dependent on the maintenance of its activity and stability during distribution and storage. Low molecular weight products such as bulk solvents, bulk organic acids, etc., are formulated as concentrated solutions after removing most of the water. When high purity is required, small molecules such as antibiotics, citric acid, sodium glutamate etc., are crystallised from solution by addition of salts once they have reached the required degree of purity. Proteins are particularly sensitive to loss of biological activity during downstream processing and subsequently during storage. This could be due to factors such as oxidation, temperature, presence of proteases etc. Protein products are formulated as solutions, suspensions or dry powders. A variety of stabilising additives are included in the formulations in order to prolong the product shelf life. The choice of a suitable stabiliser is done empirically. Among the nonspecific chemical additives used regularly as stabilisers in protein formulations are salts (ammonium sulphate or sodium chloride), sugars (sucrose, lactose etc.), polyhydric alcohols (sorbitol, glycerol etc.) or polymers (polyethylene glycol, bovine serum albumin etc.).

Bulk enzymes are commonly sold as concentrated liquid formulations. It is, however, often preferred to dry the product to decrease the volume as well as the denaturing reactions that are enhanced in aqueous solution. Bioproducts, often being sensitive to heat, require gentle drying methods. Table 5.8 lists some of the commonly used dryers. Depending on the mechanism of heat transfer, these may be broadly classified as contact-, convection-, and radiation-dryers. Batchwise drying in many contact dryers is facilitated using mechanically moved layers. The advantage is the uniform thermal stress exerted on the material being dried, high throughput and possibility for development of continuous processes. A common feature of convection dryers is that the movement of the material to be dried is promoted by a flow of gas. Drying of large streams of product is achieved in a very short time.

Spray drying essentially involves generating an aerosol of tiny droplets by passing the product containing liquid through a nozzle or a rotating atomising disc and directing it into a stream of hot gas. The water in the droplets evaporates leaving behind solid product particles. Spray-dryers are used for drying large volumes of liquids. For smaller quantities, a chamber dryer may be useful, where the product is placed on shelves and the transfer of heat takes place partly by contact and partly by convection. The material will be exposed to severe non-homogeneity in thermal stress, making such a method recommendable only at relatively low temperatures.

Table 5.8. Some of the frequently used drying equipment.

Type of dryer *Movement of the product*	*Mode of heat transfer*
Belt dryer Intensive due to gas flow	Convection
Fluidised bed dryer Intensive due to gas flow	Convection
Spray-dryer Intensive due to gas flow	Convection
Freeze-dryer None or mechanical	Contact and radiation
Drum-dryer Slight mechanical	Contact
Chamber dryer None	Convection and contact

Freeze-drying or lyophilisation represents one of the least harsh methods of protein drying. It is used for the drying of pharmaceutical products, diagnostics, foodstuffs, viruses, bacteria etc. The drying principle is based on sublimation of the liquid from a frozen material. The liquid containing the product is frozen, ideally to a temperature below its glass transition temperature, and subjected to vacuum in a freezedryer. While maintaining the internal vial temperature still below the glass transition value, the shelf temperature is increased to a temperature above zero to promote efficient sublimation of the crystallised water. After the primary drying, the protein cake still retains a significant amount of water which is removed by sublimation during secondary drying by increasing the internal vial temperature. Subsequently, the vacuum is released and the vials are sealed. Under a high vacuum, the heat transfer is solely via contact, and not by convection, therefore one needs to operate at low vacuum for an efficient drying process. Various additives are normally included in the solution prior to freezedrying. These may be compounds for maintaining the activity of the protein in the finished product, to protect the protein during the lyophilisation process (lyoprotectants), to prevent product blow out during lyophilisation (e.g., mannitol) and to enhance product solubility. All these additives influence the glass transition value of the solution and hence the freezing temperature prior to drying.

MONITORING OF DOWNSTREAM PROCESSING

There may be several reasons for monitoring a DSP step. One such reason is a desire to keep control over the presence and concentration of the target molecule which, if monitored directly e.g., by using some kind of a sensor, would enable one to take appropriate measures quickly in the event of an unexpected performance. Furthermore, when the performance of a process is well established, the signal from the sensor may be used for control purposes, thereby increasing the level of automation Continuous monitoring of DSP is also desirable in processes dealing with the production of pharmaceuticals, as a means to facilitate establishing an improved level of documentation, and hence the approval process. Furthermore, implementation of measuring and control in DSP will increase the reliability concerning reproducibility of repeated batches. From a practical point of view, fraction collection and other aspects of sampling and sample handling can be substantially facilitated by on-line monitoring combined with per-set conditions for collection/rejection of the stream. Some of the signals that are used for monitoring DSP-events are listed in Table 5.9. Important features when selecting signals for process monitoring are response-time, selectivity and sensitivity.

Table 5.9. Signals used in monitoring DSP processes.

Measuring principle	*Sensitivity*	*Selectivity*	*Response time*	*Comments*
UV-absorbance	Medium	No	Fast	Commonly used
Conductivity	Low	No	Fast	
pH	Medium	No	Fast	
Molecular size	Low	No	<10 second	
Enzyme activity	Medium/high	Yes	Medium/fast	Needs addition of substrate
Protein monitoring	High	No	Fast/medium	Needs addition of reagents
Biospecific binding reactions	High	High	Slow/medium/fast	Range of new binding reactions

In process monitoring it is important to carry out the analyses under sterile conditions so that infection of the bioprocess does not take place. An often-used method is off-line analyses. Flow injection analysis (FIA) has proven superior in this context. A small liquid sample from the medium to be analysed is introduced into a continuous flow of buffer. The sample is transported over a biosensor or a reaction site and is then analysed in a flow detector.

PROCESS INTEGRATION

Low productivity and the high cost of producing biomolecules in general have been limiting factors in the development of biotechnological processes. As the isolation and purification stage contributes substantially to this scenario, process integration has been suggested as a means to overcome these bottlenecks, wherein two or more different processing stages are integrated into one step. The outcome would be a decrease in the number of necessary steps for the complete process, leading to lower costs and higher product recovery.

Integrating a product isolation stage with the fermentation process is possible for recovery of both small molecules and proteins. *In situ* adsorption of the product is a working example. Integration of extraction in organic solvents with fermentation for in situ recovery of low molecular weight products has also been investigated; it is however limited by the toxicity of the solvent) the producing cells. Extraction using an interface such as a memberane (perstraction) has shown potential to reduce the solvent toxicity problem, but the technique is not yet used for large-scale production.

Integration of two or more separation stages into one unit operation, can lead to tightening of a purification protocol. This has been achieved for protein purification by introduction of technologies such as extraction in aqueous two-phase systems and lately, expanded bed adsorption, which can combine clarification, concentration and significant purification in one step. Direct application of an unclarified sample to a purification method is a significant simplification, especially on large scale, see Figs. 5.4 and 5.6. Another approach to decreasing the number of processing steps is to introduce specificity, e.g., by integrating affinity interactions, into techniques used for primary separation, e.g., as in affinity precipitation, membrane affinity filtration etc.

6

Production of Enzymes

INTRODUCTION

Enzymes have been used both directly and indirectly by mankind for thousands of years. Their initial discovery and use was through serendipitous observations, adoption and continued adaptation. In general the enzymes themselves were expressed through the use of live micro-organisms, for example in the leavening of bread, fermentation of fruit juices, bating of leather, or from crude tissue extracts, for example the conversion of milk to cheese. With hindsight these early discoveries are marvellous examples of the observant and creative nature of our early ancestors. Today, enzymes have many applications in a wide variety of areas of which the general consumer is unaware.

Approximately 75 per cent of industrial enzymes are used for hydrolysis and de-polymerisation of complex natural substances with proteases dominating due to their use in the detergent and dairy industries. Food applications are the largest category. These applications include conversions of starch to glucose and fructose, cheese, wine, beer, baking, flavouring, fruit juices and animal feeds.

Production of enzymes has greatly expanded since the 1960s due to the widespread introduction of fermentation technology and more recently from the introduction of genetic engineering. Recombinant micro-organisms are now becoming the dominant source for enzymes for a wide variety of types. This trend will increase in the future due to the ease of genetic manipulation and the wide variety of enzymes available from micro-organisms found in diverse and extreme environments. Many microbial enzymes have been found and developed to replace existing enzymes from animal and plant origin.

ENZYMES FROM ANIMAL AND PLANT SOURCES

The early sources of enzymes were animal or plant material. Initially their use was local. However, with the advent of slaughter houses and organised agriculture, increases in scale were possible. There still are many efficient enzyme manufacturing processes reliant on these sources. Animal tissue and organs provide excellent sources for some lipases, esterases and proteases, the most notable of which are the rennets, pepsin, trypsin and chymosin. Hen eggs continue to be a good source for lysozyme. Cultivated plants serve as excellent sources of proteases, for example bromelain from pineapple, papain from papaya and ficin from Figs. (Table 6.1).

Table 6.1. Enzymes from plant sources.

Source	*Enzyme*	*Application*
Jack bean *(Canavalia ensiformis)*	Urease	Diagnostic
Papaya *(Carica papaya)*	Papain	Baking, dairy, tanning, meat tenderises, beer haze
Fig *(Ficus carica)*	Ficin	Meat tenderiser
Pineapple *(Ananas comosus)*	Bromelain	Baking (gluten complex reduction)
Horseradise *(Armoracia rusticana)*	Peroxidase	Diagnostic
Almond *(Amygdalus communis)*	β-glucosidease	Research
Wheat *(Triticum aestivum)*	Estarase	Ester hydrolysis and synthesis
Barley *(Hordeum vulgare)*	β-amylase	Baking, maltose syrups
Soyabean *(Glycine max)*	β-amylase	Baking, maltose syrups

Animal sources tend to be more variable than plant sources (Table 6.2). There can be significant differences between the type, breed, age, condition and life history of animals prior to slaughter. Agricultural plants, on the other hand, can be cultivated specifically for the production of enzymes and some uniformity of product ensured.

Table 6.2. Enzymes from animal sources.

Source	*Enzyme*
Calf, bovine, kid, lamb, porcine	Diastase (amylase), pre-gastric esterase, lipase, pepsin, trypsin, phytase, chymosin (rennin), phospholipase
Hen eggs	Lysozyme
Human urine	Urokinase

However all crops are at the mercy of the weather and seasonal, climatic and occasionally political events. Crude tissue from animal and plant sources has to be stabilised, usually by freezing or drying, prior to transportation to enable the collection of quantities large enough for economical batch handling.

There are obvious limitations to relying solely on animal secondary products and agricultural materials. Often materials are not always available consistently and quantities can be limited. These limitations have been adequately addressed by the development of microbial enzymes with similar activities and the practice of large-scale microbial fermentations.

ENZYMES FROM MICROBIAL SOURCES

Micro-organisms are the most convenient source of enzymes. The number and diversity of enzymes is proportional to the number and diversity of micro-organisms. Microbes have been collected from environmental extremes such as hot springs, the arctic, rain forest and deserts. Each biological species has associated specific microbes and therefore the potential spectrum of enzyme activities is large. Genetic engineering techniques have enabled the enzymes industry to increase the fermentation productivity of these enzymes by many orders of magnitude. Even the properties of these enzymes can be altered and improved by protein engineering.

Many enzymes are naturally repressed and can only be expressed under certain culture conditions. These enzymes can be both intracellular, typically in the case of *E. coli* and extracellular as in *Bacillus* species. Commercial strains of non-recombinant bacilli and aspergilli are known to produce enzymes up to a concentration of 20 kg m^{-3}. Recombinant cultures can also produce enzymes at these productivities.

Species-Specific Enzymes

A number a *Aspergillus* strains are prolific producers of many types of enzymes. Due to well-developed fermentation processes, A. *niger* is also widely used as a host for expression of recombinant enzymes. This species is known to produce over 40 different commercial enzymes, the most common of which are shown in Table 6.3.

Similar enzymes can also be produced from different micro-organisms, for example there are seven different microbial sources for glucose isomerase and eight for α-amylase (Table 6.4).

Table 6.3. Bulk enzymes produced by *Aspergillus niger.*

Lipase	*Pectin esterase*
Amyloglucosidase	Cellulase
Pentosanase	Catalase
Protease	α-galactosidase
α-amylase	Insulinase
Phospholipase	β-glucanase
Phytase	Galactomannase
Glucose oxidase	Arabinase
Pectinase	

Table 6.4. Common sources of α-amylase and glucose isomerase.

α-amylase	*Glucose isomerase*
Malted cereals	*Actinoplanes missouriensis*
Animal pancreas	*Bacillus coagulans*
Aspergillus oryzae	*Mycabacterium arborescens*
Aspergillus niger	*Streptomyces murins*
Bacillus amyloliquefaciens	*Streptomyces olivaceous*
Bacillus licheniformis	*Streptomyces olivochromogens*
Bacillus stearothermophilus	*Streptomyces phoenics*
Bacillus subtilis	*Endomyces* spp.
Rhizopus oryzea	

LARGE-SCALE PRODUCTION

The cultivation of micro-organisms is economical on a large scale due to the use of inexpensive media and short fermentation cycles. The physiological state of the micro-organism in a fermentation process can be well controlled and the uniformity of each batch ensured. The harvest can be conveniently scheduled to fit in with the downstream processing. The choice of enzyme to be fermented is easy to schedule and different enzyme production campaigns can be planned and adjusted to meet sales demands. Enzyme productivity can be increased many-fold by both conventional strain improvement and fermentation process development and with the additional use of genetic engineering, several orders of magnitude of improvement can be realised within a relatively short period of time (one to two years). Recombinant DNA techniques have also opened up opportunities for the mass production of enzymes from other microbial cultures, which conventionally were fastidious growers, required expensive media or inducers or were potential pathogens. Enzymes from extremophiles (growing at the extremes of temperature, salinity, pressure, alkalinity) can now be conveniently grown in mesophilic cultures, yet produce enzymes with the benefits of temperature resistance or high salt tolerance.

Recombinant *E. coli* Fermentation

Enzymes originating from prokaryote sources can be conveniently produced on a large scale at high productivities in recombinant *E. coli* hosts. These fermentations can be carried out at a 3000 to 60,000 litre scale and do not require complex inoculum build up. A typical fermentation protocol could be as follows:

The *E. coli* host would harbour a plasmid with the DNA coding for the required enzyme, together with a suitable antibiotic resistance marker such as ampicillin or neomycin and an inducer such as TAC (codon specific to lactose or isopropylthiogalactose induction). Satisfactory high production can be achieved by a fed-batch fermentation where the initial batch medium contains components for initial growth e.g., glucose (2 per cent), yeast extract (1 per cent), phosphate (1 per cent) and other salts together with the chosen antibiotic. After the initial growth has been established, further nutrients are fed at pre-determined rates to provide a readily available supply of carbon and nitrogen. Carbon is conveniently supplied as glucose and nitrogen can be either of a complex nature e.g., yeast extract, casein hydrolysate or corn steep liquor (all providing amino acids), or as a simple ammonium salt or urea. Growth on amino acid mixtures provides faster growth and enzyme production, however, the use of simpler, more defined nitrogen, although it may require a longer fermentation period, can support similar high cell growth and enzyme producitivities. Use of defined media is a benefit to downstream processing particularly when ultrafiltration and ion exchange purification are involved. In either case the chosen antibiotic is included in the feed in order to prevent any shedding of the plasmid. Enzyme production is conveniently induced by addition of isopropylthiogalctoside to 30–300 mgl^{-1}, or lactose to 1–10 gl^{-1}. Timing and frequency of addition of the inducer are very important for maximum enzyme expression. Optical densities over 200 OD600 nm (50 gram dry cell weight $litre^{-1}$) and enzyme expression at 10 gl^{-1} (30 per cent of total protein) are desirable fermentation targets. Fermentations are normally run for 2 days at temperatures lower than the organism's optimum growth temperature, i.e., 24–28°C.

Fungal Fermentation

Fungal hosts, such as *Aspergillus* and *Fusarium* and the methylotropic yeast, *Pichia*, are suitable for the production of glycoylated enzymes i.e., originating from fungal or animal sources. The enzyme DNA is integrated directly into the chromosomal DNA together with a promoter system.

The recombinant hosts can be fermented in a similar manner to non-recombinant cultures and high cell density fermentations can easily be obtained using conventional bioengineering approaches. *Aspergillus* or *Fusarium* hosts can be grown on inexpensive raw materials such as soya flour or Pharmamedia with the controlled feeding of corn syrups. Yeasts, such as *Saccharomyces* and *Pichia*, can be grown on yeast extract or protein hydrolysates with corn syrup feeding. Cultures can be grown to high cell mass using defined media, thus providing alternate less expensive nutrients which can have benefits in down stream processing. Fermentations are typically for periods of 4–8 days. Enzyme expression in excess of 10 g l^{-1} have been reported at the industrial scale. Enzymes are usually produced extracellularly which is a benefit as mechanical cell breakage is not needed.

Microbial Enzymes Replacing Plant Enzymes

Some industrial enzymes continue to be extracted from bovine sources which pose the danger of contamination with bovine spongiform encephalopathy (BSE). For example rennin, obtained from the stomachs of newborn calves, continues to be used for cheese making. It remains to be seen whether this presents any health risk to consumers, however, recombinant calf rennin can now be produced by microbial fermentations. Alternatively, with the convenient isolation of enzymes from microbial sources and the use of appropriate screening technology, many new animal and plant-like enzymes are now available. For instance, *Mucor* protease enzymes have in many instances replaced rennin (chymosin) from calf stomach; *Aspergillus* and *Bacillus* proteases have replaced the faecal enzymes used in leather bating and a wide variety of microbial amylases have been developed to replace or supplement the plant amylases in malted barley and wheat. In addition some enzymes with mammalian origins, e.g., tissue plasminogen activator, can be produced in recombinant microbial hosts provided protein chain refolding can be attained. Glycosylated mammalian enzymes can be expressed in eukaryotic organisms such as *Saccharomyces* and *Aspergillus*. Here the host can glycosylate the enzyme to provide full activity even though the glycosylation sugars may be different from the mammalian source (also see chapter 5). Enzymes can be produced directly by mammalian cell culture, however, the manufacturing cost will be very high.

BIOCHEMICAL FUNDAMENTALS

Microbial enzyme fermentations can be improved by classical techniques similar to those used for productivity improvements in antibiotic fermentations. Improvements both at the genetic level and at the process level have led to the development of fermentations capable of producing enzyme protein up to 20 mg m^{-3}. An understanding of the genetic regulation of enzyme synthesis has been very helpful in selecting for improved strains and optimising the fermentation processes. There are many important factors that can influence the production of enzymes.

Enzyme synthesis if normally repressed, a condition that helps conserve energy from unnecessary protein synthesis, i.e., the enzyme will only be produced in the presence of an inducer, normally its substrate. The level of induction can be very strong (a more then 1000-fold increase over non-induced conditions) and acts by interfering with the controlling repressor. Many catabolic enzymes are inducible. For example;

1. Sucrose is needed for invertase production.
2. Starch for amylase production.
3. Galactosides for β-galactosidase production.

In some instances a product or intermediate can act as an inducer:

1. Phenylacetate induces penicillin G amidase.
2. Fatty acids induce lipase.
3. Xylobiose induces xylanases.

Product induction is common in the synthesis of extracellular enzymes required for the hydrolysis of large polymers that otherwise would not have the ability to enter the cell and cause the induction. Co-enzymes can act as inducers, i.e., pyruvate decarboxylase is induced by thiamine. In addition to being effective in enzyme production, induction can be useful in controlling the timing of enzyme production in the fermenter, i.e., a late rapid induction for an enzyme that is unstable under fermentation conditions. However, in practice, induction does often necessitate the handling of expensive inducer compounds, which have to be sterilised and added at specified times to established fermentations. To avoid these problems regulatory mutants can be produced in which the inducer dependence has been eliminated and are thus called constitutive mutants.

Feedback Repression

Enzyme synthesis is also controlled by feedback repression. This occurs particularly in enzymes involved in the biosynthesis of small molecules where the accumulation of the final product can cause the repression of the synthesis of particular enzymes, normally the first enzyme in the biosynthesis route. Mutants lacking feedback repression can be obtained by selecting for cultures resistant to the toxic effects of an analogue of the product or intermediate. These survivors have lost the feedback sensitivity towards the product and its toxic mimic. Similar mutants can be obtained by isolating nutritional auxotrophs where the culture cannot make the final product, but instead depends on the addition on this compound for normal growth. The controlled feeding of this nutrient will limit the intracellular concentrations to below feedback repression levels.

Nutrient Repression

Enzyme synthesis can also be controlled by nutrient repression typically by carbon, nitrogen, phosphate or sulphate. These mechanisms exist to conserve the production of unnecessary enzymes. Thus the cell only produces enzymes for the assimilation of the most easily metabolised or most readily available form of nutrient. The best known example is the control caused by the presence of glucose where this carbohydrate can effectively shut down the production of enzymes involved in the metabolism of other related and non-related compounds. Glucose catabolite repression can be very strong and can often suppress the inducer effect. Other carbon sources, such as lactate, pyruvate, succinate and citrate, are also effective repressors in some micro-organisms. Citrate can even repress the metabolism of glucose in some bacteria.

Glucose repression can be of major concern in the large-scale cultivation of cell mass as it is the most cost-effective raw material for fermentations. The effects of glucose can often be reduced by the limited feeding of the carbohydrate in such a manner that maintains practically zero concentrations of glucose in solution. If this approach fails alternative carbon sources can be used.

Glucose catabolite repression can also be solved genetically by selection of mutants resistant to this phenomenon. Mutants can easily be selected from media containing glucose and the substrate of the required enzyme, e.g., a glucose/aspartate mixture will select for aspartase producers with no glucose repression provided the aspirate is the only source of nitrogen. Penicillin G amidase production in *E. coli* has been increased many-fold by selecting for mutants capable of growth on an amide as sole nitrogen source in the presence of glucose. The resulting hyper-producers are constitutive and not subject to glucose repression. The glucose analogue, 2-deoxyglucose, is also a similar and an effective way of selecting for mutants with no glucose repression.

NH_4^+ is a much utilised source of nitrogen. It can however, have strong control over the metabolism of many amino acids and other complex sources of nitrogen. This can often present problems in large-scale fermentations where both ammonium salts, amino acids and proteins are used as nitrogen sources. Ammonium salts are very inexpensive sources of nitrogen compared to amino acid mixtures, but the presence of the latter can often lead to rapid vigorous microbial growth. Some mutants resistant to nitrogen source repression can be selected by using the ammonium anti-metabolite, methyl ammonium.

GENETIC ENGINEERING

The ready availability of recombinant techniques has had a profound efect on the microbial production of enzymes. The host *E. coli* is very suitable for high enzyme expression provided the enzyme is not glycosylated. Bacillus species are suitable for the production of extracellular non-glycosylated enzymes. *Aspergillus* species are also very prolific producers of extracellular proteins and in addition can produce glycosylated enzymes.

Several properties should be considered in the construction of recombinant bacteria suitable for large-scale enzyme production. The choice of host is critical in that it should have a known lineage, have vigorous growth, have no auxotrophy and possess no enzyme systems that would be undesirable in the final enzyme product, i.e., no β-lactamase activity in a host used for the production of penicillin G amidase. The plasmid construct should be as simple as possible.

After synthesis the recombinant enzyme can either accumulate in the cytoplasm or be transported to the periplasm. Over-production of this foreign protein in the cytoplasm can sometimes cause the formation of inclusion bodies, which generally contain the recombinant protein in an unfolded, or wrongly processed, state. This is often undesirable as it is difficult, though not impossible, to extract and refold these proteins. Extraction from inclusion bodies is only cost-effective for the production of high cost proteins and enzymes such as insulin and tissue plasminogen activator. It is often more desirable to have the protein transported into the periplasm. For this to occur, signal or secretion sequences are required.

These are located at the amino terminal end of the enzyme (upstream) and can be conveniently removed by signal peptidases. Excretion into the extracellular medium in *E. coli* is limited to certain proteins. For the production of extracellular enzymes, *Bacillus* spp., yeasts and *Aspergillus* spp. are the hosts of choice.

Plasmids should be present in high copy numbers in the host and, in order to do this, some selection pressure is necessary as a host cell without plasmid would outgrow a host with plasmid. The most convenient way of ensuring plasmid retention is the placement of an antibiotic-resistant gene on the plasmid and maintenance of the relevant antibiotic in the growth medium. Ideal systems are kanamycin phophostransferase, ampicillin β-lactamase and chloramphenicol acetyltransferase. Antibiotic concentrations of above 30 mg l^{-1} are adequate to maintain the presence of the plasmid. Expression of recombinant enzymes in fungi such as *A. niger* generally involves the integration of the gene and a fungal promoter together with an associated selection mechanism (antibiotic resistance or a wild type gene to overcome a host auxotrophic factor) directly into the chromosomal DNA. Once a stable construct has been selected it is not longer necessary to have the selective agent present (unlike the necessary plasmid retention in *E. coli*). Often multiple copies of the gene can be inserted. Constitutive promoters, such as phosphoglycerate kinase and glyceraldehyde phosphate dehydrogenase, are very powerful and are a popular choice. Promoter genes are inserted upstream of the desired enzyme gene for optimum production, e.g., a starch promoter can be used as a convenient signal for α-amylase production, or phenoxyacetic acid promoter can be used for penicillin V amidase production. Often the inducer is unrelated to the produced enzymes, e.g., methanol is used to stimulate the alcohol dehydrogenase promoter in *Pichia*, which can be coupled to many different enzymes. The inducer of choice has to be maintained above a critical level, however the timing of addition is not critical.

Site-Directed Mutagenesis

Enzymes which have known amino acid sequences and three dimension structures can be altered by site-directed mutagenesis. Amino acids in the active site or other important areas can be identified as targets for changing to other amino acids. This can be performed easily by changing the specific trinucleotide codons in the enzyme gene and expressing the altered enzyme by conventional cloning. Once the beneficial change has been selected the process can be repeated to create further amino acid changes in the enzyme. The new enzymes are easy to scale-up using the same production procedures developed for the parent wild type as these amino acid changes have no effect on the growth of the host or protein expression. Well-known such amino acid changes are increased stability, changed substrate preference, resistance to oxidation, tolerance to solvents and alkali and changes in chiral activity.

RECOVERY OF ENZYMES

Enzyme recovery and purification are as important to the economics of production as the fermentation stages. The main challenge in the recovery steps is to minimise losses in enzyme activity.

Recovery of Extracellular Enzymes

Extracellular enzymes are relatively easy to recover and purify. They can represent a major portion of the total extracellular protein and often simple cell removal and concentration of the active solution can yield enzyme preparations directly suitable for some applications. Relatively clean enzymes preparations can be obtained by cultures growing on simple defined media. The recovery and purification steps are the same as those used for intracellular enzymes, once the cells have been broken.

Recovery of Intracellular Enzymes

For preparation of intracellular enzymes from animal or plant sources, the tissue has to be disrupted to release the enzyme. Detergents or surface-active agents may have to be used to dissociate enzymes that are membrane bound. Tissue drying can be a convenient method of stabilising and disrupting animal and plant tissue. Freeze-drying is the least disruptive and avoids protein degradation, although it is prohibitively expensive on a large scale. Tissue can be air- or vacuum-dried, or subjected to water-miscible solvent precipitation. Enzyme extraction from dried materials can be as simple as rehydration in an appropriate buffered solution. Simple freezing can break some tissue although this is not a convenient method for scale-up. Some plant and animal materials require tissue homogenisation where the tissue is shredded and blended by a mechanical means to break open the cells. Once the enzyme has been solubilised, the residual cell debris can be conveniently removed by filtration. Low speed centrifugation can also be used. In any tissue with fat present, the removal of the fat layer can be problematic for centrifugation. Fatty material can be easily removed by solvent extraction; acetone precipitation is a convenient way to remove protein from lipid materials. As an alternative a water-immiscible solvent, such as hexane, can be used.

The breakage of microbial cells presents greater problems than with animal or plant cells. The cells are often tougher and sizes range from 0.2–10 microns, often necessitating specialised breakage. Many enzymes from yeast and Gram-negative bacteria can be extracted by the use of water immiscible solvents, such as toluene or chloroform, which can cause disruption of the cell membranes resulting in enzyme leakage. Water-soluble solvents, such as ethanol and 2-propanol, can be used to extract enzymes from the periplasmic space, but the use of these solvents can provide handling and safety issues on a large scale. Suitable detergents can be used in extracting small enzymes (less than 70,000 MWt). In some cases where enzymes are stable in the presence of high pH, alkali can be used to lyse bacterial cells. Cell walls of bacteria can also be lysed by the use of lysozyme, however the cost of this enzyme can be prohibitive. Similarly, yeasts can be lysed by β-glucanases. Autolysis can be effective in yeasts, however, long periods of time are required and it is difficult to control or optimise the process.

Microbial cells can be broken effectively by physical means. Sonication is convenient and effective at the laboratory scale. This process, however, cannot be scaled up. High pressure extrusion equipment such as the French Press is very effective at treating small quantities, i.e., 10–100 ml at the laboratory scale. This process of cell breakage uses high pressure to extruded material through a small orifice to impact abruptly onto a plate, thus causing very high sheer. This principle has been scaled up, e.g., Gaulin, Microfluidics, to equipment capable of breaking cell cultures at 100–1000 litres per hour. This is currently the method of choice in the industry.

Many types of cells can be broken by rapid agitation or blending with small glass or ceramic beads. At the laboratory scale this principle is available as a 'bead beater'. At a larger scale ball mills are available. With the appropriate choice of bead size and retaining mesh screens a ball mill can be a very effective, continuous process for cell disruption.

ISOLATION OF SOLUBLE ENZYMES

It is critical to separate the soluble enzyme quickly and efficiently from the remaining cell debris. Chilling, use of an appropriate buffer and the presence of enzyme protectants such as mercaptoethanol may be necessary to stabilise some enzyme preparations. Often protein inhibitors have to be added to reduce the destructive effect or proteases. Soluble enzymes can be collected by membrane filtration or by centrifugation–the latter has been used since the turn of the century. It is relatively easy to perform at a laboratory scale, where high centrifugal forces can easily be obtained. The same high *g* forces cannot be attained on large-scale equipment. Large centrifuges are often constructed from expensive titanium alloys to withstand the high *g* forces, but even then only moderately high *g* forces can be obtained. To combat this problem protein- and nucleoprotein-precipitants, such as polyethyleneimine, can be added to the cell homogenates to flocculate unwanted materials and aid the settling times.

Filtration, with the use of filtration aids, such as a diatomaceous earth, can be a convenient way to obtain soluble enzymes. Scale-up is easy. Additives can be used to speed up the filtration. Often several precipitating agents can be used together, in order to reduce subsequent processing steps. With washing, produce recoveries of 95 per cent can be obtained. Disposal of the filter aid can be an issue. Filter aids can be used for the collection of the initial whole cells provided a cell breakage technology is used that does not rely on mechanical homogenisers. For example, the cells in the filter aid mix can be lysed chemically, enzymically or physically by simple agitation due to the abrasive nature of diatomaceous earth. The released soluble enzymes can be conveniently collected by simple filtration.

Ultrafiltration is a very versatile technology, easily scaled up and, with the correct choice of membrane porosity (molecular weight cut-off), enzymes can be collected selectively according to their molecular weight range. Plate and frame and hollow fibres have been used for 20–30 years. Ceramic membranes are now becoming more popular due to their ease of cleaning and sterilisation, as they can tolerate high temperatures under high alkaline and detergent conditions.

Often the soluble enzyme solution requires no further treatment except concentration, either under reduced pressure or by membrane filtration, to produce a concentrated protein solution (10–50 per cent solids). Stabilisers such as ammonium sulphate can be added if needed. Excipients such as lactose, dextrins can be added to provide bulk and act as stabilisers. Other additives are added to reduce dust formation.

ENZYME PURIFICATION

Enzyme purification may be necessary but only if the extra cost is justified by the enzyme's application. The scale of the purification process will dictate the choice of separation technology, as some are difficult to practise on large scale. The same separation medium can often be used in different techniques, e.g., ion exchange chromatography is best practised as gradient elution from columns at the smaller scale and by batch adsorption/elution at the larger scale.

Enzyme Precipitation

It is often desirable to have an initial batch precipitation stage in order to reduce the total volume of the enzymes solution. Simple precipitation can be carried out, sequentially or in combination, with ammonium sulphate, sodium sulphate, polyethyleneimine and polyallylamine, or organic solvents such as isopropanol, ethanol and acetone. Streptomycin sulphate, polyethyleneimine and other polyamines can precipitate acidic nucleic acid and nucleoproteins. Ammonium sulphate and organic solvents can selectively precipitate the enzyme of choice with 80–90 per cent activity recovery. Simple precipitation can have the additional benefit of protease removal. The main benefit is the ability to reduce the total operational volume, often by a factor of 20. This will significantly reduce the volume of resins and the associated downstream handling used for further processing. Changes in pH and temperature can also result in the selective precipitation and removal of unwanted enzyme proteins.

Liquid-liquid partition with polyethylene glycol, dextrans or polyamines can produce separate liquid phases into which desired enzymes can be concentrated. These treatments involve the use of large quantities of additives and recovery of the polymers is desirable from a cost point of view, although recycling can reduce the initial selectivity and yield a cruder separation. These methods are easily applied to whole broths where the enzyme is soluble, or to extracellular enzymes, as the removal of the biomass is avoided.

Separation by Chromatography

Chromatographic separation techniques commonly used, include ion exchange, hydrophobic interaction, size exclusion, affinity and dye ligand chromatography. Ion exchange chromatography is the most widely used and can encompass the use of strong and weak anion and cation exchange and phosphate interaction. Weakly bound ions are more suitable for enzyme adsorption as stronger ionic charges can denature enzymes due to extremes of local pH. Enzyme proteins are generally adsorbed at low ionic strengths or at pH values where their total ionic charge is strong enough to interact with the opposite resins charge. Protein elution is easily accomplished by changing the pH or by increasing the ionic strength of the eluting solution. Such resins can be used either as a means to actively adsorb and selectively elute the desired enzyme, or by simple adsorption of unwanted protein materials.

Hydrophobic interaction chromatography uses the surface hydrophobicity of the enzymes as the selection property. Protein solutions are loaded on the resin at high salt concentrations and enzyme separations occur when decreasing salt gradient is applied to the resin. This type of separation is useful as a follow-up to ammonium sulphate precipitation as salt removal is not needed prior to this chromatographic step. Hydrophobic resins are an order or magnitude more expansive than ion exchange resins.

By using these adsorption and desorption techniques the volume of the enzyme solutions can be greatly decreased and enzymes of up to 90–95 per cent purity produced. This quality is sufficient for most specific enzyme applications. The chromatographic techniques can be chosen to remove other contaminating enzymes that would otherwise interfere with the use of the desired enzyme. Further clean up can be achieved using size exclusion chromatography. This, however, is difficult to perform on a large scale. It is often not necessary to purify an enzyme to 99 per cent plus purity. If greater than 95 per cent purity is required further adsorption/desorption chromatography can be used taking finer and more selective cuts and by using shallower eluting gradients. Such techniques do not give high activity yields, a factor that has to be taken into consideration when making very pure enzyme preparations for sale. Enzyme preparations of high purity are generally lyophilised for storage and transport. Polyhydroxy cryoprotectants, buffers, reducing agents and anti-microbial agents can be included if necessary. Alternatively enzymes can be stabilised as ammonium sulphate suspensions.

IMMOBILISED ENZYMES

Over 95 per cent of enzymes are sold in soluble forms, the majority of which are used directly on a single use basis, particularly in the food and beverage industries. For certain applications it is more desirable to have the enzyme in an immobilised form so that it can be re-used and easily removed from the reaction mixture. This is particularly important in the pharmaceutical industry where an enzyme step can be have the benefits of chiral resolution and environmental positive processing. The easy removal of the enzyme catalyst ensures better product recovery and minimum carry over of residual enzyme protein. Generally, bulk enzyme manufacturers do not manufacture immobilised enzymes. It is more cost-effective for these catalysts to be manufactured in-house using proprietary supports and immobilisation techniques.

Immobilisation Techniques

There are wide range of techniques for enzyme immobilisation. The correct choice depends upon the final application of the catalyst; whether the enzyme step is in an aqueous or solvent medium, at the beginning or at the end of a synthesis or hydrolysis process and the type of reactor. Adsorption onto an inert support is the simplest technique, however, such supports do not permit permanent adsorption when used in aqueous environments without further modification. Lipases can be effectively adsorbed onto hydrophobic supports for repeated use in non-aqueous lipid systems. Adsorbed enzymes can be permanently fixed by cross-linking reactions with glutaraldehyde treatment, which can make them re-usable in aqueous solutions. Cross-linking with glutaraldehyde onto inert supports is a common technique. The choice of inert support will dictate the final physical form of the immobilised catalyst, which in turn must be compatible with the reactor choice. Supports such as cellulose or diatomaceous earth produce a product that is suitable for stirred reactor use and recovery by filtration. Enzymes adsorbed onto beaded supports are more suitable for column reactors or stirred reactors where simple settling is sufficient to remove the final reactants from the support catalyst.

Immobilisation Supports

Many inert and activated supports are available for enzyme immobilisation. Inert supports, such as resins, porous glass beads, cellulose, gelatin, alginates or chitosans, require chemical activation or derivatisation prior to immobilisation; treatments that can add substantial costs to the final process. However some activated supports can produce high enzyme loading with a high retention of activity. Many chemically activated supports, predominately agarose-based, are available for laboratory research. Their use at the large scale is prohibitive due to very high costs, with the possible exception of antibody-immunological applications. Chemically activated supports, such as Eupergit, are available at the large scale and are priced competitively for some commercial applications.

Enzymes can also be immobilised at high activity recoveries by simple entrapment where the enzyme is mixed with a soluble material that can be easily solidified. Such reactions include cellulose acetate precipitation from an organic solvent to an aqueous change, acid to alkali changes with chitosan polymers, temperature reduction with gelatin, sodium to calcium ion change with alginates, chemical polymerisation with polyacrylamides. The resulting entrapped enzyme is in a fibre or beaded form. Enzymes in these forms have been used industrially for many years for the resolution of amino acids and chiral separations.

Legislative and Safety Aspects

Although enzymes are natural products and their past use in food processing generally proves them safe for consumer use, there is a community and government awareness of the potential hazards associated with enzymes. This perception arose from the rapid introduction of enzymes as detergent additives in the 1960s, which had the unfortunate negative consequences of allergic reactions on inhalation and skin rashes. The collapse of the detergent enzyme business left manufacturers with excess fermentation capacity, which stimulated diversification and development of other enzymes, which can now be regarded as non-traditional. Further diversification occurred with the introduction of genetic engineering techniques where large quantities of enzymes from a wide variety of sources could be easily manufactured. In response to government regulatory concerns, the industry has been proactive and has reduced and eliminated the exposure of workers and consumes to their products. Many enzymes are now sold as concentrated solutions, or as pelleted solids to avoid dust formation.

Many enzymes are GRAS (Generally Regarded as Safe) listed provided they are manufactured under Good Manufacturing Practices (GMP) and many are now made using recombinant hosts such as *Bacillus, E. coli*, yeasts and *Aspergillus*. If not GRAS, these micro-organisms are listed under the hazard group 1; most unlikely to cause human disease, the lowest category of microbial culture apart from those actually cultivated for human consumption, i.e., yeast.

Enzymes from animal and plant sources do not require toxicology studies. Enzymes from non-pathogenic micro-organisms are usually subjected to short-term toxicology, however, those from less well-known micro-organisms do require extensive testing at least in rodents. With the cultivation of any micro-organism there is always the potential danger of the inadvertent production of a toxic metabolite. The number of known mycotoxins, exotoxins and endotoxins continues to grow annually. The quantification of risk assessment associated with this potential is very difficult. However, with controlled fermentations and standardised extraction and recovery procedures, all operated under GMP, these risks can be minimised. In most food use, enzymes are only added at low levels, typically much lower than a variety of other components of biological origin.

In non-food applications such as chemical synthesis and hydrolysis, most enzymes are used in intermediary steps and the resulting products can be subjected to purification and detailed quality and impurity analysis. In the rare case where the enzyme is used as the last step, analytical procedures must ensure than any residual enzyme protein levels meet pre-determined minimum levels in the final product. All bulk enzyme products sold today come with a detailed Material Safety Data Sheet which outlines all potential dangers and describes recommended handling procedures for using the enzyme.

7

Biological Fuel Generation

INTRODUCTION

There is now a growing appreciation of biological energy systems and that biotechnological advances in this area may soon bring economic reality to selected processes. Biomass such as forest, agricultural and animal residues and industrial and domestic organic wastes can now be converted by physico-chemical and/or fermentation processes to clean fuels and petrochemical substitutes. As fossil fuel resources are depleted and become increasingly expensive, conversion of organic residues to liquid fuels becomes a more economically attractive consideration. Photosynthetically derived material is not generally in a sufficiently dry state to possess an attractive calorific value, nor in a form that is best suited to modern technology.

The total economicially recoverable world reserves of the three main fossil fuels, namely coal, natural gas and oil, applying current technology and assuming continued consumption of present-day rates, are respectively less than 1000 years, 35 years and 16 years. Modern industry is almost totally dependent on these limited supplies. Approximately 93 per cent of fossil fuel consumed throughout the world is for energy production, with only 7 per cent used by industry for the production of solvents, plastics and a host of other organic chemicals.

The continual depletion of global fossil fuel energy has generated an ever increasing need to seek out alternative sources of energy. These have so far included the harnessing of hydro, tidal, wave and wind power, the capture of solar and geothermal energy supplies and the much misunderstood, but most significant, nuclear power. With all of these systems there is yet no definitive answer on both the economic and energetic outlay necessary for successful operation. However, it cannot be doubted that fossil fuels will disappear completely in the not too distant future.

PHOTOSYNTHESIS - THE ULTIMATE ENERGY RESOURCE

Photosynthetic organisms, both terrestrial and marine, can be considered as continuous solar/energy converters and are constantly renewable. Per annum, plant photosynthesis alone fixes about 2×10^{11} tonnes of carbon with an energy content of 2×10^{21} joules, which represents about 10 times the world's annual energy use and 200 times our food energy consumption. The magnitude and role of photosynthesis has gone largely unappreciated because we use such a small proportion of the fixed carbon. Let it not be forgotten that photosynthesis in the past provided all the present fossil carbon sources. Thus photosynthetically derived biomass that exists in many available forms in the environment could well be transformed into storable fuels and chemical feedstocks such as alcohols and methane gas. The actual efficiency of solar energy capture by green plants can be as much as 3-4 per cent, the more effective photosynthetic plants such as maize, sorghum and especially sugar cane being the most productive.

The current 'energy crisis' that is reverberating throughout the world has focused attention on the finite nature of fossil fuel reserves. This, taken in association with the dramatic increase in industrialisation in many Third World countries, has generated growing economic and trade pressures for cheaper and more reliable supplies of energy. The only alternative regenerable supply of feedstocks for the chemical industry will be from the products of photosynthesis, i.e. sugar starch and lignocellulose.

Biomass can be considered as a renewable energy source, and can be converted into either direct energy or energy-carrier compounds by direct combustion, anaerobic digestion systems, destructive distillation, gasification, chemical hydrolysis and biochemical hydrolysis.

SOURCES OF BIOMASS

There are three main directions that can be followed to achieve biomass supplies (Fig. 7.1)

1. Cultivation of so-called energy crops.
2. Harvesting of natural vegetation.
3. Utilisation of agricultural and other organic wastes.

The conversion of the resulting biomass to usable fuels can be accomplished by biological or chemical means or by a combination of both. The two main end-products are methane or ethanol, although other products may arise depending on initial biomass and the processes utilised, e.g., solid fuels, hydrogen, low-energy gases, methanol and longer-chain hydrocarbons.

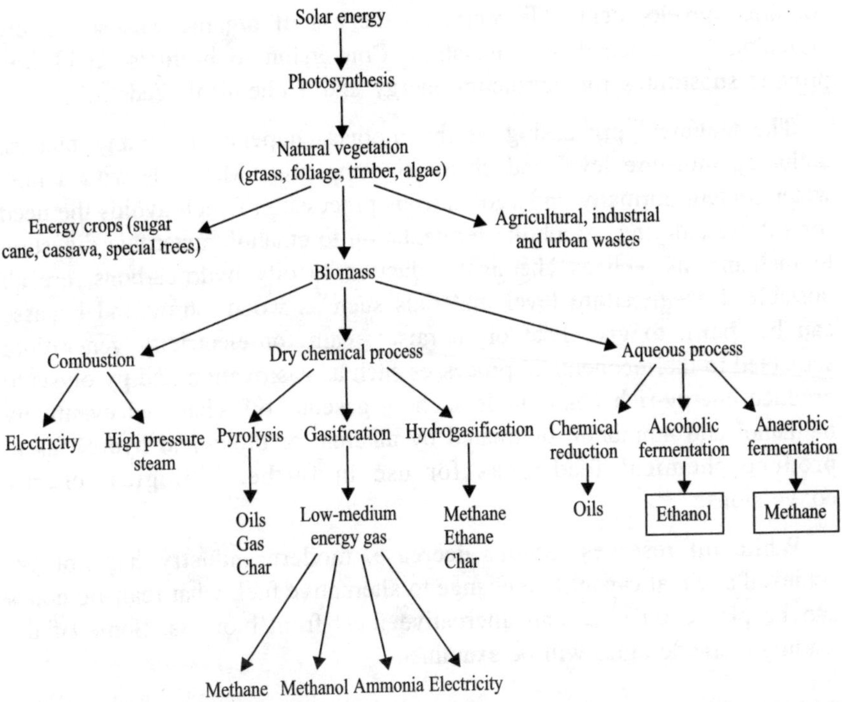

Fig. 7.1. Options for the conversion of biomass to energy.

The concept of cultivating plant biomass specifically for energy supply is based on the fact that much higher yields of fixed carbon are attainable from well-planned plantation methods than from harvesting natural vegetation or collecting agricultural or industrial wastes. Programmes of this type are now being planned and practised extensively in many countries throughout the world. Sugar cane and cassava are the two principal crops that are being developed (primarily for ethanol production) in Brazil, Australia and South Africa, whereas more lignocellulosic-based programmes are being developed in Sweden and America. In the latter case, plans are being made to grow forests for conversion into liquid fuels. Cost analysis of all of these processes offers considerable encouragement, in particular with sugar cane conversion.

Energy-crop plantations will undoubtedly supply meaningful amounts of energy in the near future. The problem of water deficiency is very real, however, and rainfall is most often the limiting factor operating in otherwise ideal conditions of solar radiation intensity, annual hours of sunshine, mild winters and an abundance of good quality land. In certain areas of the world it is possible that such plantations will rapidly become a reality, but for most countries development will centre on the use of organic wastes, namely agricultural, municipal and industrial. Conversion to biofuels could well provide substitutes for petroleum energy and a chemical feedstock.

The technical processing of the biomass depends on many factors, including moisture level and chemical complexity. Materials with a high water content normally undergo aqueous processing, which avoids the need for substrate drying. Alcoholic fermentation to ethanol, anaerobic digestion to methane, as well as chemical reduction to oily hydrocarbons, are all possible. Low mositure level materials such as wood, straw and bagasse can be: burnt to give heat or to raise steam for electricity generation; subjected to thermochemical processes such as gasification and pyrolysis to produce energy-rich compounds such as gaseous oil, char and eventually methanol and ammonia; or treated by alkaline or biological hydrolysis to produce chemical feedstocks for use in further biological energy conversions.

While oil reserves rapidly decrease, modern industry has not yet achieved any real capacity to change to alternative fuel, what realistic hopes can be placed on obtaining alternative fuels from biomass. Some of the leading methodologies will be examined.

ETHANOL FROM BIOMASS

The production of alcohol by fermentation of sugars and starch is an ancient art and is often considered to be one of the first microbial processes used by human beings.

$$C_6H_{12}O_6 \rightarrow 2CH_3CH_2OH + 2CO_2$$

Production of industrial alcohol by fermentation draws heavily upon the accumulated knowledge of the brewer and the distiller. At present, industrial alcohol production is largely synthetic, i.e. non-microbial, deriving from petrochemical processes; petrochemical ethanol is made by the hydration of ethylene, and the decline of microbial production of alcohol dates from the large-scale production of ethylene from the 1940s. Within 20 years of development of large-scale petroleum cracking, industrial production of fermentation alcohol fell below potable alcohol production in most industrialised nations. Thus, in technologically more advanced countries ethanol is produced by chemical means. In many developing countries where cheap raw materials are available, ethanol is still produced for industrial purposes using traditional fermentation techniques.

While the benefits of ethanol as a fuel are considerable, namely it is energy efficient, does not produce toxic carbon monoxide during combustion and is, therefore, much less polluting than conventional fuels, it is still cheaper to produce ethanol from oil chemically than by fermentation processes at current oil prices. In this way, ethanol usage, as with other alternative fuels, is economically hindered in industrialised countries until oil prices once again take on an upward rise in price. It is inevitable that it will happen but exactly when is hard to predict.

A dramatic change in the economics of alcohol production resulted from the massive increases in the world prices of crude oil in the 1970s. Whereas oil prices have more than quadrupled since 1975, the price of suitable cheap carbohydrates has risen far less on average.

Oil-importing nations are anxious to reduce their import costs and many now subsidise home-produced alternatives. Since alcohol can be used as a partial or complete substitute for motor fuel and can also be converted readily into ethylene and related compounds, its production from indigenous and renewable resources still seems an attractive alternative strategy.

Historically, ethanol and, to a lesser extent, methanol were used extensively as motor fuels in Europe prior to World War II and, indeed, Henry Ford's Model T car was designed to run on alcohol, petrol or any mixture in between. Nowhere has this been more actively pursued than in Brazil. Vast biotechnological processes operate throughout the country, converting sugar cane and cassava into ethanol by yeast fermentation.

Brazil's undoubted success in pioneering this production of 'green petrol' has created worldwide interest, particularly amongst poorer Third World nations with the climate and land to grow their own fuel crops but with limited currency to buy oil. Even developed nations such as Australia, the USA, Sweden and France are embarking on biological alcohol production processes utilising either large agricultural surpluses or forestry wastes.

Three billion litres of fuel alcohol were fermented in the USA from corn in 1987 and one-third of all American gasoline is 10 per cent waste-derived ethanol or gasohol. Brazilian officials now estimate that the country's entire petrol needs could be met from planting 0.3 per cent of the country's vast area with alcohol-producing crops. Over 500 fermentation and distillation plants have been built throughout the country processing the crops produced throughout the year. An additional bonus to the energy generation is the creation of over 700,000 new direct jobs and 300,000 indirect jobs in the rural areas of Brazil.

Nevertheless, there are many indirect advantages of Brazil in using ethanol instead of gasoline, While there are obvious reductions in their contributions to global warming, the addition of anhydrous ethanol to gasoline eliminates the need for tetra-ethyl lead to raise the octane rating. Studies have further shown that ethanol-powered engines produce 57 per cent less carbon monoxide, 64 per cent less hydrocarbons and 13 per cent less nitric oxides than gasoline-powered vehicles. Thus, while international oil prices continue to be low, the main justification for the continuation of bioethanol programmes must be environmental.

The world price of alcohol is still about twice that of oil, and fuel consumption is also up to 20 per cent heavier with alcohol. However, these economic considerations are constantly changing in favour of alcohol production, with future anticipated increasing oil prices and new design concepts for alcohol-based engines. Furthermore, on a global aspect, 'green' petrol production will help to take some of the pressure off oil products for the rest of the world, reducing competitive tensions and perhaps even wars.

Ethanol for fuel programmes requires a considerable capital investment and will be in keeping with large-scale needs and not small on-farm systems (where methane products are more suitable). To make available the necessary fermentable sugars (Table 7.1), most raw materials require some degree of pretreatment, depending on their chemical composition.

Table 7.1. Potential raw materials for fuel ethanol production.

Starch containing	*Cellulosics*	*Sugar-containing*	*Other*
Cereal grains	Wood	Sucrose and invert Sugar	Jerusalem artichoke
		Sweet sorghum	
Corn	Sawdust	Molasses	
Grain sorghum	Waste paper	Sugar beet	Raisins
Wheat	Forest residue	Fodder beet	Bananas
Barley	Agricultural residues	Sugar cane	
Milling products	Municipal solid wastes	Lactose	
		Whey	
Wheat flour	Intensive livestock		
Wheat mollifeeds	Production wastes	Glucose	
Corn hominy feed[a]		Sulphite wastes	
Starchy roots			
Mandioca			
Potatoes			

Notes: [a] Meal from parched corn, often with the hull and germ removed.

With sugar cane this treatment is minimal and consists mainly of the usual milling operation, whereas cassava roots (containing 25–38 per cent starch on a wet weight basis) require the action of a suitable saccharifying agent-either acid hydrolysis or enzyme hydrolysis. Cellulosic raw materials such as timber and straw require more extensive pretreatment, and this is reflected in the increased energy inputs required (Table 7.2). A flow diagram for the production of ethanol from diverse substrates is shown in Fig. 7.2.

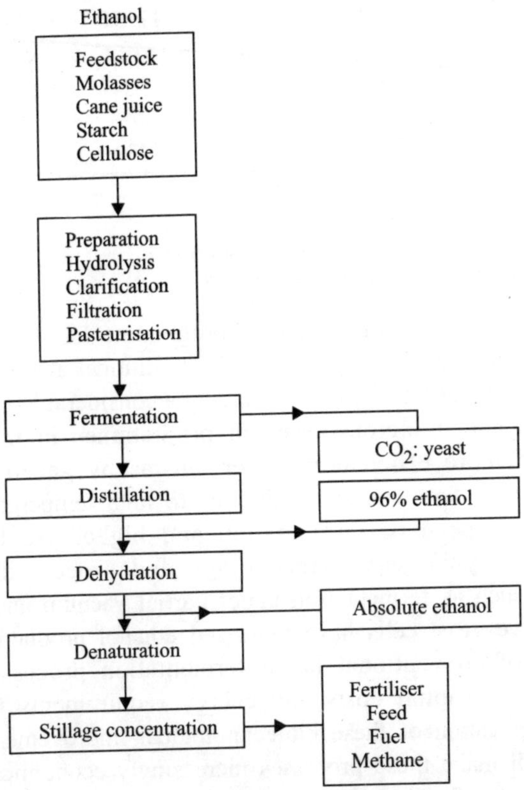

Fig. 7.2. Flow diagram for the production of ethanol.

The Brazilian programme is almost exclusively based *on batch fermentation* systems. At present the standards of these fermentations are modest and leave much room for improvement. Continuous methods of production offer many advantages but are really studied and operated only in developed nations with an interest in ethanol formation. Improvements in *continuous fermentations* have utilised many approaches, including retention of the yeast cells in the bioreactor by separation and recycling and by continuous evaporation of the fermentation broth.

Table 7.2. Gross energy requirements of ethanol produced from different substrates by microbial fermentation.

Physical inputs	*Substrates*				
	Sugar cane	*Cassava*	*Timber*[a]	*Timber*[b]	*Straw*
Substrate	7.27	19.19	12.67	20.00	4.37
Additional chemicals	0.60	0.89	4.74	6.37	4.74
Water	0.30	0.38	0.80	0.30	0.80
Electricity	7.00	10.47	175.70	7.84	166.74
Fuel oil	8.00	29.03	42.13	62.40	42.13
Capital inputs (building etc.)	0.46	1.21	3.34	0.64	3.34
Total	24	61	239	98	222

Notes:
All figures given in kJ (kg ethanol)$^{-1}$
[a] Fermentable sugars formed via enzymic hydrolysis.
[b] Fermentable sugars formed via acid hydrolysis.

So far, innovation in the Brazilian programme has been restricted to some marginal improvements in essentially traditional alcohol fermentation processes. However, biotechnology is having a considerable input, with new developments and numerous research programmes in this field, e.g., production of more efficient micro-organisms by genetic engineering (improved alcohol fermentation, resistance to high temperatures and high alcohol levels, speed of fermentation and higher yields), improved immobilised enzyme reactor technology and process design. Novel introductions such as fermentation under partial vacuum and recycling of the fermentative yeast cells have increased ethanol productivity to 10 or 12-fold that of conventional batch fermentation processes, and such increases reduce capital costs and energy requirements for fermenter operation. Application of these biotechnological improvements to ethanol production will make these processes increasingly economically attractive as a substitute for fossil fuel.

The overall economics of fermenting ethanol from specific crop cultivation in developing countries will achieve support indirectly by the expansion of agriculture, by creating more employment, because oil prices will continue to outstrip agricultural feedstocks, and because new technologies will create further economic uses for the wastes generated in fuel ethanol production.

Vast volumes of wastes or *stillage* result from the Brazilian and other alcohol programmes, and much research is in progress to seek worthwhile end-products.

Of particular significance will be:

1. Evaporation to feed or fertilisers.
2. Mineralisation to ash.
3. Anaerobic fermenation for methanol generation.
4. Conversion by micro-organisms into single cell protein.

There can be no doubt that eventually oil price rises will make gasohol programmes advantageous on simple economic terms.

Recently there has been considerable interest in Europe in the use of modified repeseed oil as a diesel substitute: *biodiesel*. This product is obtained from repeseed as the result of a reaction between the oil and methanol in the presence of a catalyst such a sodium hydroxide at 50°C producing an ester and glycerol.

Rapeseed oil +Methanol — → Diester +Glycerol
(1 tonne) (0.1 tonne) (1 tonne) (0.1 tonne)

The glycerol is allowed to settle and the biodiesel purified and used as a fuel. The purified biodiesel has physical and chemical properties close to those of diesel fuel and heating fuel oil. The use of biodiesel does not require any specific engine modifications.

What are the main advantages of biodiesel? Firstly, the energy yielded is considerably greater than that consumed during its production and will increase with improved (genetically engineered) cultivars of rapeseed. Secondly, it is non-toxic, is more than 98 per cent biodegradable, and its contribution to the greenhouse effect is three- to five-fold less than that of diesel. Above all, it is renewable. France, Italy and Germany have been the leading proponents of biodieselFrance for agricultural reasons, Italy for environmental and Germany for both.

Many public urban transport systems in these countries burn biodiesel, thus enhancing their 'environmental' image. Italy has given preference to biodiesel's use as a fuel for heating municipal buildings. In European countries a considerable amount of agricultural land must be 'set aside' or used for non-agricultural production. For this reason the growth of rapeseed as a fuel crop offers exciting commercial and environmental opportunities and almost all European countries have vested interests.

The wider application of diesel will depend on special fiscal measures. In the European Union a draft directive is under consideration with the view to harmonising the various national fiscal requirements governing biofuels, with a suggested limit of excise rate of a maximum of 10 per cent on the tax on fossil fuel. It has recently been suggested that ethanol from biomass could replace the methanol of fossil origin used in the esterification reaction. In this way bioethanol usage would be increased and the biodiesel would be derived from entirely renewable resources. The biodiesel market is real and will expand in the near future.

METHANE FROM BIOMASS

Methane gas can be used for the generation of mechanical, electrical and heat energy, and is now extensively used as a fuel source for domestic and industrial purposes through national gas pipelines or can be converted to methanol and used as fuel in internal combustion engines. Such natural gas sources were originally derived from biomass in ancient times.

Methane gas also exists in the atmosphere and is derived mainly from microbial action in natural wetlands, rice paddies and enteric fermentation in animals contributing about 20 per cent, 20 per cent and 15 per cent, respectively, to the total methane flux. Domestic cattle are the major contributors, producing about 75 per cent of all animal emissions, whereas humans produce about 0.4 per cent. After CO_2, methane is considered to be the next most important greenhouse gas and is expected to contribute 18 per cent of future warming.

The microbiology of methane production is complex (Fig. 7.3) involving mixtures of anaerobic micro-organisms. In principle, anaerobic fermentation of complex organic mixtures is believed to proceed through three main biochemical phases, each of which requires specific microbiological parameters. The initial stage requires the solubilisation of complex molecules such as cellulose, fats and proteins, which make up most raw organic matter. The resultant soluble, low molecular weight products of this stage are then converted to organic acids; in the final phase of microbial activity, these acids (primarily acetic) are specifically decomposed by the methanogenic bacteria to methane and CO_2.

The most efficient and complex methane-producing system in nature is the rumen. This anaerobic system has never been fully reproduced outside the cow, and is known to be a complex interaction of large numbers of bacteria, protozoa and fungi. All intensively studied bioreactor programmes set up to create *methanogenesis* under controlled conditions have shown that consistent high gas outputs require substantial laboratory monitoring with highly accurate control of environmental variables such as temperature, pH, moisture level, agitation and raw material input and balance. To date, most practical applications of methanogenesis have been at a very low technological level.

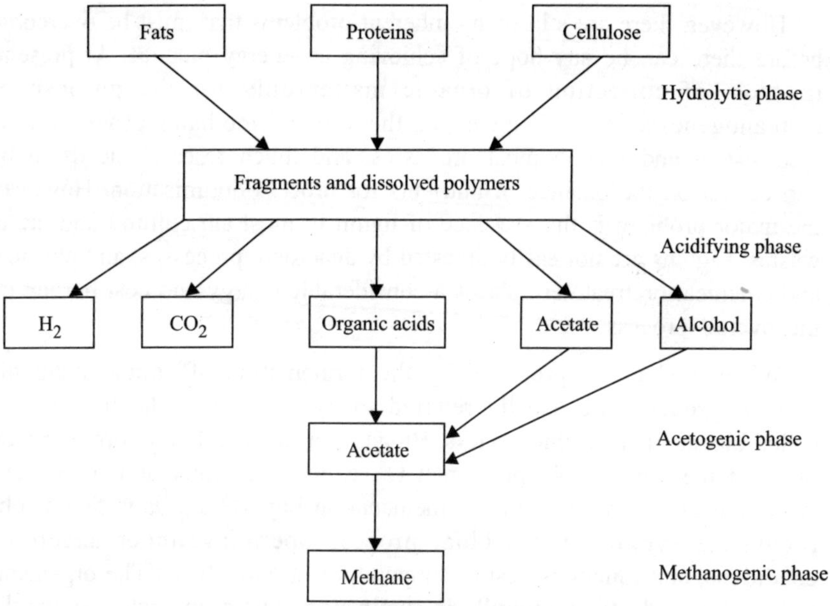

$$C_6H_{12}O_6 \rightarrow 3CH_4 + 3CO_2 + 20 \text{ kcal}$$

Fig. 7.3. The microbiology of methane generation.

There are several possible ways by which methane can be produced in a planned economy: from sewage, from agricultural and urban wastes, and in biogas reactors. The anaerobic digestion of *sewage* is a long-practised technique and many municipal systems have devised methods of capturing the methane and harnessing the energy for the needs of the sewage plant. The energy returns are modest and large-scale expansion does not seem probable.

In recent years methanogenesis of the abundantly available *agricultural and urban wastes* has appeared as an obvious and profitable way to generate energy. The energetic considerations of methane production from such wastes are complex and subject to many limitations. Using urban wastes it should be possible to convert 30-50 per cent of the combustible energy to methane, while with the use of certain other vegetable materials or forages it may be possible to achieve 70 per cent conversion. The overall economics of methane production must recognise the valuable by-products generated by the process, namely the effluent and residue rich in ammonia, phosphates and microbial cells, which may be used as fertiliser, soil conditioner or even as animal feed. Furthermore, the process can convert malodorous and pathogenic wastes into innocuous and useful materials.

However, there are still many inherent problems that must be overcome before there can be any hope of achieving an energy balance. At present, the cost of collection of organic matter only for the purpose of methanogenesis is too expensive; the rate of methane production is inconsistent and low in most processes, and much research needs to be carried out on the balance of nutrients for process optimisation. However, the major problem is the presence of lignin in most agricultural and urban wastes. Lignins are not easily digested by anaerobic processes, and physical and chemical pretreatment places a considerable energy and cost burden on the overall process.

When methane is produced by the fermentation of animal dung the gaseous products are usually referred to as *biogas* and the installations called biogas plants or bioreactors. Biogas is a flammable mixture of 50–80 per cent methane, 15–45 per cent CO_2, 5 per cent water and some trace gases. Biogas is produced via biomethanation Fig. 7.3 and is in fact a self-regulating symbiotic microbial process operating under anaerobic conditions, and functions best at temperatures around 30°C. The organisms involved are all found naturally in ruminant manures. In such systems the animal dung is mixed with water and allowed to ferment in near-anaerobic conditions. Production of biogas by such methods goes back into antiquity and is of particular importance in India, China and Pakistan. The Gobar system of biogas production in the Far East ranges from small peasant systems to quite large plants continuously producing large volumes of gas. In energy terms the simple Gobar system is very near to being a net energy producer on a small scale. Small scale plants of family, farm or village size are in operation throughout the world.

Under ideal conditions, 10 kg of dry organic matter can produce 3 m^3 of biogas, which will provide 3 hr., of cooking, 3 hr., of lighting or 24 hr., of refrigeration with suitable equipment. Biogas does, in fact, furnish a considerable part of the world's energy source; China is the largest user with over 7 million biogas units providing the equivalent energy of 22 million tonnes of coal and, with current subsidies, a biogas plant in China is cheaper than a bicycle. Larger systems of this type do not achieve a net energy balance.

Biogas combustion has been used to heat steam to drive electricity turbines and, in California, one plant provides up to 20,000 hr. of electricity by way of cow manure biomethanation. There has also been some consideration of growing crops on a large scale to provide a 'methane economy'. High-yielding crops in terms of kJ ha^{-1} hr^{-1} cultivated on massive land or water areas have been proposed. It has/been suggested that

65 per cent of the current gas consumption in the USA could be provided by an energy plantation of area 26,0000 km^2 using water hyacinth of energy content 3.8MJ (kg dry material)$^{-1}$. Marine algae have also come under special scrutiny.

Methane generated from organic materials by anaerobic fermentation offers a valuable source of energy that could be put directly to many uses. Furthermore, the associated by-products may be useful forms of fertilisers for agriculture. Yet, before the full realisation of these systems can be achieved, very considerable biotechnological studies must be undertaken. The biological aspects revolve around complex mixed cultures and it is doubtful whether the thermodynamic efficiency of the fermentation can be improved.

Thus, emphasis must be given to improving process design and to technological improvements of the control systems. New and cheap construction materials for digesters (bioreactors) and gas storage vessels will be required. In time there will be available a complete range of anaerobic technologies to deal with most kinds of biodegradable materials. Although methane will be the principal end-product, fuels such as propanol and butanol as well as fertilisers will undoubtedly add to the cost-effectiveness of the overall process.

Table 7.3. Economic arguments against large-scale methane production by microbial processes.

An abundance of methane occurs in nature, particularly in natural gas fields and oil field overlays
Methane production by gasification of coal is commercially more attractive
Microbial production of methane is more expensive than natural gas
Costs of storage, transportation and distribution of gaseous fuels is not yet economically worth while
Methane cannot be used in automobiles and is difficult and expensive to convert to liquid state

Methane as an energy source may well have economic value at local small-scale production levels, but there is considerable doubt about the future of large-scale commercial processes for methane production. Some of the more obvious considerations are shown in Table 7.3. However, anaerobic digestion of municipal, industrial and agricultural wastes can have positive environmental value, since it can combine waste removal and stabilisation with net fuel (biogas) formation. The solid or liquid residues can further be used as fertiliser, soil conditioner or animal feed. Biogas production will continue to have high priority in alternative energy research.

HYDROGEN

Consideration has been given to the use of hydrogen as a fuel or in fuel cells for the production of electricity. Hydrogen production can occur by way of photosynthetic bacteria, biophotolysis of water and by fermentation. In the first two systems, encouraging production of hydrogen has been achieved, but much research is needed to assess the significance of these methods at an applied level. It has been estimated that at least 20 to 30 years of research is needed before any type of functional system is obtained.

Although it is possible to generate hydrogen from glucose by bacterial action, the production rate is too small to make microbial genesis of hydrogen economic. The efficiency of hydrogen production by anaerobic fermentation is much less than that of methane production by the same method. Since methane also has a higher energy content it would appear that methane production by microbial processes has a much higher practical potential than hydrogen production. However, further research may well alter these considerations.

Although biomass may ultimately supply only a relatively small amount of the world's energy requirements (the estimate for the USA is approximately 5 per cent), it will nevertheless be of immense overall value. In parts of the world such as Brazil and countries of similar climatic conditions, biomass will surely attain wider exploitation and utilisation. The technical and agronomic problems are still considerable but biotechnological research is making valuable contributions to further understanding.

Although biomass may still have many disadvantages when compared with oil and coal, the very fact that it is renewable and they are not must be the spur to further research. In time, biomass will become much more easily and economically used as a source of energy.

POSTSCRIPT MICROBIAL RECOVERY OF PETROLEUM

When an oil field is opened up spontaneous flow or upwelling and/or pumping will produce approximately only about one-third of the total petroleum present. Secondary recovery techniques such as gas pressurising, water flooding, miscible flooding and thermal methods can further increase output. Tertiary oil recovery methods involving the use of solvents, surfactants and polymers able to dislodge oil from geological formations can further increase or prolong the production life of a well. Mcrobially enhanced oil recovery processes involve the use of polymers such as xanthan gum produced by large-scale fermentation of specific bacteria. Such gums have excellent viscosity and flow characteristics and can pass through small pore spaces releasing more trapped oil. Application is usually associated with water-flooding operations. A further possible approach is the use of micro-organisms *in situ* for dislodging oil by way of surfactant production, gas formation or even altering the viscosity of the oil by partial microbial degradation. At present the difficulties outweigh the advantages. Biosurfactants may also have a role in releasing oil from tar sands.

8

Single Cell Protein and Recombinant Proteins of High Value

INTRODUCTION

Protein is a complex high polymer containing carbon, hydrogen, oxygen, nitrogen and usually sulphur, and comprised of chains of amino acids (q.v.) connected by peptide linkages (—CO.NH—). Proteins occur in the cells of all living organisms and in biological fluids (blood plasma, protoplasm). They are synthesised by plants largely because of the nitrogen-fixing ability of certain soil bacteria. Their molecular weight may be as high as 40 million (tobacco mosaic virus). They have many important functional forms: enzymes, haemoglobin, hormones, viruses, genes, antibodies and nucleic acids. They also comprise the basic component of connective tissue (collagen), hair (keratin), nails, feathers, skin, etc. Some have been synthesised in the laboratory.

The sequence of amino acids in the polypeptide chain is of critical importance in genetics. Proteins can be hydrolysed to their constituent amino acids and can be broken down into simpler forms by proteolytic enzymes. They form colloidal solutions and behave chemically as both acids and bases simultaneously. They are denatured by changes in pH, by heat, ultraviolet radiation and many organic solvents.

Simple proteins contain only amino acids; conjugated proteins contain amino acids plus nucleic acids, carbohydrates, lipids, etc. On the basis of solubility they can be classified as albumins (water-soluble), globulins (insoluble in water but soluble in aqueous salt solutions) and prolamines (soluble in alcohol-water mixtures but not in alcohol or water alone). A number of proteins have been synthesised, notably the hormone insulin. Proteins are an essential component of the diet, occurring chiefly in meat, eggs, milk and fish. Edible proteins suitable for human food as well as cattle feed can be produced from micro-organisms grown in carbonaceous or nitrogenous media to form yeast-like materials. Paraffinic hydrocarbons (methane) and petroleum-derived ethyl alcohol can be used as growth media for protein biosynthesis.

Industrial applications of proteins include plastics, adhesives and fibres derived from casein and soyabean protein, but these have been declining in recent years. Special forms in which proteins are commercially available include textured proteins for food products, protein hydrolyzate and liquid predigested protein, both for medical use.

NEED FOR PROTEIN

A major problem facing the world, in particular the developing nations, is the explosive rate of population growth. The number of humans in the world now totals 6 billion, is increasing by approximately 94 million annually and could well exceed 10 billion by 2050 if left uncontrolled. Conventional agriculture may well be unable to supply sufficient food–in particular, protein–to satisfy such demands. The Food and Agriculture Organisation (FAO) already predicts a widening of the protein gap between developed and developing countries. At least 25 per cent of the world's population currently suffer from hunger and malnutrition; a disproportionate number of them live in the developing nations, where wars, arid or changing climates and infertile lands hamper productive agriculture.

However, productivity is increasing throughout the world in all branches of agriculture. Biotechnological innovations will accelerate this trend. Food surpluses are occurring in many places, particularly in North America and western Europe where there are near-static populations. Furthermore, some countries that had been net importers of major foods such as India and Indonesia are now self-sufficient. World supply of grain per head has outpaced population growth. However, there are still major imbalances in the availability of cereals; these are further disturbed by changes in global weather patterns (in particular rainfall) and by national and international warfare, with ensuing disruption of agriculture and food distribution.

The extent of the protein problem varies from country to country and must be considered within the framework of each national economy. The shift from grain to meat diets in developed and developing countries is of dramatic proportions and is leading to a much higher per capita grain consumption, since it takes 3 to 10 kg of grain to produce 1 kg of meat by animal rearing and fattening programmes.

The search for sources of protein is relentlessly pursued. New agricultural practices and widespread, high protein cereals have been developed, the cultivation of soya beans and groundnuts is ever-expanding, protein may be extracted from liquid wastes by ultrafiltration, and now the use of microbes as protein producers as gained wide experimental success. This field of study has become known as *single cell protein* or SCP production, referring to the fact that most of the micro-organisms used as producers grow as single or filamentous individuals rather than as complex multicellular organisms such as plants or animals.

Eating microbes may seem strange, but people have long recognised the nutritional value of some large fungi, namely the mushrooms. However, even here, scepticism and prejudice have influenced people's attitudes, and while in some countries mushrooms are widely consumed, in others they are avoided and neglected. Although massive production processes for mushrooms are now commercially successful throughout the world, this chapter is largely concerned with the growth of more simple organisms, namely bacteria, yeast, filamentous fungi and algae, which lend themselves to biotechnological processing. Whereas mushroom growing because of its antiquity can be considered as a conventional type of food production, the use of microbes is less appreciated and presents many problems, not all of which are technological in nature.

During the last two decades there has been a growing interest in using microbes for food production, in particular for feeding domesticated food-producing animals such as poultry. It has been argued that the use of SCP derived from low value waste materials for animal feed would improve human nutrition, by taking protein-rich vegetable foods out of the human versus animal competition and making them more freely available for human consumption in the producer countries, which are often developing countries. Many major companies throughout the world have long been actively involved in these SCP processes, and many worthwhile products are commercially available.

Protein quality and quantity are the goals of SCP production. However, the microbes also contain carbohydrates, fats, vitamins and minerals, and produce them from (in general), otherwise inedible or low quality waste materials.

SCP may be used as a protein supplement for humans and animals. With humans it has been considered as a protein supplement, as a food additive to improve flavour, fat binding, and more recently as a replacement for animal protein in the diet. Because humans have a limited capacity to degrade nucleic acids, additional processing is required before SCP can be used in human foods. Micro-organisms have high DNA and RNA contents and human metabolism of nucleic acids yields excessive amounts of uric acid, causing kidney stones and gout! In animal feeding it can serve as a replacement for such traditional protein supplements as fishmeal and soya meal. The high protein levels, bland odour and taste of SCP, together with ease of storage, confer considerable potential to SCP in food and food outlets. Its high protein content makes its use attractive in aquaculture, e.g., farming of shrimps, prawns, trout, salmon, etc.

Microorganisms produce protein much more efficiently than any farm animal (Table 8.1). The protein-producing capacities of a 250 kg cow and 250 gram of micro-organisms are often compared. Whereas the cow will put on 200 gram of protein per day, the microbes, in theory, could produce 25 tonnes in the same time under ideal growing conditions. However, the cow also has the unique ability to convert grass into protein-rich milk. After decades of research no rival method for that conversion process has been developed. The cow has recently been described as 'a live, self-reproducing and edible bioreactor'. The advantages of using microbes for SCP production are outlined in Table 8.2.

Table 8.1. Time required to double the mass of various organisms.

Organism	*Time*
Bacteria and yeasts	20–120 min.
Moulds and algae	2–6 hr.
Grass and some plants	1–2 weeks
Chickens	2–4 weeks
Pigs	4–6 weeks
Cattle (young)	1–2 months
Humans (young)	3–6 months

Table 8.2. Advantages of using microbes for SCP production.

Micro-organisms can grow at remarkably rapid rates under optimum conditions; some microbes can double their mass every 0.5 to 1 hr.
Micro-organisms are more easily modified genetically than plants and animals; they are more amenable to large-scale screening programmes to select for higher growth rate, improved acid content, etc., and can be more easily subjected to gene transfer technology.
Micro-organisms have relatively high protein content and the nutritional value of the protein is good.

(Contd...)

Micro-organisms can be grown in vast numbers in relatively small continuous fermentation processes, using relatively small land area, and growth is also independent of climate.

Micro-organisms can grow on a wide range of raw materials, in particular low value wastes and some can also use plant-derived cellulose.

ACCEPTABILITY AND TOXICOLOGY OF SCP

A unique aspect of the SCP field is the extent to which it has been influenced by factors other than purely technological or economic. Geographical, political, sociological and psychological influences have shaped the course of development to a very marked degree. In particular, a tremendous amount of attention has been given to the problems of safety, nutritional value and acceptability of the product.

The nature of the raw materials used in SCP processes represents the main safety hazard, e.g., the possibility of carcinogenic hydrocarbons in the gas oil or *n*-paraffins, of heavy metals or other contaminants in the mineral salts, of solvent residues after extraction, and of toxin (mycotoxin) production by certain fungi. The process organism must be non-pathogenic and not-toxigenic and the products of the organism's metabolism must be innocuous. Rigorous sanitation and quality control procedures must be maintained throughout the process to avoid spoilage or contamination by pathogenic or toxigenic micro-organisms.

Toxicological testing of the final product must include short-term acute toxicity testing with several different laboratory animal species, followed by extensive and detailed long-term studies, for two years or more, in both rodent and non-rodent species. All in all, it represents a major scientific and financial investment but, after the thalidomide disaster when a drug taken to suppress the nausea of 'morning sickness' caused foetal limb abnormalities, it is better to err on the side of caution.

The acceptability of SCP when presented as a human food does not depend only on its safety and nutritional value. In addition to the general reluctance of people to consume material derived from microbes, the eating of food has many subtle psychological, sociological and religious implications. In various cultures there are many specific associations related to eating, and status and symbolic values are attributed to different kinds of food. More obvious factors influencing acceptability must be considered, such as odour, colour, taste and texture. Thus, if SCP is to be used directly as food for human beings, the skills of the food technologist will be greatly challenged. Most realistic considerations would place SCP for human food at a relatively low priority. Demand will mainly be as a food source for various types of domesticated animal, poultry and fish. However, some industrial processes will be primarily directed to human consumption, e.g., Rank-Hovis-McDougall/ICI fungal protein (Quorn), which is gaining an increasing market place as a vegetable protein with meat-like texture and appearance.

What about the political and sociological influences on the production of SCP? These problems have centred mainly around the use of oil-derived materials as substrates for SCP. Such 'Petro-proteins' have been suspected of being contaminated with carcinogens and consequently massive SCP programmes in Japan, Italy and Britain were abandoned and efforts redirected to the study of alcohol/methanol-based SCP and SCP from organic wastes.

The critical parameters of SCP processes are strongly interdependent. The choice of substrate will reflect local political and economic factors and the availability of alternative outlets. The choice of organism will partly determine the process technology and the nature of the product. Acceptability of the product will depend in part on the substrate. Because of the large volumes that will be involved in SCP processes, continuous culture techniques will be widely selected for economic reasons. Most large concerns actively involved in SCP production have based their processes on such techniques.

SCP DERIVED FROM HIGH ENERGY SOURCES

Materials with a high commercial value as energy sources or derivatives of such chemicals, e.g., gas oil, methanol, ethanol, methane and *n*-alkanes, have found wide commercial interest. The microbes involved are mostly bacteria and yeasts and several processes are now in operation. As would be expected, most oil companies have been or are still involved in this field. The wisdom of using such high energy potential compounds for food production has been questioned by many scientists.

Methane as an SCP source has been extensively researched but is now considered to present too many technical difficulties to warrant exploitation. In contrast, methanol offers great economic SCP interest. A large-scale (75,000 litre) fermentation plant for producing the methanol-utilising bacterium *Methylophilus methylotrophus* was constructed by ICI, U.K. Hoechst (West Germany) and Mitsubishi (Japan) worked on a similar process, using yeast strains instead of bacteria. The ICI SCP protein (named Pruteen) was used exclusively for animal feeding. Methanol as a carbon source for SCP has many inherent advantages over *n*-paraffins, methane gas and even carbohydrates; composition is independent of seasonal fluctuations. There are no possible sources of toxicity in methanol, methanol dissolves easily in the aqueous phase in all concentrations, and no residue of carbon source remains in the harvested biomass. Several other important technical aspects are also very relevant.

The ICI Pruteen plant was the only process of its kind in the Western world but could not operate economically at present methanol prices and has ceased production. Methanol represents approximately 50 per cent of the cost of the product. In the USA the cost of SCP derived from methanol is two- to five-fold the cost of fishmeal. In the Middle East the low cost of methanol and higher costs of fishmeal coupled with a need to produce more animal products could make SCP an attractive proposition. In the former Soviet countries many methanol plants are operated. In part this is due to chronic shortages of animal feeds, excess production of methanol, lack of foreign currency to buy alternatives such as soya bean meal and above all is disregard for economic planning. However, the recent political and economic changes in Russia have led to many closures of SCP production plants.

The vast range of studies carried out in the 1980s and 1990s on the potential use of methanol and related compounds as substrates for SCP processes certainly pushed bioreactor technology to its limits for cheap bulk product formation. The aerobic process for Pruteen production was the world's largest continuous bioprocess system. The stringent economies required in these processes led to extensive use of airlift bioreactor design. Furthermore, the massive volume and expense in harvesting and preparing the final product forced many economies of scale and of downstream processing. At its peak in Russia there were several SCP plants on stream or being planned, with production capacity of between 300,000–600,000 tonnes per annum.

Ethanol is a particularly suitable source if the SCP is intended for human consumption. In the foreseeable future the comparative status of ethanol SCP will depend on local factors: over-capacity in ethylene crackers, agricultural carbohydrate surpluses and political decisions about regional economic independence and foreign trade balances.

The use of *n*-alkanes as a substrate for SCP has been extensively studied in many countries and represents a very complex biotechnological process. However, most of these processes have now ceased operation because of suspected health hazards resulting from the presence of carcinogens in the

SCP. More recently, however, the massive technology developed in this field in Japan and other Eastern countries has been turned over to the study of alcohol-based SCP and SCP from organic wastes.

SCP FROM WASTES

The materials that make up wastes should normally be recycled back into the ecosystem, e.g., straw, bagasse, citric acid, olive and date wastes, whey, molasses, animal manures and sewage. The amount of these wastes can be locally very high and may contribute to a significant level of pollution in water courses. Thus, the utilisation of such materials in SCP processes serve two functions–reduction in pollution and creation of edible protein. Of particular interest have been the extensive programmes to convert cattle and pig wastes to feed protein. However, these studies have been largely unsuccessful because of technical and sanitary difficulties.

An attractive feature of carbohydrate waste as a raw material is that, if its low cost can be coupled with suitable low process costs, an economic SCP product may be obtained from relatively small operative units. The world-wide trend towards stricter effluent control measures and parallel increases in effluent disposal charges, has led to the concept of waste as a negative cost raw material. However, the waste may not be suitable for SCP or its composition or dilution may be so dispersed that transport to a production centre may be prohibitive. Each waste material must be assessed for its suitability for conversion to SCP. In particular, the level of available technology is important. When a waste is available in large quantities and preferably over a prolonged time then a suitable method of utilisation can be planned Table 8.3.

Table 8.3. Advantages for using widely available organic wastes for SCP production.

Reduces environmental pollution.
Most organic wastes are available at low cost in most countries thus ensuring independence in supply.
The wastes are upgraded in energy and protein level.
Guards against a protein shortage in a community that may be largely dependent on imports.
Allows for technological innovation that can often be transferred to developing countries.
Many of the wastes such as cellulose and whey already form accepted parts of animal diets and will avoid the acceptability problems of other unusual wastes, e.g., human wastes and fossil fuels.

SCP processes utilising waste substrates have been carried out on a commercial scale using various yeast organisms in sophisticated bioreactor systems. Substrates used and producer organisms include molasses (*Saccharomyces cerevisiae*) and cheese whey (*Kluyveromyces fragilis*), while the Symba process developed in Sweden utilises starchy wastes by combining two yeasts, *Endomycopsis fibuligira* and *Candida utilis*.

The feed value of the yeast produced by the Symba process has been evaluated in vast feeding experiments on different types of animal, including pigs, chickens and calves. The animals grew well and no adverse effects were recorded. The Symba process can be conveniently separated into three phases:

Phase 1 Waste-water (from, for example, a potato processing plant), containing starch is fed through a heat exchanger and sterilised by steam injection.

Phase 2 Sterilised starch solution is fed through two bioreactors together with the starch hydrolysing yeast *Endomycopsis fibuligira*. The hydrolysed starch then passes into a large bioreactor with *Candida utilis* as the growing organism.

Phase 3 The harvest stream from the *Candida* bioreactor is passed through vibro-screening and hydrocycloning equipment, then centrifuged. The samples collected can be spray-dried and the dried material sifted and bagged or stored in bulk.

Pekilo was a fungal protein product that was produced by fermentation of carbohydrates derived from spent sulphite liquor, molasses, whey, waste fruits and wood or agricultural hydrolysates. It had a good amino acid composition and was rich in vitamins. Extensive animal feeding test programmes showed that Pekilo protein was a good protein source in the diet of pigs, calves, broilers, chickens and laying hens. Pekilo protein was produced by a continuous fermentation process. The organism, *Paecilomyces variotii*, a filamentous fungus, gave a good fibrous structure to the final product. However, the process does not appear to be in commercial operation at the present time.

The use of abundantly available waste starch and sugars as a source of raw material for an SCP process was considered in the early 1960s by Lord Rank to be a feasible and worthwhile project to alleviate the anticipated world protein famine. Thus, a 'starch into protein' process was commenced which ultimately was to become the only successful SCP process developed entirely with human consumption as the primary pool. Three criteria were considered essential for this new food: (a) it must be 'delicious to eat'; (b) the final product, the substrate and all possible intermediates used in the process must be safe to eat; and (c) above all, the final food presented should be highly nutritious.

Because the final product would require to be textured bacteria would be unsuitable and filamentous fungi became the obvious choice. Although starch was the initial best available substrate, it was considered that for a good controlled fermentation it would be necessary to use a soluble carbohydrate produced by the hydrolysis of starch. The fungus finally chosen was a *Fusarium* and this was followed by an extensive fermentation programme leading ultimately to a novel continuous process.

This process can be economical only if operated on a continuous basis, and is routinely utilised for periods of several weeks, without significant interruption and without contamination. The productivity, yield, composition and morphology of the organism are maintained steady by careful regulation of the operating conditions.

The final fermentation product is pale-buff in colour, bland to taste, highly nutritious and fibrous in composition. It is now texturised by food scientists into meat-like analogues and, with correct culinary ingredients, can seem remarkably similar to meat and chicken dishes. The process involves an extensive and prolonged safety and nutritional assessment and, since it was the first of its kind, this was to be the 'trend setter'.

The production process and the marketing of the mycoprotein has been highly successful and production sites should become world-wide in the very near future.

The advanced technology developed by ICI for the now uneconomical Pruteen process is used to produce large quantities of the fungal protein. The product has had excellent response from the public. A typical composition of mycoprotein compared to beef is shown in Table 8.4.

Table 8.4. Typical composition of mycoprotein in comparison to beef.

Component	*Mycoprotein*	*Raw lean beefsteak*
Protein	47	68
Fat	14	30
Dietary fibre	25	Trace
Carbohydrate	10	0
Ash	3	2
RNA	1	Trace

There can be little doubt that cellulose from agriculture and forestry sources and from wastes must constitute the future major feedstock for many biotechnological processes including SCP. Cellulose, in its more natural association with lignin, is by far the most prevalent organic material available for biotechnological conversion. Throughout the world research teams are studying ways of pretreatment to disrupt or destroy the lignin barrier and many physical and chemical methods have been examined (Table 8.5).

Table 8.5. Strategy for the utilisation of complex lignocellulose wastes for SCP production.

Substrate preparation
Pretreatment of the substrate
Physical: Milling techniques
Chemical: Short acid, alkaline or solvent treatment
Microbial: Screening for rotting organisms
Hydrolysis of substrate
Chemical: Complete saccharification, with acid or alkali
Enzymic: Production of cellulases, hemi-cellulases etc.
Microbial: Screening for good microbial enzyme producers
Selection and improvement of micro-organisms
Selection of fermentation systems
Scale-up, optimisation of equipment, process control, batch or continuous process, preservation of final product.
Nutritional and toxicological aspects
Chemical composition, animal feeding tests, acceptability, toxicology
Economics
Cost analysis, comparative considerations with other materials, energy balances.

Next to cellulose, lignin is the earth's second most abundant natural biopolymer found in plants. Approximately 30 per cent of most woody plants is composed of lignin and its catabolism and utilisation as a renewable resource are of great commercial interest. The natural abundance of lignin ensures that its biodegradation constitutes a vital part of the biophasic carbon cycle. The elucidation of the lignin biodegradation process is essential to understanding the circumstances for the recycling of carbon on earth, for establishing the technology for bioconversion of plant residues and waste lignins to useful materials (including edible biomass), and for the protection of the environment from lignin-derived pollutants.

A limited number of fungi and bacteria have then necessary enzymic activities for lignin degradation. The major group of fungi to degrade woody materials, and especially the lignin component of lignocellulosics, are the white-rot basidiomycetes. The fungus *Phanerochaete chrysosporium* is one of the most studied lignin degraders and is considered to have great potential for future biotechnological applications. As a result of these biochemical studies scientists now have a better understanding of the structures and breakdown mechanisms for lignin biodegradation.

Other basidiomycetes or mushroom-type fungi have long been shown to grow on lignocellulose materials such as wood and straws and to produce edible mushrooms. The cultivation of edible mushrooms is one of the limited examples of a microbial culture in which the cultivated organism itself–that is the macroscopic, highly developed mushroom structure–is used directly as a human food. Mushroom cultivation also provides one of the few examples of successful commercial biotechnology processes based on lignocellulose as a substrate. A wide range of edible mushrooms are now cultivated throughout the world for human consumption. This solid substrate fermentation is now one of the most challenging and technically demanding of all vegetable cultivations known to humankind. On a world-wide basis, mushroom growing is one of the fastest growing biotechnological industries and it is expected that it will expand even further with the production of enzymes and pharmaceutical compounds such as antitumour compounds and antibiotics.

The cultivation of the common white mushroom *Agaricus bisporus* has expanded world wide and accounts for over 70 per cent of all mushrooms produced and consumed. The USA continues to be the world's largest producer. However, mushrooms traditionally grown in the Far East, such as *Lentinula edodes* (the Shiitake mushroom), *Pleurotus* spp. (the oyster mushroom), and *Volvariella volvaceae* (the padi-straw mushroom) are now expanding into other areas of the world largely because of their unique flavours and textures and amenability to culinary inclusion.

Mushroom production is in principle a fermentation process. In the case of *Agaricus* the substrate for growth is straw while for *Lentinula* it is wood. For *Agaricus* cultivation the straw is composted with animal manures and other organic nitrogen compounds over a period of one to two weeks and the final product is a unique substrate suitable for the rapid growth of the *Agaricus* inoculum. There is no standard pattern for compost, formulation being based only on the availability and price of the raw materials and supplements in the particular growing region. The nature of the substrate and its pretreatment, more than all other aspects of growing determines the method by which specific mushrooms are cultivated. When the mushroom mycelium (thread-like biomass) has grown throughout the prepared compost (usually contained in large wooden boxes), the environmental conditions of temperature and humidity are altered and subsequently the large mushroom structure forms rapidly in large numbers or 'flushes'. These are then harvested by hand cutting and approximately 7–10 days later another crop will appear. Usually up to four crops are produced before the process is terminated. The cultivations of *Valvariella* and *Pleurotus* are again straw based but have a much more simple procedure.

Lentinula edodes is the second most cultivated mushroom in the world and has been farmed for over 2000 years. Currently, over 90 per cent of its production occurs in Japan but cultivation extends to China, Korea, Singapore, Taiwan, Sri Lanka and, more recently, into the USA and Europe. The principal method of cultivation has been to inoculate wooden logs (2m × 15 cm) with spore inoculum or mycelial plugs, allow the logs to stand for up to nine months to achieve colonisation by the fungus and then during subsequent early summer and autumn periods the mushrooms will grow out and be harvested. This seasonal production is augmented by drying mushrooms to achieve all-year-round consumption. The fungus derives its total nutrition from the lignocellulose of the log.

More recently a new method of cultivation has been developed in which sawdust is mixed with cereal supplements and compressed into large plastic bags–artificial log production. The bags are sterilised and then inoculated with the pure fungus and, after a period of vegetative growth, are induced to produce the mushroom. This controlled form of cultivation is leading to a wider geographical range of commercial cultivation not only for *Lentinula* but for many others.

Undoubtedly the great advantage of these mushroom fermentations is the possibility of converting industrial, urban and wood wastes into a product that is directly edible by humans. However, irrespective of public demand and relative nutritional values, the future outlook for these processes will be related largely to the economics of production methods.

Biotechnological innovation is only now seriously being involved in this field but the rewards will be great. Biotechnology need not always be high technology. To developing nations, where highly expensive systems may be impractical from the point of view of cost and lack of skilled operators, many of the new biotechnological discoveries may well lead to improvements in traditional microbial processes.

Major examples of solid substrate fermentation are the many types of oriental food fermentations. In many of these processes bland materials such as beans, bran, etc., are subjected to microbial activity, hydrolysing starch and proteins and creating products with basically enhanced flavours. Examples are traditional foods such as miso, shoyu, tempeh and many more local fermentations. Although many of these microbially-derived foods are produced on a small cottage scale, others are the basis of large industries demanding major biotechnological inputs. Such foods and flavours are slowly becoming recognised in the West and will surely become a more acceptable part of our daily food intake.

SCP FROM AGRICULTURAL CROPS

The previous section described how micro-organisms can be used to produce SCP from organic waste such as sugars, starch and cellulose. Why not grow certain plants specifically as substrates for SCP processes? The concept of plant biomass production as a feedstock for biotechnological processes is important as much higher yields of fixed carbon are attainable utilising a well-planned, preconceived plantation method than by harvesting natural vegetation or collecting crops or process wastes. At present, such programmes are practised largely for ethanol production and it is thought that cassava, sugar cane and tapioca palm represent the only crops upon which it is likely that a mainstream fermentation operation could be economically established. When lignocellulose can be economically utilised most parts of the world will have a readily renewable feedstock available for countless processes.

SCP FROM ALGAE

There has been some interest in the use of algae as SCP because they grow well in open ponds and need only CO_2 as a carbon source and sunlight as an energy source for photosynthesis. Algae such as *Chlorella* and *Senedesmus* have long been used as food in Japan, while *Spirulina* is widely used in Africa and Mexico. *Spirulina maxima* is commercially produced in Mexico as a by-product of a large solar evaporator used for production of soda lime. Up to 2 tonnes per day are produced and used as animal feed. *Chlorella* is used as a protein and vitamin supplement in some Japanese yoghurts, ice-cream and breads. In some parts of the world algae are used in ponds or lagoons to aid in the removal of organic pollution and the resultant biomass harvested, dried and the powder added to animal feed.

ECONOMIC IMPLICATIONS OF SCP

The economic feasibility of SCP, apart from mushroom production, will be dictated by possible uses in competition with comparable existing products.

SCP is rich in protein and can be stored and shipped over long distances. Its principal use will be as animal fodder, partly replacing other protein-rich materials such as soya bean meal or fishmeal. Being biological in the true sense, even though carried out industrially, these processes do not create imbalance in natural ecosystems. No novel synthetic compounds are produced and the technology, being based on recycling, is pollution free.

SCP processes are mostly capital and energy intensive, and most processes must be conducted under sterile conditions in expensive equipment that can be cleaned and sterilised. The final product must not be exposed to microbial contamination, particularly human pathogens. To achieve economies of scale, SCP processes should have an input of at least 50,000 tonnes per year unless operated as a waste-treatment facility in a food-processing plant. Thus, a considerable volume of raw materials must be close by to meet these production requirements. Water requirements for SCP production are considerable for both processing and cooling.

The world-wide, large-scale development of SCP processes has contributed greatly to the advancement of present-day biotechnology. Research and development into SCP processes has involved work in the fields of microbiology, biochemistry, genetics, chemical and process engineering, food technology, agriculture, animal nutrition, ecology, toxicology, medicine and veterinary science and economics. In developing SCP processes new technical solutions for other related technologies have been discovered, e.g., in waste-water treatment, production of alcohol and other metabolites, enzyme technology and nutritional sciences.

There is little doubt that the main short-term impetus for SCP production will come from the increasing legislative requirement for the disposal of both solid and liquid wastes in a manner that would be compatible with the preservation of the environment. The competitiveness of SCP for animal feed will improve when charges for effluent treatment are allowed for. The main reason for uncertainty with regard to profitability is undoubtedly the price of reference proteins, for example soya bean and fishmeal.

The future of SCP will be heavily dependent on reducing production costs and improving quality. This may be achieved with lower feedstock costs, improved fermentation and downstream processing and improvement in the producer organisms as a result of conventional applied genetics together with recombinant DNA technologies. However, the main limitations of SCP products for human use are sociological rather than technical, and consequently in most cases the major nutritional role of microbial biomass or SCP will be in animal feed supplements.

RECOMBINANT PROTEINS OF HIGH VALUE

Applications of High-value Proteins

Proteins used in industrial enzyme technology, e.g., detergent proteinases or enzymes applied in the food industry, are in most cases rather crude preparations and usually mixtures of different enzymes. In contrast, there are a number of commercial applications where highly purified (and therefore high-value) proteins are needed. Examples are:

1. Analytical enzymes and antibodies. For use in medical diagnostics, food analysis, as well as biochemical and molecular biological analysis.
2. Enzymes used as tools in genetic engineering technology. Gene technology has become possible through the availability of highly purified enzymes such as restriction endonucleases, DNA or RNA polymerases, nucleases and modifying enzymes. Similarly, glycohydrolases and glycosyl transferases are used increasingly in glyco-biotechnology in order to modify the sugar residues of glycoproteins.
3. Therapeutic proteins: Growth factors, antibodies or enzymes are used as the active drug ingredients for the treatment of diseases.

Furthermore, proteins with proven or supposed biological relevance in pathomechanisms are needed as targets for the search of new ligands (agonists or antagonists), inhibitors and for X-ray or NMR structural analysis in order to design novel interacting compounds by structure-based molecular modelling. This requires the production of these proteins on a relatively small scale (10 to 100 mg) but often with high purity depending on the intended use.

Analytical Enzymes

Enzymes are highly specific both in the reaction catalysed as well as in their choice of substrates. Indeed, enzymes are, besides antibodies, the most specific reagents known. The use of enzymes in analysis, therefore, offers a number of advantages compared to chemical reagents. The reactants may either become chemically transformed in the presence of an enzyme (if they are substrates), or they may modulate the enzymatic activity in a manner related to their concentration (if they act as activators or inhibitors). Enzymes also serve as 'markers' in assay techniques based on non-enzymatic interactions, such as antigen-antibody binding or DNA-oligonucleotide hybridisation.

The advances in recombinant DNA technology have made it possible to clone any gene of interest and to manipulate bacterial, fungal, insect or mammalian cells to overproduce the desired protein. Many enzymes have been expressed in manipulated micro-organisms at levels 10 to 100 times higher than in the natural host cell. This has allowed not only better economics in enzyme production, but also significant reductions of the environmental burden because of the decrease in fermentation volumes (and, thereby, in waste formation). Furthermore, the techniques of unspecific and more recently, site-directed mutagenesis have opened the way to improve relevant properties of enzymes for analytical applications, e.g., stability under the assay conditions, pH optimum, or solubility.

Enzymes in diagnostic assays

End-point assays

In this case, the compound to be determined (the analyte) takes part as the substrate in an enzyme-catalysed reaction and is converted with the simultaneous and stoichiometric production of a detectable signal (e.g., a positive or negative change in absorbance). The reaction is allowed to proceed until completion and the result can easily be calculated from known physical constants, e.g., the molar absorption coefficient in the case of light-absorbing substances. The specificity of the assay system depends on the substrate specificity of the enzyme employed. An example of this is the determination of uric acid using urate oxidase (Fig. 8.1).

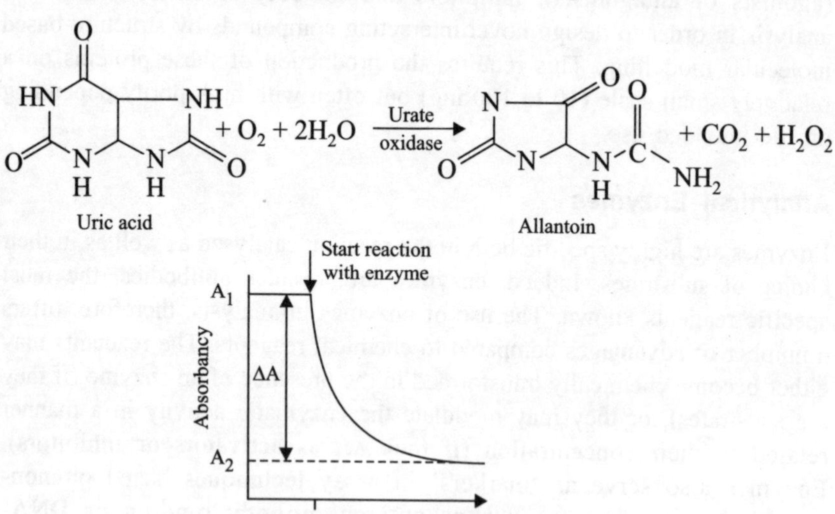

Fig. 8.1. Determination of substrate concentrations by an enzymatic end-point assay.

The same principles will, of course, apply if it is not the analyte conversion itself but the stoichiometrically linked conversion of a co-substrate or co-factor that is detected. This is the case in dehydrogenase-catalysed assays, which make use of the absorption of the reduced coenzyme NADH or NADPH at 340 nm for detection. An overview of important enzymes used in diagnostic assays is given in Table 8.6. Most of the enzymes used in enzymatic analysis are derived from bacteria or yeast and are produced using bacterial (in most cases *E. coli*) or yeast expression systems due to the lower production costs.

Table 8.6. Examples of enzymes important in diagnostics.

Enzyme	*Source (original)*	*Used for the assay of*
Cholesterol oxidase	*Nocardia erythropolis* or *Brevibacterium* sp.	Cholesterol
Creatinase	*Pseudomonas* sp.	Creatine, creatinine
Creatininase	*Pseudomonas* sp.	Creatinine
β-Galactosidase	*Escherichia coli*	Sodium ions; immunoassay marker enzyme
Glucose oxidase	*Aspergillus niger*	Glucose
Glucose-6-phosphate dehydrogenase	*Leuconostoc mesenteroides*	Glucose (indicator enzyme)
α-Glucosidase	*Yeast or Bacillus* sp.	α-Amylase activity
Glycerol-3-phosphate oxidase	*Aerococcus viridans*	Triacylglycerols
Hexokinase	Yeast	Glucose and other hexoses
Peroxidase	*Horseradish*	Indicator enzyme and immunoassay marker enzyme
Pyruvate oxidase	*Pediococcus* sp.	Pyruvate; transaminase activity
Sarcosine oxidase	*Pseudomonas* sp. *Bacillus* sp.	Creatinine
Urate oxidase (uricase)	*Arthrobacter protophormiae*	Uric acid
Urease	*Klebsiella aerogenes*	Urea

In many cases, none of the reactants or products of an enzymatic reaction lend themselves readily to physical or chemical measurement because no detectable signal is produced. In these cases, the primary (called 'auxiliary') reaction usually is coupled to a stoichiometrically linked 'indicator' reaction (mostly also enzyme-catalysed), with one of the products of the second reaction being easily detectable. In coupled systems, it is sufficient to employ an enzyme with narrow substrate specificity for only one of the reaction steps, because the specificity of the whole reaction sequence will depend on the most specific enzyme. This means that in practice, general (i.e., unspecific) indicator reactions can be coupled without individual optimisation to various specific auxiliary reactions. To reach the endpoint in an appropriate reaction time, enzymes with low K_m values for the analyte substrates are preferred to ensure that the reaction is rapid even at the low substrate concentrations reached when complete conversion is approached.

Horseradish peroxidase is a well-known example of an indicator enzyme used in a large number of commercial oxidase-based coupled assays but, in most cases, this enzyme is obtained from a non-recombinant source. An example of a coupled glucose assay which uses recombinant enzymes as the indicator enzyme is shown in Fig. 8.2.

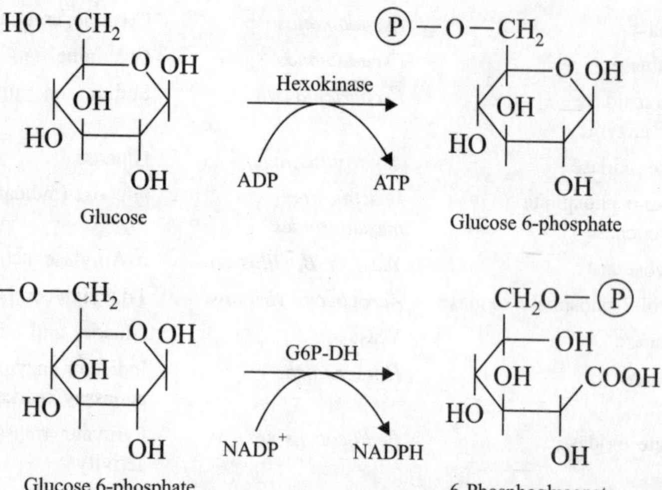

Fig. 8.2. An example of a coupled enzymatic assay system using an indicator enzyme: glucose assay with hexokinase and glucose-6-phosphate dehydrogenase.

Kinetic assays

If the substrate concentration [S] is much smaller than the Michaelis-Menten constant K_m of the enzyme, it follows from the Michaelis-Menten equation that the observed reaction rate becomes linearly proportional to [S]:

$$v = \frac{V \times [S]}{K_m + [S]}$$

where v = observed reaction rate; V = theoretical maximal rate with fixed amount of enzyme; [S] = substrate concentration; K_m = Michaelis-Menten constant.

By monitoring the reaction kinetics, either by the rate of disappearance of substrate –d[S]/dt, or the rate of formation of product, d[P]/dt, catalysed by a fixed amount of enzyme, the analyte concentrations can be determined by measuring the reaction rate. Kinetic assays allow a drastic reduction of the time required for analysis and are less sensitive to interferences than end-point assays. However, for rate measurements the reaction conditions, such as temperature, pH amount of enzyme, etc., must be held constant. Therefore, kinetic assays are best performed with automated analysers rather than carried out manually.

Substrate concentration determination by rate measurement is only possible if the initial substrate concentration is low compared to the K_m value of the enzyme. Therefore, enzymes with high K_m values (i.e., low substrate affinity) are used in kinetic assays to increase the dynamic concentration range of the assay.

In contrast, when [S] >> K_m, it follows from the Michaelis-Menten equation that the observed reaction rate now becomes independent of the substrate concentration (zero-order kinetics). Under these conditions, the enzyme activity cannot be used to determine the substrate concentration. However, in zero-order reactions a direct correlation exists between the reaction rates v and V, which is proportional to the total enzyme activity present in the assay mixture. Therefore, if a substance exerts an activating or inhibiting effect on the catalyst in a concentration-dependent manner, it will be possible to determine its concentration indirectly by kinetic measurement of the enzyme activity under substrate saturation conditions. This principle has been used to design kinetic substrate concentration assays for a number of effectors, e.g., heparin (based on inhibition of thrombin), insecticides (inhibition of acetylcholinesterase), theophyllin (inhibition of alkaline phosphatase) and several ions able to enhance the activity of apo-enzymes, e.g., Na^+, K^+, Mg^{2+}, Cu^{2+} or Cl^-.

Test strip systems and biosensors

In order to facilitate handling of the reagents used for determination of analyte concentration or enzyme activity, attempts have been made to use immobilised enzymes for analytical purposes. An early and still useful approach is to bind the enzyme non-covalently to paper. This has been used in the design of the test strips for use in biological samples such as urine or blood. In biosensors, the enzyme (usually an oxidoreductase) serves as the 'specifier' mediating analyte recognition. This biological interaction is then transformed through a 'transducer' component into an electrical signal which can be amplified and processed electronically (Fig. 8.3). A number of commercial test strip and biosensor systems for use in clinical chemistry have been developed.

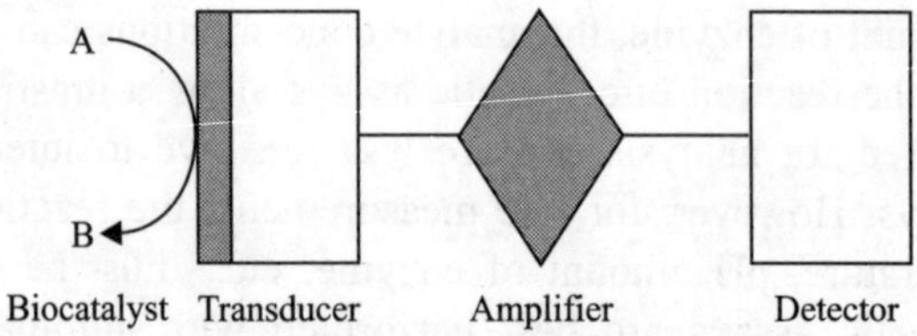

Fig. 8.3. General concept of a biosensor.

Enzymes as tools in biochemical analysis

Purified enzymes are widely used for analytical problems outside medical diagnostics and food analysis, e.g., in molecular biology and gene technology and in protein and glycoconjugate analysis.

In molecular biology, enzymes are used for molecular analysis of chromosome and genome structure, for the characterisation of genetic defects on the DNA level, for taxonomy of viruses and other organisms by correlation of characteristic fragment patterns, as well as for elucidating phylogenetic relationships. In addition, DNA and RNA modifying enzymes as well as polymerases (e.g., *Taq* DNA polymerase) are essential tools for all cloning techniques. Highly purified enzymes are also important tools in analysis of protein structure and modification (including glycosylation). Specific proteases are used for protein fragmentation and fingerprinting as well as for C-terminal protein sequencing (an example is given in Fig. 8.4).

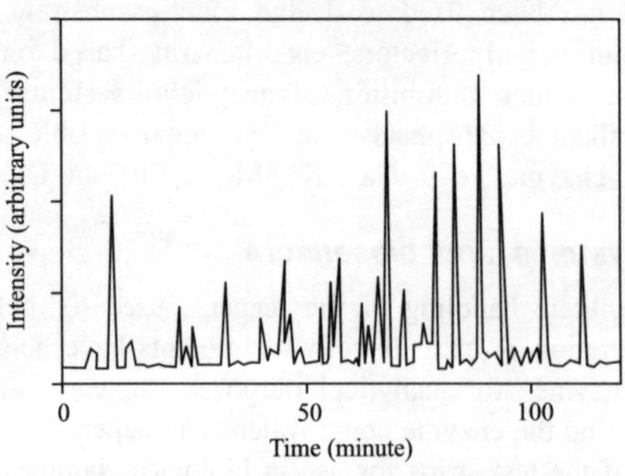

Fig. 8.4. Peptide map of a therapeutic protein.

Similarly, a set of glycohydrolases is used to analyse the carbohydrate residue structures of glycoproteins. A number of enzyme used for this purposes are listed in Table 8.7.

Table 8.7. Some enzymes used in biochemical analysis.

Enzyme (source)	*Use*
Proteases Trypsin (bovine) Chymotrypsin (bovine)	Protein fragmentation for sequence analysis, peptide fingerprinting, limited proteolysis of enzymes or receptors to study structure-function relationships
Endoproteinase Lys-C from *Lysobacter enzymogenes*	
Endoproteinase Glu-C (V-8 protease) from *Staphylococcus aureus V8*	
Carboxypeptidases A, B, C, Y	C-terminal protein sequencing
Restriction proteases	
Factor Xa (bovine or human)	Processing of recombinant fusion proteins
Enterokinase (bovine)	
IgA protease from *Neisseria gonorrhoe*	
Glycosidases	
Endoglycosidase D and *O*-Glycosidase from *Diplococcus pneumoniae*	Carbohydrate and glycoprotein analysis
Endoglycosidase F and *N*-Glycosidase F from *Flavobacterium meningosepticum*	
Endoglycosidase H from *Steptomyces plicatus*	
Many exoglycosidases	

Special requirements for analytical enzymes

Enzymes to be used for analytical applications have to satisfy a number of quality criteria concerning.

1. Specificity - absence of side activities towards other substances which may be present in the sample or the reaction mixture.
2. Purity - absence of contaminating activities or other contaminants interfering with the analytical and detection systems (and not necessarily purity with respect to the absence of other inactive proteins unless the enzyme is used in protein analysis).
3. Stability - in the reaction mixture during the reaction as well as in long-term storage.

4. Kinetic properties - suitable K_m and k_{cat} values and no inhibition by substances present in the sample.
5. pH optimum - suitable for the required experimental conditions.
6. Solubility and surface properties - no interference by adsorption or aggregation effects.
7. Cost.

The criteria are dependent on each other. Therefore, the choice and quality of an enzyme must be optimised in each case with regard to the particular analytical application.

Therapeutic Proteins

Proteins are part of numerous traditional medicines, such as snake and bee venom and enzyme preparations but, in most cases, these are ill-defined mixtures. Furthermore, naturally occurring proteins obtained from animals, plants or micro-organisms as well as from the human body (blood, urine, placenta or adenohypophysis) have long been used as drug ingredients. Typical examples are porcine insulin, blood coagulation factors VIII and IX from human blood fractionation, pancreatin as a digestive aid, or the Ancord and Batroxobin proteases obtained from snake venoms. However, foreign proteins are immunogenic for humans and will give rise to immune responses if injected into the bloodstream. This may lead to rapid inactivation and may prevent repeated application of the protein drug. Furthermore, the isolation of therapeutic proteins from human (and animal) fluids and tissues poses a potential risk of infecting the patients with other diseases, for example with HIV.

The concentration of physiologically active proteins in human body fluids or organs is very low, necessitating the processing of much material for low amounts of product. Chemical synthesis of proteins, although feasible in principle, is not normally an economically viable production method for protein drugs. So, only through the advent of gene technology has it become possible to produce human proteins in large amounts and high purity. In many cases, recombinant human proteins allowed, for the first time, a rational therapy with the body's own substances based on knowledge of the causes of disease and pathobiological mechanisms. Therapeutic proteins such as hormones, growth and differentiation factors, act in signalling while others function as biocatalysts or inhibitors. They are used for substitution, amplification, or inhibition of physiological processes.

At present more than 60 recombinant proteins are used in therapy (a selection is given in Table 8.8), and more than 200 others are in development. Erythro (EPO), insulin, somatotropin (human growth hormone), granulocyte-colony stimulating factor (G-CSF) and alpha-interferon are among the most successful drugs and are placed in the 'top twenty' on the list of worldwide sales of pharmaceuticals. Analysts expect that in the near future, more than 100 pharmaceuticals based on gene technology will be on the market.

Table 8.8. Selection of recombinant protein drugs.

Protein	*Diseases treated (selection)*	*Year of first approval*
Insulin	Diabetes mellitus	1982
Human growth hormone (hGH)	Short stature (hGH deficiency)	1985
Interferon-alpha	Hepatitis, cancers, genital warts	1986
Hepatitis B vaccine	Hepatitis B prevention	1986
Plasminogen activators (alteplase, reteplase)	Acute myocardial infarction	1987/1996
Erythropoietin	Anaemia of chronic renal failure	1989
Interferon-gamma	Chronic granulomatous disease	1990
Granulocyte colony stimulating factor (G-CSF)	Neutropenia, bone marrow transplantation	1991
Granulocyte–macrophage colony stimulating factor (GM-CSF)	Bone marrow transplantation	1991
Interleukin-2	Renal cell carcinoma	1992
Coagulation factor VIII	Haemophilia A	1992
Interferon-beta	Relapsing multiple sclerosis	1993
DNase	Cystic fibrosis	1994
Glucocerebrosidase	Gaucher's disease	1994
Coagulation factor IX	Christmas disease	1997
Interleukin-10	Prevention of thrombocytopenia	1997

Notes: Many of the proteins listed are commercialised as various brands by the same or different pharmaceutical companies.

Choice of expression system

Bacteria, yeast, insect and mammalian cells are the most commonly used hosts for heterologous protein expression. A brief overview of the main advantages and disadvantages of the various host systems referring especially to the use of the expression systems for therapeutic proteins will be given below.

Mammalian cells

In mammalian host cells, such as Chinese hamster ovary (CHO) or baby hamster kidney (BHK) cells, high-level expression (from 10 to more than 100 picogram per day per cell) of the recombinant protein can be achieved. Usually, the proteins are secreted into the fermentation medium in properly folded, active form and in most cases, glycosylation and other post-translational modifications occur in a more or less 'human-like' manner, although minor differences may exist. However, development of a stable cell line that expresses the therapeutic protein at a high level may take many months and the manufacturing costs are high. Therefore, the developed technology using CHO, or BHK, cells is today the systems of choice for large-scale production of modified, e.g., glycosylated, therapeutic proteins, especially if correct protein modification is crucial for the therapeutic effect.

Human cells

One way to ensure the identity of the recombinant protein product with the original human protein is to use human cell lines as the expression system. Instead of cloning a gene or DNA sequence coding for the desired protein into the host cell, it is an interesting alternative to manipulate not the gene itself, but its promoter, in order to activate expression of the endogenous human gene. This 'gene activation' technology may eventually become a commercially advantageous way for the production of many therapeutic proteins.

Insect cells

The gene coding for a recombinant protein can be inserted into the genome of the baculovirus, *Autographa californica.* It will very efficiently infect insect cells and uses their protein synthesis machinery to produce large amounts of protein (up to 500 mg of protein per litre of culture medium) within two to three days after infection. Therefore, this system is very suitable to obtain protein rapidly for feasibility studies. However, post-translational processing differs from the mammalian cell systems and the system is not very suitable for scaling-up.

Yeast

Yeast and other fungi are eukaryotic micro-organisms that are routinely cultivated on a large scale. Recombinant proteins are usually located inside the yeast cell, however, it is also possible to attach a leader sequence in order to induce protein secretion. Whereas intracellular heterologous proteins may accumulate to g l^{-1} level, secreted proteins usually reach titres of about 10 to 100 mgl^{-1}. When expressed in yeast, human proteins are correctly folded and disulphide-bridged, but glycosylation differs significantly from the mammalian pattern. In most cases, the well-known baker's yeast (*Saccharomyces cerevisiae*) is used, but for the production of large amounts of recombinant proteins, *Pichia pastoris* is the system of choice in many cases.

Bacteria

Bacteria host systems offer fast development, high efficiency and relatively inexpensive production as the main advantages. Recombinant proteins can be accumulated intracellularly or secreted into the peri-plasmic space (*Escherichia coli*) or into the fermentation medium (*Bacillus* species). However, post-translational processing of complex eukaryotic proteins, such as by glycosylation or disulphide bond formation, does not occur in the bacterial cell. Protein expressed in bacteria may also differ from their natural form concerning their amino terminus where an *N*-formylmethionine residue may be present. A correct *N*-terminus can however be obtained when the recombinant protein is expressed as a fusion construct with an *N*-terminal extension that is then cleaved off with a suitable protease. In many cases, the sequence of this fusion tail is chosen so that it can be exploited to facilitate purification, e.g., poly(His) tails, which mediate binding to metal chelate chromatography materials.

In addition, upon high-level expression in bacterial cells, whether cytosolic or periplasmic, many eukaryotic proteins are accumulated as misfolded, aggregated and insoluble particles (termed 'inclusion bodies') that have to be refolded *in vitro* into their native conformations. If this refolding process can be designed to offer a fair yield and good economics, the bacterial (especially *E. coli*) host system is very will suited for the production of all the therapeutic proteins that do not require post-translational processing for *in vivo* bioactivity.

Transgenic animals and plants

Transgenic manipulation means that a gene from one species (the transgene) is introduced into the germ line of another species, either plant or animal. Milk, blood as well as urine have been proposed for transgenic protein production and a number of different proteins have already been produced in this way; some proteins, e.g., milk-produced antithrombin-III and α_1-antitrypsin, are currently in clinical trails. Expression levels up to 35 g l^{-1} milk have been reported, suggesting that transgenic dairy animals may provide a cost-effective route to the large-scale manufacture of biotherapeutics. However, development times are long because the gestation period and the onset of sexual maturity of the animal are rate limiting and there are still a number of concerns with respect to the consistency of protein production from different animals. Nevertheless, transgenic technology represents a real challenge for biotechnology.

Production from transgenic plants is potentially a more economically attractive system for large-scale production of recombinant proteins, offering advantages in the low cost of growing plants on large acreage, the availability of natural protein-storage organs and the established practices for harvesting, transporting, storing and processing. At present, the main disadvantages are low accumulation levels of recombinant proteins, insufficient information on post-translational events and limited knowledge of relevant downstream processing technology.

Protein folding from inclusion bodies

Protein folding *in vitro* has often been compared to the task of unboiling an egg; to reform the biologically active, native protein conformation from insoluble and inactive aggregates. This 'naturation' process is usually done in several steps, as shown in Fig. 8.5.

In unfolded proteins, hydrophobic regions that would be buried within the native globular protein structure are exposed. These parts of the polypeptide chain tend to induce unspecific aggregation, thereby decreasing the refolding yield (Fig. 8.6).

Kinetically, aggregation is a bimolecular reaction and therefore concentration dependent. As a consequence, protein naturation usually has to be performed at high dilution and the resulting low protein concentrations and large reaction volumes lead to unfavourable economics for large-scale production. However, it has been demonstrated that a correctly folded (thus, hydrophilic) protein does not interfere with folding of further portions of the same, still unfolded protein. Therefore, if one starts refolding at a low protein concentration and adds further portions of unfolded protein continuously or discontinuously to the same mixture only after the initial amount has already found its correct conformation, the total protein concentration can be increased stepwise up to economically attractive levels. This process of 'pulse naturation' is commercially used in the production of a plasminogen activator.

Application, delivery and targeting of therapeutic proteins

Because of their typical substance class properties, proteins generally would be no means be considered 'ideal' therapeutic agents for reasons related to stability and application:

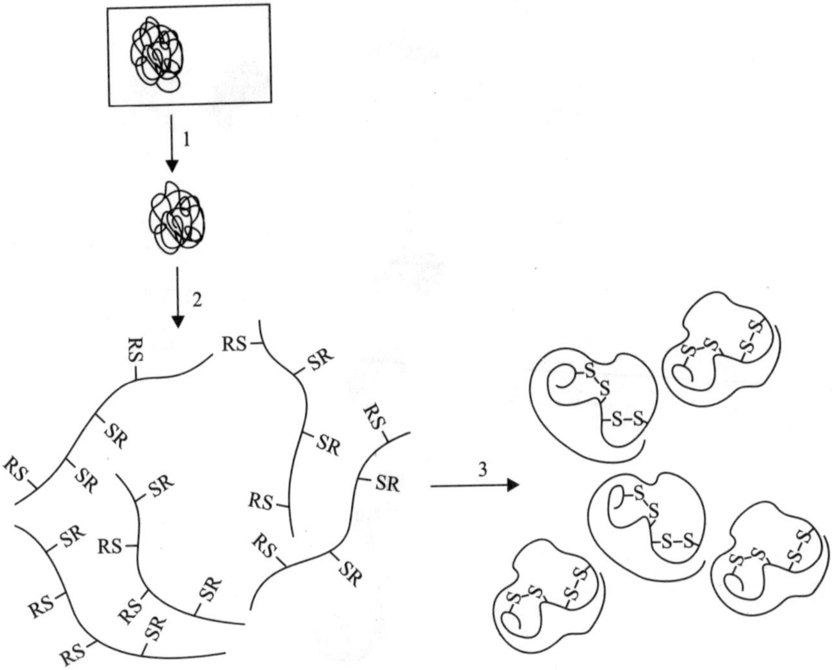

Fig. 8.5. Protein 'naturation' from inclusion bodies. Recombinant proteins overexpressed in bacterial cells often are formed as insoluble and misfolded 'inclusion bodies'. (1) After cell lysis, the inclusion bodies are collected by centrifugation, washed with buffer to remove soluble cell components (which may already lead to ≥ 90% purity of the desired protein) and are then dissolved in a concentrated solution of a strong denaturant, e.g., 6–8 M urea or 5–6 M guanidinium. HCl; (2) If the recombinant protein contains cysteine residues, a redox buffer system such as a mixture of reduced and oxidised glutathione (GSH/GSSG) at a slightly alkaline pH is added. In some cases, it has proven advantageous to modify the sulphydryl groups reversibly, e.g., by formation of mixed disulphides with glutathione, to increase solubility of the denatured polypeptide chain (R). The protein is then allowed to refold by slow removal of the denaturing agent, usually by dilution or dialysis, with concomitant formation of the correct disulphide bonds; (3) In many cases, the addition of additives such as arginine, tris(hydroxymethyl)amino methane, or alkylurea derivatives has been shown to improve the refolding yields considerably.

Stability

Proteins are polypeptides and therefore labile against heat, extreme pH values and biological degradation. This may lead to limited shelf-life as well as short half-lives in the human body, e.g., due to proteolysis in the stomach and intestine and to receptor mediated clearance from the blood followed by proteolytic degradation in the liver.

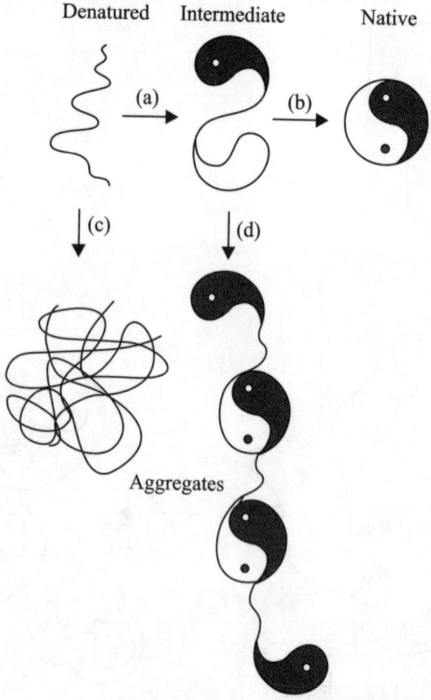

Fig. 8.6. Kinetic competition between protein folding and association. Steps (a) and (b) are productive first-order folding steps whereas steps (c) and (d) are unproductive second or higher order association processes.

In addition to the risk of proteolytic cleavage of the polypeptide chain, amino acid side chains of proteins may also be modified during storage, e.g., by oxidation or isopeptide bond formation. Whereas shelf-life can be improved by suitable additives such as sugars or amino acids, attempts to prolong the biological *in vivo* half-life of protein drugs have met limited success until now. Microencapsulation into biodegradable polymers, mostly polylactide-polyglycolide co-polymers, appears to offer an interesting approach in this respect.

Ways of application

The molecular surface of soluble proteins is hydrolphilic. Therefore, as a rule, proteins cannot pass through biological membranes and will not enter into tissue through the intestinal wall or into human cells from the bloodstream. Proteolytic degradation in the stomach could be prevented by encapsulation, giving the protein resistance against the acid pH and proteases. Therefore, oral application of protein drugs would not result in sufficient bioavailability unless the protein is intended to act in the oral cavity itself (as for example lysozyme, used to inhibit bacterial infections in the mouth cavity) or in the gastorintestinal tract (e.g., lipases and amylases, used to support food digestion). Protein therapeutics therefore cannot be given orally but have to be injected or infused into the bloodstream.

Immunogenicity of foreign proteins

Proteins that are foreign to the human body are immunogenic. When injected into the bloodstream, they may induce the formation of antibodies and cellular immune response. Furthermore, proteins obtained from natural sources may contain immunogenic contaminants. This may prevent repeated or prolonged application of the same protein drug. (The immunogenicity is desired when proteins are used as vaccines).

One way to decrease immunogenicity of proteins is chemical coupling to water-soluble polymers, especially to polyethylene glycol. Such 'pegylated' proteins are in use as therapeutics, for example PEG-adenosine deaminase (PEG-ADA) for treatment of ADA deficiency (SCID-severe combined immunodeficiency disease) by substitution of the missing enzyme, as well as PEG-asparaginase in tumour therapy. It is, however, not easy to ensure product homogeneity after chemical modification and of course production costs are increased by the additional chemical modification step. On the other hand recombinant human proteins are expected not to be immunogenic. Depending on the expression system used, however, proteins may differ from original human proteins in their post-translational modification (e.g., glycosylation, processing of *N*-terminus, etc.).

First-generation therapeutic proteins

The first-generation of recombinant therapeutic proteins are protein drugs made with the aid of gene technology. These are identical to natural human proteins. A few typical examples are listed below.

Insulin

This is a pancreatic hormone which has been used for treatment of type I diabetes since 1922 because of its effect in lowering blood glucose levels. Insulin consists of two polypeptide chains connected by disulphide bonds. The A chain has 21 amino acid residues and the B chain has 30 amino acid residues. Insulin biosynthesis involves proteolytic processing from the single-chain precursor molecule proinsulin, with release of a connecting C-peptide, as illustrated in Fig. 8.7.

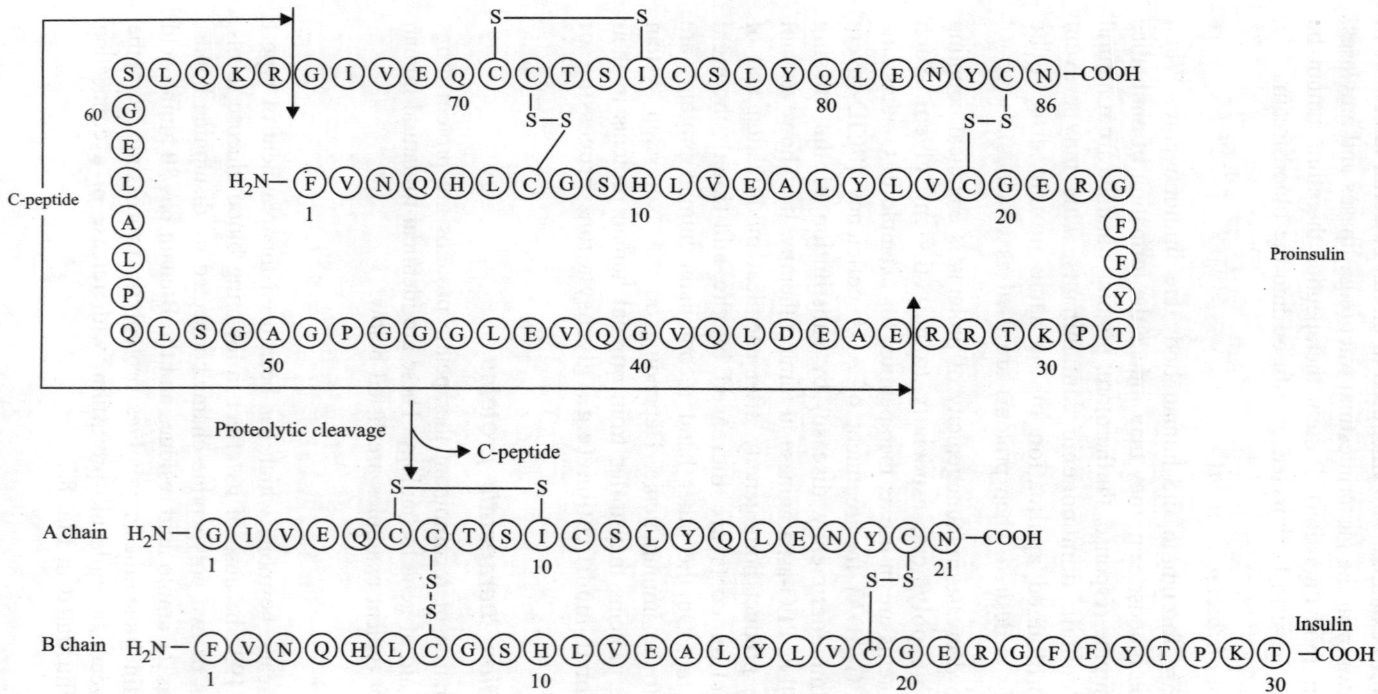

Fig. 8.7. Biosynthesis and amino acid sequence of human proinsulin and insulin. Proteolytic cleavage removes the connecting C-peptide from the single-chain precursor, proinsulin, to release the active two-chain, disulphide-linked insulin. In the analogue 'insulin lispro', the order of residues B28 and B29 is changed (Lys-Pro instead of Pro-Lys). B30 is the only amino acid residue which is different between human (B30 = Thr) and porcine (B30 = Ala) insulin.

During the first decades of insulin therapy, bovine or porcine insulin had to be used. In these animal proteins, there are some amino acid sequence differences from human insulin that may lead to formation of insulin antibodies during long-term application. In the 1970s it became possible to replace the alanine residue, B30, of porcine insulin with a threonine residue by protease-catalysed semisynthesis and, thus, insulin identical to the human molecule could now be produced. However, due to the growing population of patients needing insulin (about 1 in 1000), there were concerns that the supply of porcine insulin might become limited and the porcine or semisynthetic human material has been replaced by recombinant production of human insulin. Several strategies have been developed to produce recombinant insulin. In the original process described by Genentech, Inc. and Eli Lilly, the A and B chains are expressed separately in *E. coli* as fusion proteins with tryptophan synthease or β-galactosidase and, after processing by cleavage with cyanogen bromide, the two chains are connected by chemical reoxidation. In an alternative process, the physiological biosynthetic intermediate proinsulin (Fig. 8.7) or analogues with shortened connecting peptide sequence are expressed in *E. coli* or yeast and the connecting peptide is removed enzymatically.

Erythropoietin

Erythropoietin (Epoietin alpha and beta, EPO) is a glycoprotein of 165 amino acid residues. It is formed in the foetal liver and in the kidneys of adults. The EPO hormone belongs to the haematopoietic growth factors and induces the formation of erythrocytes from precursor cells (termed BFU-E und CFU-E) in the bone narrow.

Recombinant erythropoietin has to be produced in mammalian cell systems due to the necessity of glycosylation (Chinese hamster ovary (CHO) cells are used in the commercial processes) and is used therapeutically mainly in renal anaemia, but also in other indications, e.g., in tumour anaemia.

Granulocyte-colony stimulating factor (G-CSF)

G-CSF belongs, as EPO, to the class of haematopoietic growth factors. G-CSF stimulates proliferation and differentiation of neutrophil precursor cells to mature granulocytes. It is therefore used as an adjunct in chemotherapy of cancer to treat neutropenia caused by the destruction of white blood cells by the cytotoxic agent. Furthermore, G-CSF is also used in the treatment of myelosuppression after bone marrow transplantation, chronic neutropenia, acute leukaemia, aplastic anaemia, as well as to mobilise haematopoietic precursor cells from peripheral blood.

G-CSF is a glycoprotein containing 174 amino acid residues. Products have been launched which contain either the glycosylated molecule produced from recombinant CHO cells (Lenograstim) or alternatively an unglycosylated, but therapeutically equally effective, form produced from recombinant *E. coli* (Filgrastim) which additionally possesses an *N*-terminal methionine residue.

Among the first-generation of recombinant therapeutic proteins, there are also various antibodies, enzymes, such as glucocerebrosidase, used for treatment of glucocerebrosidase deficiency (Gaucher's disease) as well as coagulation factors, Factor VIIa, Factor VIII and Factor IX, used for substitution therapy of haemophilia. Tissue plasminogen activator (t-PA) is described below. Additional examples are listed in Table 8.3.

Second-generation therapeutic proteins (muteins)

DNA sequences coding for proteins can nowadays be modified by site-directed mutagenesis so that the amino acid sequence of recombinant proteins can be designed as desired. This is known as protein engineering. Mutated proteins obtained in this way are called muteins. The changes may be restricted to isolated amino acid residues (point mutations), but may also involve the deletion or insertion of larger sequence regions, or newly introduced connection of originally unrelated sequences (protein fusions). These technologies may offer a strategic approach to modify protein properties in a rational way, such as stability, solubility, substrate or receptor binding specificity, or pharmacokinetics. Muteins obtained by rational design have been described as the second generation of therapeutic proteins.

However, present knowledge of structure-function relationships in proteins is far from complete. Only in some simple cases has it been possible to predict the effects of sequence changes on observable protein properties but in most cases, the assumptions leading to the introduction of mutations have to be verified empirically. Nevertherless, in several cases recombinant proteins have been successfully designed for use as therapeutic agents. Protein engineering has been used in these cases for quite diverse purposes.

Insulin lispro

After an insulin injection, the plasma concentration of insulin rises so slowly that the injection should be done at least 15 minutes before a meal. Similarly, the plasma level insulin also decreases more slowly than physiologically required, so there is the danger of hyperinsulinaemia. The slower increase in concentration is due to the time needed for dissociation of insulin hexamer to the pharmacologically active dimers and monomers. To accelerate this process, a large number of insulin muteins have been constructed which still are biological active but show faster dissociation of hexamers in solution. One of these fast-acting insulin analogues is *insulin lispro* where, in analogy to the naturally occurring insulin homologue insulin-like growth factor-I (IGF-I), the order of the amino acid residues B28 and B29 was changed, (Fig. 8.7). It was reported that insulin lispro reaches pharmacologically efficient levels faster and therefore could be injected immediately before a meal.

Tissue plasminogen activators

Acute myocardial infarction (AMI) is the principal cause of deaths in most Western hemisphere countries. One approach to improve treatment of AMI is the use of thrombolytic enzymes. Plasminogen activators catalyse the proteolytic processing of the inactive proenzyme plasminogen, which circulates in the bloodstream, into the active protease plasmin. Plasmin is able to cleave the insoluble fibrin of blood clots into soluble fibrin fragment peptides so that the clot is dissolved and the blood vessel is opened. The reaction scheme is outlined in Fig. 8.8.

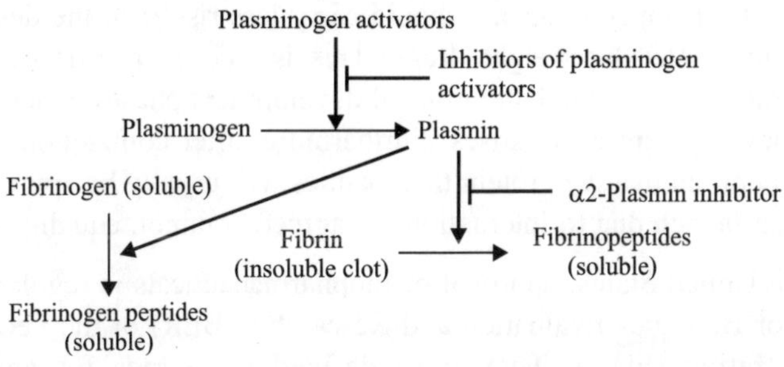

Fig. 8.8. General scheme of fibrinolysis. Thick arrows (↓) designated catalytic activation by proteolysis which is under control of plasma inhibitors (⊥).

Plasminogen activators (e.g., Alteplase and Reteplase, a mutein with an increased *in vivo* half-life) are used increasingly as thrombolytic agents in the treatment of AMI and are also used in studies on related diseases such as stroke or deep vein thrombosis.

Other second generation recombinant protein drugs

In the near future, much progress is expected in the field of recombinant immunotoxins used in experimental treatment of various cancers. These are chemical conjugates of recombinant fusion proteins constructed from a cell-binding part (mostly the antigen binding parts of an antibody), a translocation domain mediating transfer through the cell membrane and a cytotoxic portion, e.g., protein domains from bacterial toxins (such as *Diphtheria* toxin or *Pseudomonas* exotoxin) or a chemical cytotoxic agent. The idea is that the toxic agent should be targeted to the selected cancer cell population by the antibody domain directed against specific surface antigens of these cells, which then should be killed after internalisation of the toxin part. In clinical experiments, encouraging results have been obtained with immunotoxins targeting the transferring or interleukin-2 receptors or the erb B2 or Lewis-Y antigens. However, problems concerning in particular the immunogenicity of the bacterial toxin sequences still have to be solved.

Regulatory Aspects of Therapeutic Proteins

Development and approval risk

The toxicity of proteins is usually less severe than with chemically synthesised substances since they have fewer undesirable side-effects associated with them, due to their specific physiological roles. As natural substances, proteins are neither carcinogenic nor teratogenic. If the principle of action has been identified, development of recombinant human proteins into therapeutic agents should carry less risk than the development of new low-molecular weight drugs. This is true once efficacy has been demonstrated, e.g., in the latter clinical development phases, where the main part of development costs arises. Furthermore, after completion of clinical development, innovative protein therapeutics will usually be approved faster for market launch due to internationally agreed common equality standards.

In the United States, approval of biopharmaceuticals is regulated by the Centre for Biologics Evaluation and Research (CBER) of the Federal Drug Administration (FDA). They have defined guidelines for approval of therapeutic proteins as 'well characterised biotechnology pharmaceuticals'. In Europe, the approval procedure for biotherpeutics has been centralised through the European Medicines Evaluation Agency (EMEA) which follows similar guidelines.

Safety

In contrast to proteins isolated from human or animal, including transgenic, sources or from pathogenic organisms, e.g., vaccines obtained from bacteria or viruses, highly purified and carefully analysed recombinant proteins do not bear the risk of contamination with allergenic substances, pathogenic viruses, e.g., HIV, or prions from cattle or humans causing new variant Creutzfeldt-Jakob disease. For this reason, products such as coagulation factors (formerly produced from human blood or plasma), human growth hormone (in the past obtained from adenohypophysis extracts), or hepatitis B vaccines are today manufactured from recombinant systems.

Outlook to the Future of Protein Therapies

Considering the general advantages and disadvantages of protein therapeutics, it can be concluded that they are not equally attractive in all therapeutic areas and indications when compared with competing approaches such as low-molecular weight chemical substances on the one hand and gene therapy on the other. Protein drugs would be especially useful in the following cases:

1. In indications where no alternative therapy is available, particularly for potentially live-threatening diseases such as acute myocardial infarction, cancer or viral infections.
2. For substitution therapy if essential human proteins are missing or inactive, e.g., in ADA deficiency or in coagulation factor deficiencies.
3. To modulate the regulation of biological processes such as metabolism, cell growth, wound healing, etc., or to influence the immune system by proteins acting as hormones, growth factors, or cytokines (e.g., insulin, erythropoietin, G-CSF, somatotropin, interactions or interleukins). In these cases, protein-protein interactions have to be modulated. This may be more effective with therapeutic proteins as 'nature's own ligands' optimised in the course of evolution, than with small chemical substances.
4. As vaccines, especially against viral infectious diseases.

Human proteins identical to the body's own substances have become available through the advent of gene technology. Besides the first-generation biotherapeutics, an increasing number of redesigned, second-generation protein muteins with improved properties are being introduced to the marketplace. Once the present problems of low transfection and expression efficiency have been solved, it may be possible to substitute defect genes, or add therapeutic genes, to human cells *in vivo* so that the patient's body itself will act as the manufacturing facility where the synthesis of therapeutic proteins occurs. In this sense, gene therapy may represent the future third-generation of therapeutic proteins and may help to approach the final goal to cure, rather than treat, disease.

PURIFICATION AND CHARACTERISATION OF PROTEINS

The goals of purification vary with the intended use of the purified protein. The amount of protein needed can range from micrograms for microsequencing or the raising of antiserum, through milligrams for activity characterisation or structural studies, to grams or more for pharmaceutical products. Purity is defined by the general level of other protein contaminants, but also by the absence of contaminants of special interest, which can include endotoxins, highly antigenic proteins, high levels of salts, viruses, and competing enzymatic or biological activities.

Protein purification can generally be divided into five broad stages, which need not all occur sequentially: preparation of the source, gathering of all available information about the protein's properties, development of an assay, primary isolation, and final purification.

Preparation of the Source

Preparation begins with selection of the raw material from which the protein will be isolated. The source is here assumed to be a microbial or cultured metazoan cell line, although transgenic animals and plants will be of increasing importance in the foreseeable future.

It is important to start with ample quantities of target protein in the raw material. Overall purification yields for optimised pharmaceutical protein processes often exceed 80 per cent, but the first few developmental purifications of a poorly characterised target may achieve levels closer to 5 per cent, or even less. Make sure there is at least 10 to 20 times as much protein present in the raw material as the amount required for the intended use; there have been attempts in which even a 100 per cent yield would not have sufficed! An abundance of raw material permits greater exploration of alternatives and allows each step to achieve a greater degree of purification, at the expense of yield.

Protein supplies can be increased by increasing the cultivation volume, by growing more cells per unit volume (e.g., by computer-controlled fermentation with pH control, enhanced oxygen transfer, and regulated feeding of substrate), or by producing more of the desired protein per cell. Expression levels can be increased by screening of high-producing cell lines; cloning and overexpression; use of antibiotic selection pressures other than ampicillin, which is usually ineffective at high cell densities; use of tightly regulated, strong promoters such as T7 or alkaline phosphatase; and often by expression at lower temperatures, e.g., as low as 10°C for *Escherichia coli*.

In addition to its quantity, the quality of the raw material is subject to improvement. These improvements can include elimination of key contaminants (e.g., proteases, competing enzymatic activities), optimisation of posttranslational modifications, maximisation of extracellular secretion of correct folding, and the genetic addition of purification 'handles' such or protein fusions, as discussed below.

Properties of Proteins

Knowledge of almost any characteristic of the protein may find application in the development of a purification process; properties of interest are summarised as follows.

1. Source: cell type, expression level, intra/extracellular location, folding state, presence of proteases/glycosidases.
2. Stability: To endogenous proteases, temperature range, pH range, ionic strength, hydrophobic surfaces, aggregation tendency, cofactor or metal ion loss/requirement, freeze-thaw stability (for storage during and after purification).
3. Size: multimeric state, molecular weight, peptide chain(s), hydrodynamic radius.
4. Charge: isoelectric point, titration curve, electrophoretic mobility.
5. Binding partners: substrates and cofactors (and analogs), screening-derived binding agents (antibodies, peptides, aptamers), metal affinity. Binding to any of these in immobilised forms for affinity chromatography. Stabilising effects of binding partners.

Assay

An assay for the desired activity or protein is required. Assays for use in purification must be convenient and rapid; extreme precision is less important. Typical methods include UV/visible absorbance, dot-blot and gel-shift binding assays, chromogenic enzyme activity assays (e.g., in microtiter plates), and biosensor (BIAcore) testing of fractions.

Initial Isolation

Initial isolation steps separate the product from the majority of the water in the cultivation medium and from the majority of the host cell components. Recovery of both extracellular and intracellular proteins usually begins with cell separation by centrifugation or filtration.

Concentration

Extracellular (secreted) proteins are usually concentrated from the cell-free broth by ultrafiltration or adsorption on chromatographic media. Ultrafiltration can be carried out at small scale in centrifugally driven cells, or at larger scale in stirred cells or tangential-flow units. Adsorption can employ any of the chemistries noted below, although expensive and/or delicate adsorbents (e.g., immobilised antibodies) are normally used for initial capture only on the small scale, or for unstable proteins. An increasingly popular approach at larger scale is direct protein capture in fluidised beds of dense adsorbent particles from culture media and cell lysates, thus reducing the number of early process steps.

Sometimes a secreted protein is adsorbed to the outside of the cells and can be concentrated along with them, then liberated by washing, often with a high-salt buffer.

Cell lysis

Intracellular proteins are liberated by cell lysis. Total lysis of cells is required for liberation of cytoplasmic proteins. Microbial lysis often employs chemical agents such as chloroform, toluene, sodium dodecyl sulphate (SDS), EDTA, lysozyme (especially for gram-positive organisms), freeze-thaw cycling, and mechanical stress. Chemical, enzymatic, and freeze-thaw methods are often less than completely effective and may best be used on smaller scales. Mechanical methods include sonication (for small scale), grinding, glass bead mills, the French press, and the Manton-Gaulin homogeniser. Each of these methods destroys proteins as well as cells and requires optimisation of the release of active protein, but each can give reliable, satisfactory results. The French press in particular is a standard tool for lysis of some tens of grams of wet cell paste by forcing cells through a small orifice at high pressure. Primary isolation yields may benefit from cooling, or from addition of protease or glycosidase inhibitors. These measures can be particularly valuable with cell lysis using mechanical methods, which tend to heat samples.

Periplasmic proteins can be selectively liberated by osmotic shock, which ruptures only the outer membrane by sudden transfer from hyper- to hypotonic medium. Successful performance and scale-up of this operation can be tricky and may benefit from measurement of release of enzyme activities known to be cytoplasmic (e.g., β-galactosidase) or periplasmic (e.g., alkaline phosphatase).

Cell lysates are unattractive substances, rich in degradative activities, sticky lipids, and viscous nucleic acids along with thousands of protein contaminants. Soluble proteins are often recovered from cell lysates using precipitation or (less commonly) liquid partitioning, as these methods are tolerant of viscosity, particulates, and fouling. Ribosomes and nucleic acids can first be precipitated with streptomycin, protamine, or polyethyleneimine. The desired protein is often concentrated and partially purified by precipitation with ammonium sulphate or polyethylene glycol. Stepwise addition of optimised levels of precipitant can first precipitate some contaminants, and then (at a higher concentration) the desired protein, leaving other contaminants in the final supernatant. Exact results depend on the rate of precipitant addition, mixing, and ageing for precipitate formation, but these are generally robust methods from which proteins can be recovered in high yields.

Refolding

Recombinant proteins often misfold to form dense, insoluble aggregates of inactive protein. (Such aggregation can be detected by SDS-polyacrylamide gel electrophoresis [SDS-PAGE] analysis of the soluble and insoluble fractions of a cell lysate, or inclusion bodies can be directly observed by microscopy.) Proteins in inclusion bodies can be harvested by centrifugation at purities exceeding 70 per cent, but they are valueless unless the native tertiary structure can be restored. The first step in the renaturation process is dissolution of the inclusion bodies (sometimes even whole cells) in a strong chaotrope solution such as 6 M urea or guanidine hydrochloride. Dissolution in denaturant is rapid and reliable. The dissolved, denatured protein is then (hopefully) allowed to renature to its native conformation by removing the denaturant through dialysis, dilution, or chromatographic separation. In the renaturation step, unimolecular refolding processes must compete with multimolecular reaggregation reactions, so operation with immobilised or highly dilute protein molecules is required. Development of a process that recovers the protein in native form while losing as little as possible to reaggregation can be difficult and is always unpredictable. In many cases, a low refolding yield is acceptable because the protein is abundantly available from inclusion bodies.

Several rules of thumb for refolding have emerged.

1. Small, disulphide-free, single-peptide-chain proteins are easiest to refold, but proteins of all types have been refolded.
2. Refold at low protein concentrations.
3. If possible, immobilise the refolding protein by adsorption to prevent multimolecular reactions, which promote reaggregation.

4. Control redox potential (through dithiothreitol and β-mercaptoethanol addition) and oxygen supply.
5. Add arginine, polyethylene glycol, or cyclodextrins.
6. Reduce denaturant concentration in two or more steps, rather than attempting a drastic reduction in a single step.
7. Allow up to 7 to 10 days for refolding.
8. Recycle the reaggregated protein back to the refolding process.

High-Resolution Purification

Chromatography is the usual method of preparing highly purified, active proteins (3, 5, 7, 9–12, 25, 28, 32). Most forms of chromatography involve selective adsorption of proteins on the surface of porous particles, through interactions that can be broadly classified as ion-exchange, hydrophobic, and affinity. Chromatographic operations are also classified as low-pressure, medium-pressure (including Pharmacia's popular FPLC) and high-pressure (HPLC), depending on the pressure used to force liquid through the packed bed of adsorbent particles (Table 8.9). In general, high pressures imply higher costs but also finer adsorbent particles and better resolving power. Samples loaded onto high- and medium-pressure columns must be rigorously free of particulates.

Table 8.9. Types of chromatography.

Feature	*Low-pressure*	*Medium-pressure*	*High-pressure (HPLC)*
Particle size	40–150 μm	10–75 μm	2–15 μm
Flow driver	Gravity, peristaltic	Piston or syringe	Positive displacement
Run time	40–1,000 min.	15–16 min.	0.5–30 min.
Apparatus cost	Low	Medium high	High
Resolving power	Lowest	Intermediate	Highest
Particulate tolerance	Low	Very low	Lowest

In addition to the adsorbent matrix, apparatus for chromatography includes a column into which the particles are packed (glass, plastic, or steel, depending on operating pressure), a pump or height differential to drive liquid flow, some method for introducing the sample into the flow before the column (a switching valve, or manual pipetting onto the top of the packed bed) and a fraction collector that deposits the emerging, separated proteins into different vessels. Possible enhancements include UV/visible absorbance detectors for monitoring the emergence of the separates proteins from the column, a conductivity monitor for tracking changes in salt concentration used to elute proteins from the column, and a computer for control and for storage of the results.

Not all adsorptive separations involve particles in a column. Particles can be used for batch adsorption by gently mixing them into a solution containing proteins to be captured. The resolution obtained in this manner is low, but the method is tolerant of fouling and particulates and can easily be applied to large volumes of solution. Nonparticulate supports, notably membranes, are available in many of the surface chemistries discussed below. These give rapid adsorption and desorption kinetics, although with some what reduced capacity.

There are a variety of excellent references for the methodological details of chromatographic separation of proteins. The present discussion will address general characteristic, along with some guidelines for method development.

Ion-exchange chromatography

Ion-exchange chromatography is the most common high-resolution method for preparative separation of proteins and is used in most protocols (Table 8.10). It will be discussed in particular detail because of its importance, and also to illustrate the general issues involved in development of chromatographic separations. Adsorbents with fixed positive charges (from immobilised amines, e.g., DEAE, QMA) are called anion exchangers and are employed at pHs above the isoelectric point of the protein to be adsorbed so that the protein's net charge will be of sign opposite to that of the adsorbent.

Table 8.10. Modes of adsorptive chromatography.

Feature	*Ion-exchange*	*Hydrophobic*	*Metal chelate*	*Biospecific*
Adsorbent	Carboxyl, amine, sulphonate	Porpyl, butyl, phenyl	Chelator, loaded with Ni^{2+}, Zn^{2+}	Antibody, cofactor, receptor
Selectivity	Moderate-high	Moderate-high	Moderate-high	High-very high
Capacity	High	High	Moderate-high	Moderate-high
Matrix cost	Low	Low	Moderate	High
Elution	High salt, pH	Low salt	pH, imidazole	pH, chaotrope
Initial salt	Low	High	Indifferent	Often indifferent

Matrices bearing negative charges (carboxylates and sulphonates, e.g., CM, SP) are called cation exchangers and are used at lower pHs. In practice, the protein of interest is usually adsorbed rather than passed through the column, as higher resolution and a degree of concentration can be achieved. Ion-exchange adsorption depends strongly on the ionic strength, equal to half the sum of the concentrations of all ions present, each multiplied by the square of the ion's charge. For 1 M NaCl, I = 1 M; for 1 M $(NH_4)_2SO_4$, I = 3 M. Initial trials might employ ionic strengths of 10 to 50 mM for loading and 1 M for elution.

The combinations of pH and matrix type (anion or cation exchange) at which the protein will be adsorbed can be predicted with some confidence from its pI, although the detailed distribution charges on the protein's surface plays a role, as do nonspecific interactions with the adsorbent backbone. If the pI is not known, it can sometimes be measured through isoelectric focusing or can be estimated from the sequence. Otherwise, a sample containing the protein is loaded (at low ionic strength) on an anion-exchange column at the lowest pH at which it is known to be stable, or on a cation-exchange column at its highest stable pH. The column is then washed with 2 to 3 volumes of the starting buffer, or until the initial peak of unretained protein has fallen to baseline. Most proteins will be retained on the column and can be eluted with a gradient of increasing salt concentration up to 1 to 2 M applied over a volume 5 times that of the column packing. Failure to bind can be due to overloading of the column (5 to 10 mg of total protein per ml of column packing is a conservative starting point); high loading ionic strength; failure to regenerate the column sufficiently at low ionic strength after the previous elution (this can require more than 5 column volumes (this can improve at lower temperatures, or with slightly higher loading ionic strengths); or instability at too extreme a loading pH.

Optimisation of the method involves increasing the fraction of the protein recovered in active form, the purity of the recovered material, and the rate at which purified protein is isolated (through loading of larger samples or shorter separation times). Steps in process development often include loading under less permissive conditions (higher salt, pH closer to neutrality) to cause more contaminants to wash through unadsorbed, higher operating temperatures, faster liquid flow rates (to the limits specified by the adsorbent manufacturer), and testing of shorter gradients, or gradients tailored to the shallow only near the ionic strength at which the protein of interest is eluted.

Hydrophobic interaction chromatography

Hydrophobic interaction chromatography (HIC) exploits the presence of exposed hydrophobic groups on the surfaces of proteins, which can interact with immobilised nonpolar moieties such as short alkyl chains and phenyl rings (i.e., immobilised propane and benzene). Adsorption is promoted by high ionic strength, and elution often employs a salt gradient of decreasing concentration. As burial of nonpolar groups is a major contributor to protein stability, there is reason for concern that exposure to hydrophobic surface may be denaturing. Many very hydrophobic polymer surfaces are in fact rapidly coated with denatured protein, and this is a major constraint on the selection of support materials for chromatographic matrices. Isolated hydrophobic groups sparsely immobilised on polar supports such as polysaccharides and acrylates, however, are not nearly as denaturing as bulk materials, and recoveries of many proteins from optimised hydrophobic separations can approach 100 per cent. The cost and capacity of HIC matrices are attractive, as is the methods's tolerance of high salt concentrations, which can allow immediate processing of proteins eluted in high salt from a previous ion-exchange chromatography or ammonium sulphate precipitation step. HIC is widely used in industrial processes and is probably under utilised in academic laboratories.

Affinity Chromatography

The remaining adsorptive techniques are affinity methods, which exploit selective interactions characteristic of biological systems. The normal biological functions of many proteins depend on their ability to associate with substrates, cofactors, etc., and proteins often will bind selectively to immobilised forms of these partners. Alternatively, an affinity agent such as an antibody can be used to selectively capture the protein of interest. Finally, it has become increasingly common to append DNA encoding a 'purification handle' protein with affinity for an available ligand to the gene encoding a recombinant protein of interest. The resulting fusion protein can be purified by its affinity for this ligand. The purification handle sequence can be removed by proteolysis, or left in place for many applications. Typical purification handles include cationic polyarginine, the immunoglobulin-binding domain of protein A of a *Staphylococcus aureus*, cellulose- and maltose-binding proteins, and the increasingly popular immobilised-metal-binding hexahistidine (His_6) described in the next section.

Immobilised-metal affinity chromatography (IMAC) exploits the affinity of certain chelated metal ions (e.g., Ni^{2+} and Zn^{2+}) for the amino acid side chains of cysteine and especially histidine. Elution is achieved by pH gradient or the addition of competitive metal-chelating compounds such as imidazole. IMAC adsorbent matrices are relatively inexpensive and robust, and selectivity can be quite high. The method is also tolerant of variations in ionic strength. The popularity of IMAC has increased dramatically with the widespread use of genetically added chelating-peptide purification handles on recombinant proteins, especially His_6.

Biospecific affinity chromatography describes separations base on molecular recognition interactions normally found in nature, such as antibody/antigen, enzyme/cofactor, and hormone/receptor affinity. Many widely applicable premade affinity matrices (at least 50 types) are commercially available. Among the most popular affinity methods are Cibacron Blue dye affinity for purification of enzymes recognising ATP or NAD(P)(H), to which this textile dye bears a coincidental structural resemblance, and chromatography on protein A of S *aureus* for purification of antibodies. There are also available many 'activate' matrices that have been chemically derivitised for facile coupling of affinity ligands through surface amines, carbohydrate groups, etc.

The selectivity of affinity separations can be enormously high. This advantage is balanced in some cases by the cost and delay of obtaining and coupling suitable ligands. Also, the harsh conditions that may be required for elution of the bound protein can damage both target protein and affinity column. An active area of development in purification technology is the screening of large numbers of potential ligands, both phage-displayed peptides, aptamers, and small organic molecules, for use as improved affinity separation ligands.

Size exclusion chromatography

Size exclusion chromatography (SEC, also known as gel filtration), differs from the adsorptive methods in that protein interactions with the chromatographic matrix are not exploited, but minimised. SEC media display pores of varying sizes, and differential access to these pores is the basis of separation. Large molecules run faster, as they enter fewer pores and flow through an effectively smaller volume. All molecules above a certain size run together, 'excluded,' as they enter very few pores. Small molecules below the 'included' size limit rung together, exploring nearly all pores. Useful resolving power is confined to a range of hydrodynamic radii (not molecular weights) between the effective sizes of the largest and smallest pores; matrices of differing average pore size are used to purify proteins of different sizes. The inclusion and exclusion limits of a given matrix are given by the manufacturer. It is important to remember that larger molecules run faster (the opposite of a sieve or filter), and that separation is on the basis of molecular radius, not molecular weight. This implies that an elongated protein runs more rapidly than a compact globular protein of the same molecular weight, and both run much more slowly than a nucleic acid or random coil carbohydrate of the same molecular weight as the proteins.

SEC has relatively low sample capacity, and the packing and sample loading of long, narrow SEC columns is more demanding than preparation of relatively squat columns of adsorptive media. It is well worth testing the quality of a newly packed SEC column using a tracer such as acetone or blue dextran. SEC is a gentle, nondenaturing method, and in contrast to most methods is capable of removing aggregated forms from nearly pure protein samples. An SEC column can be 'calibrated' by running a set of proteins differing in molecular weight, but all of similar shape (in practice, all globular). The retention times will ideally give a linear dependence on the log of the molecular weight of the protein (deviations often indicate interactions with the matrix, which can be suppressed by adding salts or traces of organic solvents to the running buffer). The retention time of the unknown protein indicates its largest dimension. This information is especially valuable if it can be compared with the peptide molecular weight from SDS-PAGE, as large differences in the results of the two methods can often be the first sign of aggregation or multimerisation.

A few general rules for protein purification are given here.

1. Tailor the process to the intended purpose.
2. Obtain a convenient assay.
3. Select the right source of raw material.
4. Work quickly, or in the cold, or both.
5. Eliminate proteases early, through host selection, separation, or inhibition.
6. Identify stable intermediate storage forms.
7. Avoid freeze-thaw cycles.
8. Use columns of appropriate capacity.
9. Build on previous work.

Characterisation

Complete characterisation of the purified product involves study of the idiosyncratic activities of each protein. There are also, however, general structural properties that are subject to determination by generic methods.

Electrophoresis

Probably the most common method of protein characterisation is SDS-PAGE, which gives the molecular weight(s) of the peptide chain(s) of the protein, and also a good indication of the purity of the sample. SDS converts nearly all proteins into regular rodlike forms of constant charge density per unit mass, so that electrophoretic mobility is independent of amino acid composition. The gel serves to suppress mixing induced by temperature gradients, which would reduce resolution. SDS-PAGE can use commercial pre-cast gels (e.g., Pharmacia's pHAST system), or gels can be cast at the time of use. Related electrophoretic methods include isoelectric focusing (IEF; for measuring isoelectric point), native gel electrophoresis (in which proteins are not denatured with SDS), and two-dimensional gel electrophoresis, in which IEF separation by charge is crossed with orthogonal separation by mass using SDS-PAGE. Native electrophoresis and IEF also show promise as preparative methods.

Capillary electrophoresis (CE) is conducted in tubing of very small diameter (<100 μm) using high voltages (e.g., 30 kV). CE can achieve very high resolution in separations requiring only a few minutes and is widely regarded as an extremely promising method. The methods required for CE analysis of protein samples are still relatively sophisticated, however, and this technique is not yet recommended for the beginner.

Peptide sequencing

Amino-terminal sequencing is used to identify the first few amino acids of the protein. Most workers will choose to have this procedure performed by a specialist core facility. This sequence information can be used to confirm the identity of the protein, to establish the degree of removal of *N*-terminal methionines, purification handles, or secretion leader peptides, and to detect exoproteolytic degradation. Protein sequencing depends on sequential, stepwise removal of *N*-terminal amino acids in the form of labeled derivatives that are separated by HPLC and identified by their characteristic retention times. Blocked *N*-terminal are not uncommon and require removal of the blockage, isolation and sequencing of internal peptides, or use of mass spectrometric sequencing as discussed below. Carboxy-terminal sequencing, while successfully demonstrated, is now yet readily available in most core facilities.

Tryptic mapping

Tryptic mapping, in which small peptides derived from the protein by endoprotease action are separated by high-resolution reverse-phase HPLC, allows sensitive detection of variations in protein (especially pharmaceutical) preparations. Individual peptides can be subjected to sequencing as described above (e.g., avoiding problems with blocked *N*-terminal) to confirm the identity of a given peptide, to develop information of DNA probe development, or even to establish the entire sequence of the protein.

Analytical ultracentrifugation

Analytical ultracentrifugation has undergone a renaissance with the recent introduction of the first commercial instrument in many years (Beckman's XL-A). This technique, which requires considerable investment in equipment and training, allows measurement of a variety of properties of a protein sample, including solution molecular weight, self-association, interaction with other molecules, and sample homogeneity. Sedimentation field-flow fractionation is useful for characterising similar properties of larger particles.

Spectroscopy

Circular dichroism gives an indication of the fraction of the polypeptide that is composed of specific secondary structural features, such as α-helix and β-sheet. Raman spectroscopy has been used (less widely) for similar purposes, and especially for characterising metal-containing protein cofactors.

Biosensors

Biosensor devices such as Pharmacia's BIAcore, Fison's IASys, and ASI's Integrated Optics biosensor are now widely used for detecting and characterising intermolecular interactions. These devices share the property of continuous detection of proteins, cells, or nucleic acids present in a narrow region adjacent to the detector surface. If, for example, an antibody to a particular protein is immobilised on the sensor surface, addition of a sample containing that protein will produce an immediate signal. Because of their sensitivity, low material consumption, and convenience, these devices are very useful for tracking the desired protein in the course of purification. They also have a role to play in characterising the purified protein, as they can readily answer questions such as: Does this protein bind to this immobilised ligand/antibody/oligonucleotide? Does this compound inhibit binding of this pair of molecules? What is the (approximate) concentration of the molecule of interest in this preparation? Do the binding/desorption kinetics of this batch of this material match those of the last batch? It is necessary, however, to guard against the (often severe) effects of mass-transfer limitation, rebinding, flow, multivalency, etc., in attempting to derive quantitative biophysical information from these devices; errors have been published as users have gradually become familiar with these devices.

Use of a biosensor device requires covalent coupling of one member of an interacting pair to the surface of the detector, in active form. The chemistries involved are derived from those of affinity chromatography and are now generally well characterised by the instrument manufacturers. For a signal to be produced, a meaningful amount of mass must be captured by the interaction. For example, in studying the interaction of an antibody with a small hapten, it would be better to immobilise the hapten and capture the much larger antibody, as capturing the hapten might not produce a useful signal.

Analysis of glycosylation

Analysis of glycosylation is an actively developing field, in part because the characterisation of glycoproteins remains quite challenging. Many eukaryotic proteins contain sugars, which are sometimes essential to the protein's function. Available methods include stains and blots for detection of (particular classes of) glycosylation, deglycosylating enzymes, sugar-specific lectins for blotting and separations, and nuclear magnetic resonance and mass spectrometric methods of characterising individual side chains. An emerging area is the controlled creation and remodeling of glycosylation to create what are called 'neoglycoproteins'.

Mass spectrometry

Mass spectrometry (MS) has blossomed in recent years with continuing advances in instrumentation and methods of sample introduction. MS services are now commonly available in core facilities and are probably somewhat underutilised by potential users, considering the amount of information MS can provide for a small investment of sample and time. While there exists a family of related MS techniques, the present discussion will introduce the uses of MS as a whole, all of which are driven by the ability to determine the mass of proteins and peptides with remarkable precision, of the order of a few hydrogen atoms in a typical protein. The technique, therefore, can detect and characterise almost any important modification or variation in a protein's structure. These include post-translational modifications, *N*-terminal methionine addition or removal, methionine oxidation, proteolytic trimming, and many sequence mutations and deamidation events. Because these modifications are detected without being specifically looked for, MS is an excellent way to characterise a newly purified protein.

Mass spectrometry is advancing into new applications. These include protein sequencing, on-line coupling to HPLC and CE, and identification of 'hits' from combinatorial libraries. A new application driven by the availability of genome sequences and cDNA sequence banks is the identification of unknown proteins separated by electrophoresis or HPLC from a diseased tissue, or a cell line expose to an agent thought to alter expression of one or more proteins. MS determination of the molecular weight of a protein showing interesting behaviour (or, if digestion can be done, of one or more peptides derived from the protein) gives mass values that can then be searched against the (large) database of all proteins/peptides encoded by that genome or cDNA bank. Remarkably, this type of search can often be successful and in many cases will avoid the need to purify the protein in question at all.

Emerging Trends

1. Automation: Computer control of chromatographic processes is increasingly common. Several manufacturers have introduced software capable of testing a large number of gradient designs, pHs, etc., to minimise the human effort required to develop optimal methods.
2. Purification handles: Genetic addition of biotinylation signals, or proteins or domains having affinity for particular ligands, continues to be popular. Probably the newest development is the use of self-removing intein fusions designed to recover the native protein with no added sequences.
3. Combinatorially derived separation ligands: Phage display, screening of libraries of small molecules, and nucleic-acid-base ligand (aptamer) libraries are all promising methods of identifying optimised ligands for affinity separations.
4. CE and MS: Each of these methods is advancing rapidly. MS should and will be more widely used, largely through core facilities, in the near term. CE will likely replace SDS-PAGE in a few more years.

9

Biotechnology and Medicine

INTRODUCTION

The impact of pharmaceuticals on human health care is an area where biotechnological innovations are likely to have the earliest commercial realisation. The long-standing awareness within the health-related industries of biological and biochemical innovations has led to these industries being heavily involved in biotechnological research, particularly molecular biology. Furthermore, since health-related products are generally of high value, the financial return warrants extensive research investment. However, the considerable time required to develop a modern pharmaceutical product must not be underestimated, and long periods of toxicological testing are necessary before the national regulatory bodies will grant approval for marketing. The cost of achieving this approval can be many millions of dollars, and the product must have a high sales potential to warrant this investment. Many potentially worthwhile products will not appear on the market because it is not in the financial interest of the producing companies to meet such vast costs of gaining official approval.

New medical treatments based on biotechnology are appearing almost daily in the marketplace. These include (a) therapeutic products (hormones, regulatory proteins, antibiotics), (b) prenatal diagnosis of genetic diseases, (c) vaccines, (d) immunodiagnostic and DNA probes for disease identification, and (e) genetic therapy. This is the largest commercially developed area of new biotechnology, with massive present and future markets and can only be selectively examined here.

PHARMACEUTICALS AND BIOPHARMACEUTICALS

The vast bulk of pharmaceutical drugs presently on sale are synthetic chemicals derived either directly by chemical synthesis or by chemically modifying molecules derived from biological sources. Biopharmaceuticals are considered to be recombinant protein drugs, recombinant vaccines and monoclonal antibodies (for therapeutic roles). Biopharmaceuticals are becoming increasingly relevant in biological applications but are still only a small part of the pharmaceutical industry. However, there can be little doubt that the techniques of molecular biology/genetic engineering will become a dominating factor of drug discovery, design and development. Biotechnology will also accelerate screening, speed bioassays and the production of new drugs, and also explain more accurately how drugs act in the human system. Biotechnology will almost certainly vastly reduce the huge costs presently incurred in product development of new drugs (e.g., costs of discovery, development, scale-up, clinical trials and regulatory paperwork).

ANTIBIOTICS

The discovery in 1928 by Alexander Fleming that a fungus called *Penicillium notatum* could produce a compound selectively able to inactivate a wide range of bacteria, without unduly influencing the host, set in motion scientific studies that profoundly altered the relationship of humans to the controlling influence of bacterial diseases. From these studies emerged the fungal antibiotics penicillin and cephalosporin, and the actinomycete antibiotics streptomycin aureomycin, chloramphenicol, tetracyclines and many others. Many bacterial diseases have largely been brought under control by the use of antibiotics. Pneumonia, tuberculosis, cholera and leprosy, to mention only a few, no longer dominate society and, at least in the developed parts of the world, have been relegated to minor diseases. Griseofulvin, an antibiotic active against fungi, has brought great relief to those infected with debilitating fungal skin diseases such as ringworm.

Antibiotics are antimicrobial compounds produced by living microorganisms, and are used therapeutically and sometimes prophylactically in the control of infectious diseases. Over 4000 antibiotics have been isolated but only about 50 have achieved wide usage (Table 9.1). The other antibiotic compounds failed to achieve commercial importance for reasons such as toxicity to humans or animals ineffectiveness or high production costs. Antibiotics were extensively used in medicine from about 1945 with the arrival of penicillin. New antibiotics soon extended the range of antimicrobial control, and antibiotics are now widely used in human and veterinary medicine and (to a lesser extent) in animal farming, where some antibiotics have been shown to increase the weight of livestock and poultry.

Antibiotics can also be used to a limited extent to control plant diseases and as insecticides.

Table 9.1. Some economically important antibiotics

Antibiotic compound	*Producer microorganism*	*Activity spectrum*
Actinomycin D	*Streptomyces sp.*	Antitumour
Bacitracin	*Bacillus sp.*	Antibacterial
Bleomycin	*Streptomyces sp.*	Anticancer
Cephalosporin	*Acremonium sp.*	Antibacterial
Chloramphenicol	*Cephalosporium sp.*	Antibacterial
Daunorubicin	*Streptomyces sp.*	Antiprotozoal
Fumagillin	*Aspergillus sp.*	Amoebicidal
Griseofulvin	*Penicillium sp.*	Antifungal
Mitomycin C	*Streptomyces sp.*	Antitumour
Natamycin	*Streptomyces sp.*	Food preservative
Nisin	*Streptococcus sp.*	Food preservative
Penicillin G	*Penicillium sp.*	Antibacterial
Rifamycin	*Nocardia sp.*	Antituberculosis
Streptomycin	*Streptomyces sp.*	Antibacterial
Tetracycline	*Streptomyces sp.*	Antibacterial, antiamoebic

Antibiotics that affect a wide range of micro-organisms are termed *broad spectrum*, for example, choramphenicol and the tetracyclines, which can control such unrelated organisms as *Rickettsia*, *Chlamydia* and *Mycolplasma* species. In contrast, streptomycin and penicillin are examples of *narrow-spectrum* antibiotics, being effective against only a few bacterial species. Most antibiotics have been derived from the actinomycetes and the mould fungi.

The production of antibiotics has undoubtedly been a highly profitable part of the pharmaceutical industries in the industrialised world.

In 1992, the cephalosporins (products derived from cephalosporin C and penicillin G or V) were one of the largest business sectors in the global pharmaceutical market sales. The present processes are highly efficient and have been achieved with little knowledge about the genetics of the producing organisms. This was, in part, due to the lack of an obvious sexual cycle, which limited cross-breeding experiments. However, new techniques such as protoplast fusion and gene transfer technologies are leading to the development of new strains with higher productivity, improved stability and possible new products. These improvements have all resulted in continued decreases in overall costs of production. At present all antibiotic fermentations involve centrally stirred tank reactors run under aerobic batch conditions. Modifications in production processes may well follow on from the novel fermenter designs that are gaining wider industrial acceptance.

It is regrettable to note that most studies on antibiotics have been concerned with diseases prevalent in the developed nations. Many diseases of developing countries, including many major tropical diseases, have received little attention from the major pharmaceutical industries. In part this may be due to the high level of technology, including specially trained personnel, normally associated with antibiotic research and development. More probably, the reason lies with the economics of developing new drugs for countries with limited financial resources. Let it be hoped that the advances in biotechnology may make it possible to follow a more enlightened pathway to develop the antibiotics necessary to combat the massive specific disease problems of the developing nations. Biotechnology may well make it possible to economically produce *orphan drugs*–drugs with specific needs but small profit return.

A disquieting observation has been the gradual evolution of drug resistance in many bacteria. The possibility of this acquired resistance being transmitted to another species of bacterium is now real. For example, gonorrhoea (a venereal disease) resistant to treatment with penicillin, is now present in 19 countries. It is recognised that the resistance factors are located on plasmids within the bacterium and because of this can be more easily transmitted between organisms. The very core of gene transfer technology derives from this phenomenon. Antibiotic resistance in many well-known diseases is steadily increasing and must cause serious concern in our society. For example, tuberculosis, a former scourge of nations, has been nearly eradicated but is now increasingly being diagnosed in Western countries!

The very large market for antibiotics in animal feeds and food preservation is now under considerable reappraisal. Without doubt, the addition of relatively small amounts of certain antibiotics (e.g., bacitracin, chlortetracycline, procaine, penicillin) in the feed of livestock and poultry led to the production of animals that were healthier, grew more rapidly and achieved marketable weight faster. However, there is now little doubt that the incorporation of medically important antibiotics into feed has led to increased spread of drug-resistant micro-organisms, increase shedding of dangerous *Salmonella* bacteria in animal dung, and the transfer of antibiotic residues into human food.

As a consequence of the dangers of using antibiotics of human relevance in animal feed, there has been a massive effort to produce antibiotics specifically for animal feed incorporation, so replacing the medically used antibiotics. Thus, antibiotics of low therapeutic potency in humans or with an insufficient spectrum of activity are now more regularly used in animal nutrition.

The fight to produce food for the world's ever-increasing population goes on relentlessly. A major part of this strategy involves the battle against microbes. Biotechnology will play a major part in developing and producing new and effective antibiotic compounds.

The combination of new and traditional technology in the pharmaceutical industry holds outstanding potential for improving micro-organisms used in antibiotic production and the isolation of new antibiotic products. The high values of antibiotics and the proportionately low cost of raw materials is a further incentive to research. There is also no competition from chemical processes.

VACCINES AND MONOCLONAL ANTIBODIES

During the last 15 years we have witnessed the unravelling of the bewildering processes of immune response in human and animal systems. When a foreign substance (e.g., a micro-organism) enters an animal system a remarkable chain of reactions is set in motion, which, if successful, will result in the inactivation and exclusion of the invading organism. This molecular response can in some cases remain in the animal system for many years, giving complete or partial immunity against that type of micro-organism. As the foreign molecule is the *antigen*, which can elicit a counteracting response, the *antibody*, from the host system.

In general, antigens are proteins, or proteins combined with other substances such as sugars, though polysaccharides and other complex molecules may also act as antigens. In the disease process, antigens usually reside on the surface of the invading micro-organism and trigger the body's defences against it. In this way antibodies are the essence of immunity against disease.

Antibodies are made by special cells throughout the body and it is now recognised that individual animal species, including humans, can produce unbelievable numbers of different antibodies. The antibody-producing cells recognise the shape of particular determinant groups of the antigen and produce specific antibodies in order to neutralise and eliminate the foreign substance. Thus, an animal has sufficient antibodies to combat not only the vast array of microbial invasions that can occur but also an unlimited range of synthetic chemicals. In short, the mammalian system can bind and inactivate almost any foreign molecule that gets in. However, should a particular antigen challenge not be adequately dealt with, then the invading micro-organism can rapidly multiply and create imbalance, illness and perhaps death in the susceptible host.

The ability to stimulate the natural antibodies by vaccines has long been known. Vaccines are preparations of dead micro-organisms (or fractions of them), or living attenuated or weakened micro-organisms that can be given to humans or animals to stimulate their immunity to infection. In this way they mimic infectious agents without the pathogenic consequences and elicit in the body, protective immune responses. When used on a large scale, vaccines have been a major force in the control of microbial diseases within communities. The major goals of vaccine research are to identify and characterise the individual antigens of infectious agents that elicit protective immune responses and to define the components in the immune response that induce protection.

Vaccines have been developed against many microbial diseases. However, the success and persistence of the antimicrobial effect varies widely between types of vaccine. Thus, a vaccine for poliomyelitis has almost eliminated this disease on a world scale, while vaccines against typhoid and cholera are still unsatisfactory. It has not yet been possible to produce any fully effective vaccines against such world-wide diseases as gonorrhoea, syphilis, serum hepatitis, malaria and many others. A massive global effort is now in progress to develop a vaccine against human immune deficiency virus (HIV), which is responsible for acquired immunodeficiency syndrome (AIDS).

Vaccine production is a high cost, low volume production system that encompasses many basic principles of biotechnology. Scale-up, in particular, is a constant problem with virus diseases that need to be produced from animal or human cell cultures. New advances in fermenter technology are rapidly revolutionising this work and should greatly increase vaccine production in the near future.

New methods of antibody production are now being considered. In current practice antibodies are obtained from immunised animals, but this is usually a tedious and time-consuming operation. At the end of the various extraction and purification stages the antibodies are usually weakly specific, available only is small batches and of variable activity. Attempts to culture antibody-secreting cells have been unsuccessful, since such cells neither survive long enough nor produce enough antibodies in culture to become worthwhile sources of antibodies. Furthermore, such systems normally produce mixtures of different antibodies (polyclonal antibodies).

The production of vaccines to combat human and animal diseases represents an immense market that has been extensively developed by the pharmaceutical industries. At present, vaccine quality and efficacy range from excellent to unsatisfactory. In virally derived diseases, vaccines are being developed by recombinant DNA technology against the influenza virus, polio virus, hepatitis B virus, herpes virus and more recently HIV. Successful vaccines in these important world disease could mean massive commercial gains for the producing companies. The biggest opportunities exist for diseases such as AIDS or herpes.

Extensive studies are also in progress with certain bacterial vaccines, as well as vaccines for parasitic diseases. Malaria still remains the most prevalent infectious disease in the world; this complex and demanding problem could well be overcome in the very near future.

A significant new development in medically related biotechnology has been the ability to produce monoclonal antibodies. A major advance of this technique is that when antibody producing cells are immortalised and stabilised the secreted antibodies will always be the same from that particular cell line and can be fully characterised to assess their suitability for different applications. In this way suitable antibodies can be produce, scaled-up (either as ascites tumours in mice or by various forms of fermentation technology), in large quantities allowing much greater standardisation for diagnostic applications. Monoclonal antibodies are now finding wide applications in diagnostic techniques requiring highly specific reagents for the detection and measurement of soluble proteins and cell surface markers in blood transfusions, haematology, histology, microbiology and clinical chemistry, as well as in other, non-medical, areas (Table 9.2).

Table 9.2. Monoclonal antibody markets.

Cancer diagnosis and therapy
Diagnosis of pregnancy
Diagnosis of sexually transmitted diseases
Prevention of immune rejection of organ implants
Purification of industrial products
Detection of trace molecules in food, agriculture and industry

Monoclonal antibodies may also be used in the treatment of tumours and possibly to carry cytotoxic drugs directly to the tumour site. In particular, monoclonal antibodies have found ready application in *in vitro* diagnostic products that do not need such rigorous safety testing. Diagnosis can be achieved for many diseases, including human venereal diseases, hepatitis B and some bacterial diseases, Such antibodies can also be used in pregnancy testing.

On a commercial scale monoclonal antibodies are being produced in 100 litre airlift fermenters, by encapsulation in 100 litre fermenters, and in perfusion chambers using lymph from live cattle. Commercially, monoclonal antibodies have been one of the most rewarding areas of new biotechnology.

BIOPHARMACEUTICALS

The vast majority of pharmaceutical products are compounds derived either from synthetic chemical process, from naturally occurring sources (plants, microorganisms), or combinations of both. Such compounds are used to regulate essential bodily functions or to combat disease-causing micro-organisms. Increasing attention is now being directed to the body's own regulatory molecules, which occur normally only in very small concentrations and have predominantly defied modern methods of extraction or synthesis. Limited quantities of some of these compounds have historically been derived from organs of cadavers and from blood banks. Genetic engineering is now increasingly being recognised as a practical means of providing some of these scarce molecules in unrestricted quantities. In practice, this involves inserting the necessary human-derived gene construct into suitable host micro-organisms that will produce the therapeutic protein (biopharmaceutical) in quantities related to the scale of operation. Not only is it now possible to produce these biopharmaceuticals in a form identical to that normally occurring in the human body but also to design meaningful improvements in activity, stability or bioavailability. Such products will also be free from the dangerous contaminants that have occasionally arisen from extraction of cadavers, e.g. the degenerative brain disease Creutzfelt-Jakob disease has been associated with the administration of human growth hormone from early extractions.

The successful development of biopharmaceuticals requires:

1. Advanced biochemical/biomedical research to identify and characterise the native compounds.
2. Skilled molecular biology and cloning technology to identify the relevant gene sequences and insert them into a mammalian or microbiological production host.

3. Bioprocess technology to grow the organisms and to isolate, concentrate and purify the chosen compounds.
4. Clinical and marketing expertise.

Table 9.3 indicates some of the main biopharmaceuticals approved for marketing world wide between 1982 and mid-1999. It is probable that between 50 and 100 new biopharmaceuticals are currently undergoing final clinical trials.

Table 9.3. Biopharmaceuticals approved for marketing between 1982 and 1999.

Product	*Broad approved medical use*
Insulin	Diabetes
Human growth hormone	Growth deficiency
α-Interferon	Cancer, viral infections
Anti-T-cell	Organ transplantation
Hepatitis B vaccine	Hepatitis B prevention
Tissue plasminogen activator	Cardiovascular disease
Erythropoietin	Anaemia
Interleukin-2	Cancer

However, the use of these protein-type pharmaceuticals has several restrictions that will limit their use and size of market. As proteins they are unstable and poorly absorbed from the gastrointestinal tract-consequently they have to be given by a medically trained person. Thus, their main use will be for acute rather than chronic conditions. With long-term therapy, some proteins may also cause allergic reactions in the patient.

Important Therapeutic Proteins, Cloned into Micro-organisms

The first human gene sequences encoding important therapeutic proteins, cloned into micro-organisms, were insulin, human growth hormone (somatostatin) and the interferons.

Insulin

Throughout the world there are millions of people who need regular intakes of insulin to overcome the lethal effects of diabetes. Insulin extracted from pigs and cattle has long been the source of world-wide usage and it is now believed that some of the unfortunate side-effects that have occurred with continued long-term use of insulin could be due to additional contaminating compounds present in the animal insulin. Recombinant human insulin appears not to have such problems and is increasingly having the largest market share of sales. Production is unlimited and free from market shortage of animals and all the problems associated with previous production methods.

Somatostatin

The growth hormone, somatostatin, has been extremely difficult to isolate from animals; half a million sheep brains were required to yield 0.005 g of pure somatostatin. By cloning the human gene for somatostatin into a bacterium, this same amount of hormone can be produced from 9 litres of a transgenic bacterial fermentation. One child in 5000 suffers from hypopituitary dwarfism resulting from growth hormone deficiency and the easy availability of this biopharmaceutical will be of immense benefit to these child sufferers.

However, a potential massive market could arise from the increasing evidence that this growth hormone can increase muscle formation in normal individuals and is now being exploited by some athletes. There are also claims that regular administration of the hormone can improve quality of life in the aged!

Interferons

In 1957 two British researchers discovered substances produced within the body that could act against viruses by making cells resistant to virus attack. Most vertebrate animals can produce these substances, known as interferons, and many animal viruses can induce their *in vitro* synthesis and become sensitive to them. Why, then, have the interferons not become the 'penicillins' of virus infections? Primarily, this is because only minute amounts of interferon are produced within cells, and it has proved unbelievably complicated to extract and separate them from other cellular proteins.

Human interferons are glycoproteins (proteins with attached sugar molecules) and are believed to play a part in controlling many types of viral infection, including the common cold, as well as having potential in controlling cancer. However, the scarcity of these compounds has consistently hampered efforts to understand the extent of their effectiveness.

There are many different types of interferon characteristic of individual species of animals; mouse interferon will respond to mouse cells but not human cells, and vice versa. Furthermore, different tissues from the same species appear to produce different interferons. Thus, interferon for human studies must be derived from human cells, and it has been here that the blockage to production has occurred. Most early human interferon production was carried out in Finland using leukocytes from blood, and the small amounts of interferon produced this way were used for limited clinical tests throughout the world.

So far, studies have shown that interferons can confer resistance to some virus infections, and are involved in the body's natural immune reactions even in the absence of viruses. However, much of the current interest in interferons arises from their ability to inhibit cancer in experimental animals. Interferons present a new approach to cancer therapy because they appear to attack the cancer cells by inhibiting their growth, and that of any viruses involved in the cancer process, and they can also stimulate the body's natural immune defences against the cancer cell. Although the limited clinical studies of these compounds have indicated considerable potential in cancer therapy, the restricted supplies have severely hampered conclusive experimentation; this must await greater availability of interferons.

Two sources of interferon are currently available. The first is from human diploid fibroblasts growing attached to a suitable surface and the interferon produced is widely considered to be the safest available. The second source is from bacteria in which the gene for human fibroblast interferon has been inserted into a plasmid in such a manner that interferon is synthesised and can then be extracted and purified.

Lymphokines

Lymphokines are proteins produced by lymphocytes (part of the body's immune system) and are considered to be crucially important to immune reactions. They appear to have the capability of enhancing or restoring the capacity of the immune system to fight infectious diseases or cancer. Interleukin-2 at present offers the greatest potential and is now produced by genetic engineering and consequently more readily available on the market. With each of these important compounds it has been possible to achieve a level of realistic pharmaceutical drug delivery only because recombinant DNA technology enabled the synthesis of large quantities of the product.

The production of human vaccines by recombinant methods has been quite successful and should allow for new approaches to diseases without existing remedial treatments. Recombinant Hepatitis B virus has gained regulatory approval and has high market sales.

At present, all biopharmaceuticals are produced by way of genetically engineered mammalian cell or microbial fermentations. However, with the development of transgenic animals it has become possible to produce certain human proteins of biopharmaceutical potential, including tissue plasminogen activator, blood clotting factors, etc., in the lactating glands of several animal species, such as mouse, sheep, cow and pig, and to be able to express the products in the milk of the animal. These products can then be more easily extracted and purified. A further feature of this process is that it is accomplished in a mammalian system that can confer on certain human proteins the complex structural modifications required for full biological activity. Such modifications cannot be achieved in microbial systems. As yet no commercial production is in operation and awaits improvements in yields and final regulatory approval. Undoubtedly this will become a major source of production of certain complex human proteins. There is no apparent adverse effect on the animal, which continues to produce milk in the normal way.

An American company can now produce human haemoglobin in the blood transgenic pigs that could therefore serve as a human blood substitute. Such transgenic haemoglobin could capture a massive market. This transgenic haemoglobin would be free from human pathogens such as HIV and would not need typing or matching before transfusion because it is not composed of red blood cells. Much needs to be done before this becomes a reality.

Pharming – producing human pharmaceuticals in transgenic animals – is rapidly becoming a reality. The animals can be considered in biotechnology terms as bioreactors operating on a continuous basis. Overall this could be a massive global market. At present the major limiting factor is achieving the number of transgenic animals.

GENE THERAPY

Undoubtedly, the most far reaching and controversial area of genetic engineering of humans is gene therapy. This is the treatment of disease by the transfer and expression of genetic material in a patient's cells in order to restore normal cellular function. It is, however, essential to distinguish between *germ cell gene therapy* and *somatic cell gene therapy*. In germ cell gene therapy changes are directed at the individual's genetic make-up and can be passed on to the offspring. Ethics and practical wisdom ensures that this type of therapy will *not* be permitted in any country, in the foreseeable future. In contrast, in somatic cell gene therapy, functioning genes are introduced into body cells that lack them. The effects of the therapy are confined to the person undergoing the treatment and are *not* passed on to the offspring.

The main thrust of gene therapy has been directed at correcting single-gene defects (mutations), such as cystic fibrosis and haemophilia (Table 9.4), that have been observed in families by their Mendelian pattern of inheritance. It is believed by some that many hundreds of such diseases could be treated by this process and the next decade should see significant progress if the many technical problems can be readily overcome. At present most genetic diseases have no effective treatment so gene therapy could offer hope for so many people. Somatic cell gene therapy for complex multifactorial diseases, e.g., Parkinson's disease, cancer, must be a long way off. In many of these diseases there can be many genes involved as well as an interaction with environmental factors.

Table 9.4. Selected human genetic diseases for possible single gene therapy.

Disease	*Target tissue*	*Incidence*
Thalassaemia	Bone marrow	1:600 is some populations
Cystic fibrosis	Liver	1:500
Duchenne muscular dystrophy	Muscle/brain	1:300 males
Haemophilia A	Liver	1:6000 males
Haemophilia B	Fibroblasts	1:30,000 males

The human genome is believed to contain up to 100,000 genes involving 3000 million nucleotides per haploid chromosome. In America and Europe there are now human genome mapping projects with aims: (i) to provide a genetic map for the relative position of the genes; (ii) to provide a physical map for the actual gene positions; and (iii) to determine the sequence of bases in the DNA.

This is a vastly expensive programme and the current goal is to have the total human genome sequenced by the year 2007. It should also be noted that such ambitious projects do divert limited resources from other *equally* important areas of research. World wide, healthcare systems are threatening to break down under the escalating costs of treating ageing populations. Undoubtedly there will be increasing resort to cost–benefit analysis and many of these treatments will not be universally available.

Gene therapy is a complex series of events relying heavily on new biotechnological techniques. Therapy will require a full understanding of the mechanism by which the defective or unusual gene exerts its effect on the individual, an ability to switch off the defective gene and to substitute a healthy gene copy (Fig. 9.1). It is truly a multidisciplinary activity involving skills in molecular biology, cell biology, virology, pharmacology, clinical application and patient interaction. The USA is the undoubted world leader in gene therapy research and application. Some of the protocols are now proceeding through regulatory processes and already patent coverage for gene construction, delivery systems and supporting technology has been granted. Private medical systems internationally must see gene therapy as a very lucrative market in the affluent nations.

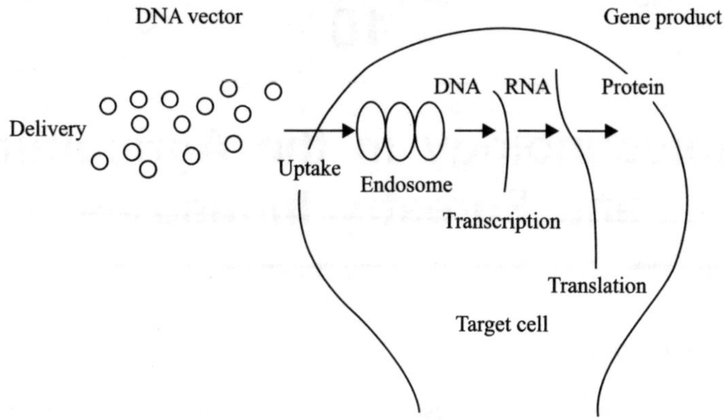

Fig. 9.1. Enctional steps in gene transfer and expression. Gene therapy involves the delivery of DNA vectors to specific target cells within the body, uptake (commonly by endocytosis), transcription and translation of sequences within the DNA vector, and production of therapeutic gene product.

10

Biotechnology in the Agricultural and Forestry Industries

INTRODUCTION

Agriculture is the world's largest, single industry and in advanced societies such as the USA agriculture contributes over 20 per cent of gross national product. In developed economies agriculture relies heavily on technology to achieve productivity and profitability. Many aspects of modern biotechnology are now being applied increasingly to agriculture. Agriculture in many parts of the world is undergoing a major strategic restructuring to achieve vertical integration between production and ultimate utilisation. Whereas the food-processing company would formerly buy raw agricultural products on the commodity market breeding programmes are increasingly becoming established to create the desired raw materials with the specific traits required for higher value processing. Genetic engineering is creating a revolution in agriculture allowing an ever-increasing range of plants and animals. There will be increased stability in the marketplace and much less wastage. Agricultural biotechnology will allow higher quality standards with lower costs of production.

The human race is totally dependent on agriculture and as world populations continue to expand there must be continuous reassessment of agricultural practices to optimise their efficiency. However, agriculture is a politically sensitive area, with many selective trade barriers and protectionist policies. In many parts of the world such as Europe and the USA agriculture is a highly efficient industry and continues to demonstrate annual increases in productivity. In contrast, many countries are still not self-sufficient in food production due to many reasons such as lack of good agricultural practices, hostile or changing climate, or political instability. Many countries are intrinsically poor and lack the ability to take advantage of new biotechnology. This can only be remedied when the advanced nations make available the training programmes and financial investments required. However, much is already taking place through agencies such as the World Bank and the EU.

PLANT CELL CULTURE

Many plant cell cultures can regenerate entire plants, making it possible to exploit this property for introducing large scale cloning in horticulture cell also provides a good way to extend studies in plant pathology to casual relationships in tumor formation, host-parasite interactions and in sanitations of pathogens–infected stock of crop plants.

Plant cell culture has found wide applications ranging from studies of basic plant biochemistry and molecular biology to mass propagation and genetic engineering of crop species. The media, conditions, culture vessels, and other critical parameters vary widely depending on the intended use of the culture. This chapter covers the basic requirements for common procedures in plant cell culture. For work with a particular species, it is recommended that one survey the litreature to see if procedures and media for the chosen species or a close relative have already been developed, since optimum conditions can be species or genus dependent.

In general, plant tissue culture begins with the initiation of callus cultures, i.e., dedifferentiated masses of rapidly dividing cell cultures, on solid, agar-based nutrient media. The callus is generated by exposing sterile pieces of a plant to plant growth regulators. Normally, cell division in an intact plant is restricted to various meristematic regions (for example, root and shoot tip meristems) by the plant regulating its natural hormone distribution. The plant growth regulators are either natural plant hormones or synthetic analogs which bind to receptors in the plants cells, causing all cells to resume division and growth.

Callus cultures can be maintained indefinitely by passing clumps of cells to fresh agar media at regular intervals. Alternatively, callus pieces can be placed in shake flasks containing liquid media to eventually form suspension cultures. These grow faster than callus cultures, probably due to better nutrient exchange with the medium. Plant cell suspensions consist of some single cells and many clumps of a few to hundreds of cells; the size of the clumps depends upon the species, the medium, and the length of time since the culture was initiated. Suspension cultures can be used as uniform, year-round source of plant cells for biochemical and plant physiological studies or as a source of valuable plant secondary metabolites following appropriate treatments.

The callus cultures can also be allowed to differentiate (regenerate) back into plants; this process is often accelerated by the omission of or changes in the types of plant growth regulators. Each piece of callus can give rise to one to hundreds of plantlets, each with the same genome as the parent, giving rise to the industry of micropropagation. Either early or late in the callus stage, foreign DNA can be introduced into the dividing plant cells. If the cells are then allowed to regenerate, a portion of the plantlets will be transgenic and can be selected for in a number of ways. At one time, the main way of introducing foreign DNA and organelles into plant cells involved the production of protoplasts (wall-less plant cells released by the action of cellulases and pectinases). Currently, several methods of DNA delivery are used, including delivery by the 'tamed' plant pathogen *Agrobacterium tumefaciens* or delivery by direct shooting of DNA-coated particles into the cells.

Given good sterile technique and some patience for the comparatively slow growth, much can be accomplished today with plant cell culture.

Culture Media

The composition of a plant cell culture medium resembles that of an all-purpose plant fertiliser, with the addition of a carbon source, since most cultures cannot photosynthesise enough sugars for growth, and plant growth regulators to stimulate cell growth. Most plant cell lines can be cultured in completely defined media, although for some applications complex ingredients such as coconut milk were used in the early formulations and are still used today. The compositions of two of the most commonly used media are shown in Table 10.1.

Table 10.1. Composition of common plant cell culture media.

Component	*Concentration in:*			
	MS		*B5*	
	mM	*mg/litre*	*mM*	*mg/litre*
Micronutrients				
NH_4NO_3	20.6	1,650	–	–
KNO_3	18.8	1,900	25	2,500
$CaCl_2$-$2H_2O$	3.0	440	1.0	150
$MgSO_4$-7 H_2O	1.5	370	1.0	250
KH_2PO_4	1.25	170	–	–
$(NH_4)_2$ SO_4	–	–	1.0	134
NaH_2PO_4-H_2O	–	–	1.1	150

(Contd...)

Component	*Concentration in:*			
	MS		*B5*	
	mM	*mg/litre*	*mM*	*mg/litre*
Macronutrients				
KI	0.005	0.83	0.0045	0.75
H_3BO_3	0.100	6.3	0.050	3.0
$MnSO_4$-4 H_2O	0.100	22.3	–	–
$MnSO_4$-H_2O	–	–	0.060	10.0
$ZnSO_4$-7 H_2O	0.030	8.6	0.007	2.0
Na_2MoO_4-2 H_2O	0.001	0.25	0.001	0.25
$CuSO_4$-5 H_2O	0.0001	0.025	0.0001	0.025
$CoSO_4$-6 H_2O	0.0001	0.025	0.0001	0.025
Na_2EDTA	0.100	37.3	0.100	37.3
$FeSO_4$-7 H_2O	0.100	27.8	0.100	27.8
Vitamins/amino acid				
Inositol	0.49	100	0.49	100
Nicotinic acid	0.0047	0.5	0.0094	1.0
Pyridoxine HCI	0.0024	0.5	0.0048	1.0
Thiamine HCI	0.0003	0.1	0.0300	10.0
Glycine	0.005	2.0	–	–
Carbon source				
Sucrose	30/g/litre		20 g/litre	
pH	5.8		5.5	

Murashige and Skoog medium (MS) was one of the first developed for the growth of plant cells, using tobacco as the test species; it is still used today for the growth of callus and suspension cultures, as well as the growth of axenic cultures of plants. Gamborg's B5 medium (B5) was an adaptation of MS, optimised for the growth of soyabean cells, and is used for many species. There is some confusion in the terminology used in the litreature, but generally a medium is defined by its formulation of inorganic salts and organic compounds (the macro- and micronutrients), often excluding the major carbon sources (usually sucrose). Some authors include either the carbon source or growth regulators or both in their definitions, so it is important to check the formulations in the litreature carefully.

Macronutrients

The macronutrients provide the sources of N, P, K, S, and Ca, all required for healthy plant growth. The Na and Cl included as counterions are not essential; MS contains no Na source other than a trace amount included with the Fe source. Na, K, and Cl ions can be tolerated at levels up to 50 to 60 mM and can be used as counterions for additives or for pH adjustment without concern. Nitrogen is provided mainly as nitrate, although some media contain both ammonium and nitrate. Utilisation of ammonium causes the culture pH to drop, while utilisation of nitrate causes the pH to rise. Some species have a requirement for ammonium or greatly benefit from the inclusion of ammonium or glutamine, but high concentrations of ammonium can be toxic either directly or by causing extremely low pH shifts. Phosphate salts provide both a source of P and the majority of the buffer capacity of the media.

Micronutrients

The micronutrients are sources of various trace elements required for growth. These are similar to those found microbial defined media, with the addition of iodide. Studies have indicated that iodide is not essential but greatly stimulates cell growth. Iron is always added in chelated form, either as an EDTA salt or as agricultural chelates (e.g., Sequestrene 330 Fe).

Carbon sources

The most common carbon source is sucrose, generally in the range of 20 to 30 g/litre; the original formulation for MS called for 30 g/litre, while B5 called for 20 g/litre. Glucose is used less often, and fructose is used only occasionally. In many cultures, much of the sucrose is rapidly taken up by the cells in the first 1 or 2 days of culture and converted to starch granules inside the cells. If different carbon sources are to be tested for growth, one must take into account that the cells previously cultured on sucrose may have substantial starch reserves, allowing growth for several days. Cultures of other species often hydrolyse the sucrose to fructose and glucose and utilise the glucose first and then the fructose. Very high sugar concentrations (40 to 100 g/litre) have been used in specialised secondary-metabolite production media and to adjust the osmotic potential of the media in short-term treatments for regeneration.

Vitamins

Although intact, soil-grown plants normally synthesise all of the vitamins required for growth, vitamins are generally added to all plant culture media. Very early work indicated an absolute requirement for thiamine (vitamin B_1) in the growth of tomato cells, but this has not been determined for all species. Pyridoxine, nicotinic acid, and *myo*-inositol have each been shown to increase the growth of certain species and should also be included, especially when establishing new callus cultures. In many cases, however, these may not be required for growth and some species have been reported to be unable to take up exogenous *myo*-inositol. All of the common vitamins listed here (thiamine, pyridoxine, nicotinic acid and myo-inositol) are sufficiently stable during autoclaving at common medium pH values. Other vitamins have been added (1 mg/litre or less) in some instances, especially when cells or protoplasts are cultured at very low densities; there include biotin, pantothenic acid, folic acid, choline chloride, p-aminobenzoic acid, riboflavin, vitamin B_{12}, and ascorbic acid. Ascorbic acid and cysteine HCI are often added at very high concentrations (25 to 100 mg/litre) as antioxidants to prevent browning of cultures.

pH

Most cell culture media are adjusted to pH 5.5 to 5.8 before autoclaving. Following inoculation, the cells may rapidly shift the pH of the media by the release of stored ions or by the uptake and utilisation of nutrients such as ammonium. After a week or more of growth, suspension cell culture media may exceed pH 6.0. Rapid alkalinisation of the medium is also common after elicitation, with values reaching as high as pH 7.0.

Plant growth regulators

Plant growth regulators include both the naturally occurring plant hormones (phytohormones) and synthetic compounds, most of which are structural analogs of the natural hormones and bind to the same receptors. Commonly used regulators are listed in Table 10.2, along with their abbreviations. One class of regulators are called auxins, which cause plant cells to grow, the naturally occurring plant auxin being indoleacetic acid (IAA). (IAA is unstable solution and is easily oxidised and conjugated to inactive forms by plant cells. However, IAA and the slightly more stable indolebutyric acid (IBA) are still used when low doses or short pulses of hormone are needed, as in certain regeneration systems or in rooting micropropagated plants or cuttings. Naphthaleneacetic acid (NAA) and 2,4-dichlorophenoxyacetic acid (2,4-D) are the most frequently used auxins. 2,4-D is more potent than NAA, followed by IBA, then IAA. IAA and IBA must be added after autoclaving, and the media must be used immediately thereafter; stock solutions should be frozen or made weekly and refrigerated. 2,4-D and NAA can be added before autoclaving and will not break down in the media during storage; stock solutions can be stored indefinitely at 4°C. Many callus cultures can grow with auxins alone; the original formulation for B5 contained 1 ppm of 2,4-D (i.e., 1 mg/litre, or 4.5 μM).

Table 10.2. Growth regulators and additives commonly used in plant tissue cultures.

Abbreviation	*Full name*
Auxins	
IAA	Indole-3-acetic acid
IBA	Indole-3-butyric acid
N AA	1-Naphthaleneacetic acid
2,4-D	(2,4-Dichlorophenoxy) acetic acid
2,4, 5-T	(2, 4, 5-Dichlorophenoxy) acetic acid
PCPA or CPA	(para- or 4-Chlorophenoxy) acetic acid
PIC	Picloram (4-amino-3, 5, 6-trichloropicolinic acid)
NOA	2-Naphthoxyacetic acid
Cytokinins	
BA or BAP	6-Benzylaminopurine
ZEA	Zeatin
KIN	Kinetin
2iP	(2-Isopentenyl) adenine
ADE	Adenine
Other hormones	
GA or GA3	Gibberellic acid (gribberellin A3)
ABA	Abscisic acid
Other additives:	
CH or C	Casein hydrolysate
CW	Coconut water
EDTA	Ethylenediaminetetraacetic acid

A second class of regulators, the cytokinins, promote cell division and are sometimes required for culture initiation or long-term growth. Cytokinins are often added to induce shoot initiation during plant regeneration from callus, at which time the auxins may be decreased or omitted from the medium. The first naturally occurring cytokinin to be identified was zeatin, a purine derivative. Zeatin is occasionally used in culture media, but the most common cytokinins in use today are kinetin (6-furfurylaminopurine) and 6-benzylaminopurine (BAP; also known as N6-benzyladenine BA); adenine and adenosine can also have cytokinin activity, particularly at high levels. Most cytokinins, except for zeatin and its derivatives, can be sterilised by autoclaving.

Gibberellins (usually GA3, gibberellic acid) and abscisic acid are sometimes used to stimulate embryo or shoot development. Other plant growth regulators such as the recently discovered brassinosteroids are still being evaluated in culture systems.

Auxins and cytokinins are used in the range of 1 to 50 M. Most plant growth regulators can be dissolved in a small volume of reagent grade ethanol and then diluted in warm water to the final stock concentration; the ethanol will evaporate during autoclaving. Alternatively, NaOH is used to dissolve most auxins and HCI can be used to dissolve cytokinins, but these should be kept to a minimum to avoid altering the pH of the media.

Other additives

Complex ingredients such as coconut milk, yeast extract, fruit juices, and casein hydrolysates (such as NZ Amine) were used in many early culture experiments and are still used today. They provide plant growth regulators, vitamins, and amino acids which can enhance plant cell growth rates and survival of some cultures. However, they are expensive, sometimes variable in quality, and sometimes detrimental, such as when yeast extract induces stress responses in dilute cell suspensions.

Mannitol, sorbitol, or combinations of these two sugar alcohols are routinely added as an osmoticum (i.e., a supplemental agent increasing osmolarity). A concentration of 100 g of sugar alcohol per litre (0.6 M) is used in most media for the isolation of plant protoplasts, to prevent the cells from bursting once the rigid cellulose cell walls are removed. The addition of 0.2 to 0.5 M total osmoticum (300 to 900 mOsm/kg of H_2O, added as an equimolar mixture of sorbitol and mannitol) to the culture medium increases the number of transformants produced by bombardment techniques.

Organic acids such as citric, fumaric, malic, or succinic acids, or synthetic buffers such as Tris, MES, or HEPES have been used to help maintain culture pH, especially when ammonium salts are used as the nitrogen source. Neutralised activated charcoal is occasionally added to young regenerating cultures to remove toxic phenolics realised by the stressed plant cells, or to help remove plant growth regulators introduced at an earlier stage. Silver thiosulphate serves as an ethylene biosynthesis inhibitor, preventing this gaseous plant hormone from accumulating to detrimental concentrations.

'Nurse' medium or 'conditioned' medium is the liquid medium removed from a suspension of fast-growing cells. It contains uncharacterised growth factors released by growing cells, or it may be depleted of growth-inhibiting substances, such as high levels of ammonium ions. Conditioned medium is usually removed aseptically from a culture and never autoclaved. A nurse culture or 'feeder' layer is usually a thin layer of fine, fast-growing cells spread on the agar surface and covered with a moist sterile filter paper. Conditioned media and nurse cultures are sometimes used in the culture of regenerating protoplasts, single cells, or very dilute cell suspensions. The nurse media or cultures are often not from the same species as the species of interest; e.g., a tobacco cell culture may be used as the nurse culture for carrot cells.

Medium Preparation

Only high-purity (reagent grade) chemicals should be used for plant cell culture, especially when new cultures are being initiated or sensitive cell lines are being used. Water should be double-distilled (glass stills preferred) or deionised, such as 18 MΩ water produced by MilliQ devices (Millipore, Bedford, Mass.). Glassware must be free of residual detergents. If glassware is cleaned with chromic acid (or any solution containing heavy metal ions), it must be treated with EDTA or another chelator to remove all traces of chromium since some cultures are very sensitive. In fermentors or bioreactors, all parts in contact with the medium must be made of high-quality stainless steel, glass, or plastic for work with sensitive cell lines. In studies with opium poppy cells (*Papaver somniferum*), the use of bioreactors containing even a small amount of silver solder or brass screws results in the browning and death of the cultures, while periwinkle (*Catharanthus roseus*) and tobacco (*Nicotiana tobacum*) cells grow well (personal observation).

The options for gelling agents for plate cultures include bacteriological grade agar (such as Bacto Agar), Noble agar (agar reduced in ionic impurities, tissue culture grade Phytagar (specially prepared agar, agarose (highly purified agar), and Phytagel (Gellam gum agar substitute or Gelrite. For routine work, some laboratories prefers Phytagar at 0.6 to 0.8 per cent or agarose at 0.55 per cent, although many cultures can tolerate the impurities in bacteriological agar. Carageenan (vegetable gelatin) and calcium alginate have been used for cell immobilisation, particularly in bioreactor studies.

A number of strategies have been used to handle medium preparation. One traditional method is to prepare four liquid stock solutions, containing (i) the five or six macroelements (20 × stock), (ii) the microelements omitting the iron source (200 ×), (iii) the iron chelate (200 ×), and (iv) the vitamins (200 to 1,000 ×). The first three are stored at 4°C, while the vitamins are stored frozen. The carbon source is added as a solid. If hormones are to be included (Table 10.2), heat-stable hormones are predissolved and added as a 200 to 1,000× stock; otherwise, they are filter-sterilised and added after autoclaving. Agar, if required, is added after pH adjustment. The use of multiple concentrated stocks of N and P sources make medium variations easier.

Once an acceptable medium is identified and large quantities are needed, a convenient practice is to prepare a 10× concentrated stock and freeze it in appropriately sized aliquots. An example of this strategy for the preparation of an SH stock (supplemented with three growth regulators and 30 g of sucrose per litre for use with alfalfa cell cultures is provided in Table 10.3. A 4-litre volume of 10× concentrate, including three growth regulators but excluding sucrose, is dissolved, distributed into plastic Whirlpak bags in 100-ml aliquots and frozen at –20°C or lower.

Table 10.3. Preparation of SH frozen stock.

Component	*Concentration (mM) in final medium*	*Amount (g/40 litres) per batch of 10 × stock*
KNO_3	25	101
$MgSO_4$-$7H_2O$	1.5	14.8
$NH_4H_2PO_4$	2.5	11.6
$CaCl_2$-$2H_2O$	1.5	8.8
myo-Inositol	5.6	40
$MnSO_4$	0.059	0.358
H_3BO_3	0.13	0.200
$ZnSO_4$-$7H_2O$	0.0035	0.040
Kl	0.0060	0.040
$CuSO_4$-$5H_2O$	0.0008	0.008
$FeSO_4$-$7H_2O$	0.0054	0.600
Na_2EDTA	0.0054	0.800
Thiamine HCI	0.0150	0.200
Nicotinic acid	0.0410	0.200
Pyridoxine HCI	0.0024	0.020

For 10 × concentrate without sugar or pH adjustment. The method is as follows. Dissolve all of the above solids in 2.5 litres of distilled H_2O. Add 8 ml of Mo/Co. microstock (50 mg of Na_2MoO_4 and 50 mg of $CoCl_2$-$6H_2O$ dissolved together in 100 ml and stored at 4°C). Add 20 ml of kinetin stock (43 mg/200 ml). Add 80 ml of 2,4-D stock (225 mg/litre). Add 400 ml of PCPA stock (186.6 mg/litre). Adjust the volume to 4 litres. Mix well. Divide into 100-ml lots in Whirlpak bags. Store at –20 to –40°C. Note: Each 100-ml bag is equivalent to 1 litre of SH. To use, thaw 1 bag/litre needed, add 30 g of sucrose litre, adjust the volume, and titrate to pH 5.7 with 1 N NaOH. For plates, add 8 g of agar/litre.

To prepare the medium, on, bag of concentrate is thawed for every litre of liquid or agar needed, and the contents of the bag are transferred with rinsing to a large flask or beaker and dissolved by stirring for 30 to 60 minute at room temperature. Sucrose (30 g/litre) is added, the pH is adjusted to 5.7, and the medium is distributed into flasks (with or without agar) for autoclaving. We find it advantageous to add the sugar at the time the concentrate is thawed for use, because otherwise the stock would contain 30 per cent sucrose, which is difficult to dissolve, distribute, and freeze. The pH of the stock is slightly acidic and should not be adjusted prior to freezing. A slight precipitate may form in the bags during freezing, but it will redissolve with extended stirring.

Various companies now produce a variety of premixed plant cell culture media. Usually, the medium is divided into 'basal salts' (the macro- and micro-nutrients), sold as a dried powder, and vitamins, sold separately as a solution. These components may also be sold as a combined dried powder ('basal salts with minimal organics'), with or without sugar included. The medium is usually dispensed into preweighed packets such that each makes 1 to 10 litres of medium; one can also purchase a jar of well-blended powder that makes 50 litres and weigh the powder as needed. The plant growth regulators and agar are rarely included. Due to inconsistencies in formulations and terminology in the litreature, manufacturers usually provide very detailed tables of their formulations; one manufacturer supplied 15 variations of MS salts. If one is careful to select the correct preparation and labour is limited, these commercial powders offer a slightly expensive but reliable option.

For many liquid media including MS, a small amount of white precipitate may be visible in shake flasks, especially after extended autoclaving or storage. For routine culture maintenance, this medium will work well, since the precipitate will gradually be used by the cells. Media should be autoclaved at 121°C for a minimum of 15 minutes for small volumes and for longer times for larger volumes or thick-walled vessels. The required times may need to be evaluated on-site; 30 minutes for 500 ml of agar or 1 hr. for a 2-litre bioreactor vessel is not unusual.

Facilities

Successful plant cell culture requires extremely close attention to proper sterile technique and temperature control. The growth of plant cell cultures is relatively slow compared to common microbial cultures, necessitating very long culture times for experiments; a low level of contamination which would be outgrown and perhaps unnoticed in a bacterial culture would have time to overwhelm plant cultures. It may take months to establish a high-producing cell line or the regenerate transgenic plants from the original explants.

Sterile transfer facilities

Laminar-flow hoods are strongly recommended. An even flow of HEPA (high-efficiency particulate air) filter-sterilised air comes from the back of the hood, protecting the cultures when the vessels are opened. Unlike primary animal cell cultures, no biohazard containment of the cultures is required. Some laboratories prefer to use containment hoods when introducing microbes into plant cell cultures, such as during genetic transformation or inoculation with plant pathogens. Each hood should be equipped with a natural gas outlet for a Bunsen burner and lighting (usually overhead); A UV germicidal lamp is strongly recommended for use in decontaminating the hood work surface after the HEPA filter is replaced, if the hood is switched off for more than 30 minute, or if the hood is accidentally contaminated. A selection of stainless-steel forceps, spatulas, and scalpels with replaceable blades are required for culture initiation, callus subculture, and transformation and regeneration. For quantitative transfers, an electronic (digital) top-loading balance (400-g to 4-kg capacity) with auto-tare feature is essential.

Temperature

Most cell cultures grow well only in a narrow temperature range. The most common growth temperature is 25°C. Some plant species, especially heat-tolerant plants such as tobacco or *Catharanthus roseus*, can continue to grow at temperatures of 30°C and above, while heat-sensitive cultures like opium poppy will cease to grow above 27°C and will die at higher temperatures.

Cyclic temperature changes may occur during the day-night cycles used for some culture procedures. Localised heating of shelves occurs during the day, and they then cool during the night. Although not directly lethal to cultures, this contributes to the formation of condensation inside the lids of petri dishes and other culture vessels, which can dry out the cultures or contribute to contamination. Stacking an uninoculated petri dish containing water agar on top of the important cultures will greatly reduce condensation yet still allow sufficient light for regeneration.

Light

Callus and shake flask cultures generally do not require light for growth. Usually they are grown in dark rooms, equipped with lights which are turned on when inspecting or subculturing the cultures. Absolute darkness is not usually required. High light intensities or low levels of continuous illumination can cause cultures to turn green (from small numbers of chloroplasts) or brown (accumulate polymerised phenolics). Enough light to induce either colour change can be sufficient to inhibit secondary-metabolite accumulation, and browning may greatly slow the growth of cultures. If callus and shake flasks must be incubated in lighted rooms (such as rooms shared with regenerating cultures or aseptically grown plants), they can be shielded with foil or boxes.

For regeneration, cool white fluorescent lights are recommended. Interspersing incandescent bulbs will greatly increase the quality of the light, providing a spectrum close to that of sunlight. Lights are generally suspended 12 to 14 inch above the culture vessels. Sufficient distance must be maintained between the lights and vessels to allow cooler air to circulate and avoid overheating the plates. Wire shelving, positioned a few inches from the culture room walls, are often preferred, providing good air circulation at a reasonable cost. The shelves can be partially covered with aluminium foil or Plexiglas to provide a more stable, easily cleaned surface. If several layer of closely lit shelves are used, heat from a lower shelf may warm the shelf above; plates can be partially insulated by placing them on foil-covered sheets of styrofoam, or lights should be suspended by chains to allow heat dissipation.

Too much light can actually be detrimental to culture growth and regeneration. One indication of a high light level is the accumulation of anthocyanins (red to purple pigments) in callus or differentiated tissues. Continuous light is sometimes detrimental to whole-plant growth and is rarely used in plant cell cultures. The use of household timers is adequate for generating day-night cycles (16 to 18 hr. of light, 6 to 8 hr. of dark). One exception is the culture of photoautotrophic cells, which require high-intensity and high-quality light to enable the cells to photosynthesise sufficient sugars from CO_2.

Aeration

Under most conditions, shake flask cultures of plant suspension have a lower volumetric oxygen demand than do microbial cultures. The respiration rate (Q_{o2}) reported for plant cell cultures range from 0.2 to 3.6 mmol of O_2/hr./g (dry weight), compared with 2 to 16 mmol/hr./g for *Escherichia coli*, 39 to 570 mmol/hr./g for *Bacillus* spp., and 1.9 to 2.8 mmol/hr./g for *Saccharomyces* spp. Shaker speeds (100 to 150 rpm with a 1-inch throw) are generally lower than those used for microbial cultures, and the fill ratios in the flasks (volume of medium/volume of flask generally 0.2 to 0.4) are generally higher. Baffled flasks are rarely used to increase aeration but are occasionally used for breaking up clumps of cells, especially when a suspension is first being initiated.

Generally, shake flask cultures can be removed from a shaker for subculturing and observation for 1 hr. or more without killing the cells. However, this procedure will affect the growth rate and can affect the ability of the culture to produce secondary metabolites. High oxygen demands have been reported immediately following elicitation, since a short-term burst of O_2 utilisation occurs as a result of activation of many defense processes. If elicitation is used to induce secondary-metabolite accumulation, one should prevent even brief interruptions in aeration of shake flask cultures, and it may be to necessary to increase the aeration of production bioreactors.

A wide variety of closures have been used for shake flask cultures. Traditional cotton plugs in Erlenmeyer flasks can work well, particularly if the cotton plugs 'mushroom' over the rim of the flask, keeping it free of microbial contamination. Since the cultures may be on the shakers for weeks, particulates in the culture room air will tend to collect around the rim if unprotected, making it impossible to open the flask for subculture without contamination. A preferred option is straight-walled Delong-style flasks with metal culture tube closures; a small piece of folded cheese-cloth or gauze is placed in the top of the closure to act as a sterile filter, and the prongs on the closures hold it securely on the flask. These caps have the advantage of providing the same sterility barrier and good aeration as cotton plugs while covering the rim and upper edge of the flasks. Any dust from the shaker rooms or medium storage areas can be easily removed by a brief flaming before the flask is opened.

Petri dishes containing agar media, and callus cultures or regenerating plants are incubated agar side down, not inverted like microbial cultures. Plates are always sealed with Parafilm or porous medical tape (such as Urgopore tape) or can be placed inside a sterilised box or new plastic self-sealing freezer bag for short-term incubation. Otherwise, contaminants will enter the plate during the long cultures or the agar media will dehydrate.

Closures on culture vessels must not only allow oxygen and CO_2 to pass but, in sensitive species, must also allow ethylene and other volatiles to escape. Ethylene, a gas, is a plant hormone, often released by plant cells in response to wounding and infection or during senescence. It can have a number of effects on axenically grown plants and cell cultures. In very sensitive species, such as *Arabidopsis* and potato, the addition of ethylene biosynthesis inhibitors such as silver ions (as silver thiosulphate) greatly increases the yield of regenerated transgenic plants. Without these, the wounded tissue produces enough ethylene to inhibit the development or lead to browning and death of nonwounded cells.

Dedicated growth rooms provide the most efficient and flexible use of space. Commercial growth chambers equipped with lights, temperature control, and (optional) humidity control are available; several growth chambers allow many different conditions to be maintained, which may be important if diverse species are to be cultivated. For important cultures, it is recommended that several separate cultures be maintained, preferably in separate culture rooms or incubators. Temperature recorders are strongly recommended to diagnose possible problems in temperature control. Motion sensors and temperature monitors with audible alarms can also be used to minimise the damage caused by heating and cooling and by shaker equipment breakdowns. Even brief overheating to 35°C or more can greatly reduce regeneration and secondary-metabolite production if it occurs at critical times, while regenerated plantlets are relatively resistant. Long-term exposures (hours) to high temperatures can stop the growth or rapidly kill suspension cultures.

Culture Initiation

Rapidly growing plant tissues are generally best for rapidly producing callus cultures, although any living plant cell can in theory be induced to divide. While authors often indicate the plant organ from which a callus was derived (i.e., root, leaf, or hypocotyl), after several generations the cultures are fully dedifferentiated and there is little evidence that the original tissue source influences the behaviour of the culture, such as the products it will accumulate. In contrast, for short-term cultures such as for transformation and quick regeneration, the explant source can be critical; for wheat and some recalcitrant legumes, immature seeds or cotyledonary nodes will produce easily regenerable callus while leaf or root callus cultures will not.

Sterile explants

There are two basic sources for obtaining surface-sterilised plant tissues. One is to surface sterilise seeds, which are dormant and often dried and can therefore tolerate very harsh sterilisation conditions. The seeds are then germinated under sterile conditions, and the various tissues are placed on callus-inducing media. The other is to surface sterilise undamaged tissue from a growing plant, which may be the only option of vegetatively propagated species or tree species which are slow to set seed. The same sterilisation strategies can be used in each case. Most commonly, a brief ethanol wash serves to remove and kill spores and bacteria on the plant surface and to thoroughly wet the tissue; it is followed by a hypochlorite soak to kill any remaining microbes.

Seeds

The following protocol works well for soyabeans and other medium-sized to large seeds. The seeds are placed in a clean plastic petri dish (25 by 100 preferred) or beaker and covered with 70 per cent ethanol for 5 minute. The dish is stirred or gently swirled occasionally to ensure that the ethanol contacts all the seed surfaces. The seeds are transferred with a sterile (flamed) spoon-shaped spatula to a sterile dish with enough bleach solution (20 per cent Clorox in distilled water with 1 drop of Tween 20 or 2 drops of 10 per cent sodium dodecyl sulphate solution per 100 ml; equivalent to 1 per cent hypochlorite (final concentration) to cover the seeds. They are soaked for 15 to 20 minute with stirring. Any cracked, floating, or noticeably damaged seeds are removed. The seeds may begin to swell during this time.

The seeds are transferred with a sterile spatula to a sterile dish containing distilled water and are rinsed two or three more times by transferring to new dishes of sterile water. They are then transferred to sugar-water agar plates (5 gram of sucrose and 8 gram of Phytagar/litre; 50 ml/25-mm-deep plate) and sealed with Parafilm. If possible, the seeds are spread out so that they do not touch each other, and several plates are used, so that unsterilised seeds will not contaminate clean seeds. The seeds are incubated at approximately 25°C.

Germinating the seeds on media containing sugar or other nutrients helps locate contaminants by stimulating their growth on the agar surface. Seeds may also be germinated on damp filter paper or cotton. Tiny seeds, such as those of tobacco, lettuce, and *Arabidopsis* spp., may be more easily sterilised by being treated in sterile 1.5 -ml plastic snap-cap tubes, briefly centrifuging them to the bottom of the tubes to remove solutions. The seeds can be transferred to new tubes by being suspended and pipetted. To make tiny seeds easier to distribute, after sterilisation they can be suspended in 0.1 per cent agarose and dripped onto the agar surface. If a special treatment is recommended before planting in soil, the same treatment should be applied before sterilisation. Some species require extended cold treatments, exposure to light, or scarification (scratching the seed surface with sandpaper or nicking the seed coat with a razor blade).

Leaves

Young leaves, free of insect and microbial damage, should be selected. The are rinsed with distilled water to remove dirt and kept in a moist container to prevent wilting. For leaves which are thick or have a thick waxy cuticle, the above protocol should work well. For very thin leaves or for delicate leaves such as those grown in a very humid environment, less harsh conditions may be required. For example, for tender alfalfa leaves, reduce the ethanol concentration to 50 per cent and the exposure time to 1 to 2 minute to avoid killing the leaf cells.

Other tissues

Depending upon the species and environment, other plant parts may be available. For species which make tubers or rhizomes, these organs should be first scrubbed free of soil and may be surface sterilised, peeled, then sterilised briefly again. Young stems and petioles are often more resistant to damage during sterilisation than are leaves.

Other sterilising agents have been used, such as $HgCl_2$ or antibiotics, especially for seeds which have natural bacterial population under their seed coats. A procedure involving chlorine gas is recommended for dry seeds such as soyabeans. The seeds are placed in a dish in a larger, sealable glass container such as a desiccator, together with a beaker containing bleach. The bleach is acidified (3 to 5 ml of concentrated HCI per 100 ml of Clorox), which releases chlorine gas, and the container is quickly sealed. After 6 hr. to overnight, the container is opened and the seeds are rinsed with sterile water and transferred to plates for germination. Procedures have also recommended acidifying the bleach solution in the leaf and seed procedures described above to cause less damage to the tissues and to 'activate' the chlorine. Any procedures in which bleach and acid are mixed should be carried out in a chemical fume hood, not sterile transfer hood, due to human toxicity of the chlorine gas.

Callus culture initiation

Once surface-sterilised seedlings or explants are available, these are cut into small pieces (0.5- to 1-cm segments or squares) with a sterile scalpel and placed on callus-inducing media. Hypocotyl and root segments are most commonly used. The edges of leaves sand any cut surface that was exposed to bleach should be trimmed away and discarded, since these will not yield callus. The tissues will increase in size and should show visible callus formation within 1 month. Some tissues, such as older leaf and stem segments, may show callus development only at the cut edges, while younger tissues, such as roots, hypocotyls, and very young leaves, may completely dedifferentiate. If tissues curl and lose contact with the medium, they should be cut and pushed gently into the agar surface.

Much time can be saved by consulting the litreature and finding medium formulations which have been previously developed for the species of interest. If this information is not available, one should try media which have been successful with closely related species or try a number of different basal salt and hormone combinations. MS and B5 salts with 1 mg of 2,4-D per litre and SH medium with the two auxins and one cytokinin described above (Table 10.3) provide a wide range of media for dicots. While callus may form on several media, the texture and growth rate may vary greatly. For example, peanut callus on B5 salts with 1 mg of 2,4-D per litre is very soft and friable, while the callus after several subcultures on SH medium (Table 10.3) is extremely dense and hard. Monocots often require very high concentrations of auxins (up to 20 mg of 2,4-D per litre) for callus initiation but require reduced levels for extended rapid callus growth. If a particular cultivar or variety of a species is not required, several should be tried, since callus formation may vary with the genotype. Several plates of each type should be initiated, due to variations in callus formation response between individual explants and to allow for possible losses to contamination.

The callus should be subcultured every 2 weeks to 1 month by simply moving clumps of rapidly dividing cells to fresh plates with a cooled sterilised spatula. During the first transfer, it may be beneficial to cut any remaining intact explants to expose fresh tissue to the medium, or to break up large callus pieces. If sufficient material is available, the softest, most friable calli should be selected for subculturing.

Suspension culture initiation

Suspension cultures are initiated by inoculating liquid media in shake flasks with callus clumps. Approximately 1 to 2 inch^3 of callus (5 to 15 gram fresh weight) should be suspended in 50 ml of medium. Generally, the same medium used for callus culture is used for suspension cultures. Callus from a single explant may be used, or calli from several plates of the same type can be combined. Large clumps can be gently broken or cut into smaller pieces. Softer, friable, rapidly growing callus works best, since it will more quickly disperse into a fine suspension, whereas hard callus clumps may simply enlarge without dispersing. If only a small amount of callus is available, less medium should be used, since high liquid-to-cell ratios may inhibit growth.

The growth rate of suspension cultures will gradually increase, as faster-growing cells take over the culture. Initially, growth may be slow and it may take weeks for the culture to double or triple in biomass. At this time, the culture should be divided into two new medium flasks. This is often done by simply swirling the culture and pouring cells and old medium into the new flasks. If the major cell aggregates are white but the medium is brown and contains much cell debris, it may be beneficial to pour off the old medium and transfer the cells with a spatula. As the growth rate increases, the cultures will have to be divided more frequently. Also, the aggregate size should decrease, and the ratio of old culture to new medium can be reduced.

A portion of the original callus should be reserved and subcultured. If the initial suspension does not grow well, after additional months of subculturing, a callus which as too hard to form good suspensions may improve in texture and can be tried again. After a good suspension culture is established, a portion of the cells may be returned to agar to corm a callus culture again; this callus will usually grow faster than the original. If the shaker stops and/or the shaker room overheats, suspension cultures can be killed or may lose desirable properties such as the ability to produce secondary metabolites; healthy callus cultures serve as a back-up in case such problems arise.

Bioreactors and scale-up

The same principles regarding scale-up that apply to microbial or mammalian cell cultures also apply to the scale-up of plant cell cultures. Because aeration is usually not limiting for growth in shake flasks, scale-up to larger shake flasks (up to 4 litres) is often easily accomplished for biomass production, keeping the ratio of medium volume to flask volume constant.

Plant cell suspension cultures have been grown successfully in a wide range of bioreactor designs, but there are two important design considerations. First, plant cells tend to be very shear sensitive, and impeller damage can greatly impair the growth of a culture. Strategies to decrease impeller damage include decreasing the angle of the impeller, changing to a marine-style impeller, or using air-lift fermentors. By manipulating the hormones or medium components, the culture may be induced to form larger aggregates, which can be less shear-sensitive. Second, the cultures can be very heterogeneous. Large cell aggregates tend to settle to the bottom of the reactor vessels, while small aggregates may be carried out of the medium to stick to the reactor walls or float on foam, forming a 'meringue' layer above the useful area of the reactor.

To reduce the problems of shear and cell floating or settling, many immobilisation matrices have been tested. Cells have been embedded in beads of calcium alginate and carrageenan or entrapped in porous matrices such as foam rubber or polyester fabric. For cells which release products into the medium or carry out biotransformation reactions, immobilisation allows the cells to be recycled. Unfortunately, new problems arise, such as decreased mass transfer to and from the cells and decreased volumetric productivity of the bioreactor, plus the added cost associated with immobilisation.

Plant cell suspension cultures have been grown in industrial-scale fermentors. An extremely fast-growing cell line of tobacco (TBY-2) was grown in a 369-litre, then a 1,500-litre, and finally a specially designed 20,000-litre culture tank. For the commercial production of shikonin, cells are first cultured in a 250-litre fermentor in a growth medium and then transferred to a production medium in a 750-litre fermentor.

Growth Quantitation

Many methods have been used to estimate the growth of plant cell cultures both directly and indirectly. The method chosen will depend on the requirements for maintaining the sterility of the culture and the speed and accuracy of the measurement. One must keep in mind that the doubling time for most plant cell cultures exceeds 12 hr. and that a typical culture experiment will continue for 1 to 4 weeks.

Fresh weight

The growth of callus cultures can be measured by simply moving the callus mass to a preweighed sterile container and measuring the weight of the callus at regular intervals. The new agar medium plates can be weighed before and after inoculation at each subculture.

Cells can be collected from suspension cultures by filtration by using Miracloth, nylon mesh filters (40- to 200-μm mesh), or sintered glass filters. Filter paper may also be used, but it may clog quickly due to the abundant cell debris or polysaccharides present in most cultures. Due to their extremely high water content, the cells will quickly become too dry, so the length of time the vacuum is applied should be closely monitored, either by timing the filtration or by stopping it as soon as the liquid stops dripping from the filter. The weight of cells collected in this manner is often referred to as 'filtered fresh weight.'

An alternative measurement is 'wet fresh weight' or 'drained fresh wet.' By placing a pipette loosely against the bottom of a flask, much of the liquid medium can be withdrawn from a suspension culture while the majority of the cells will remain in the flask. The cell mass will retain a high percentage of medium (often about 50 per cent), but this percentage will be fairly constant if the cell aggregate size remains constant. The wet fresh weight is calculated by subtracting the weight of the culture flask. This procedure can be carried out aseptically with cotton-plugged pipettes, preweighed flasks, and a clean top-loading digital balance in a laminar-flow transfer hood. These cells can then be weighed into new culture flasks, giving fairly reproducible inoculum levels without risking damage to the cells by filtering them dry. This measurement has also been adapted to immobilised cells, for which the weight of the medium soaked immobilisation matrix is subtracted from the weight of the inoculated, drained matrix.

Dry weight

Collected fresh cells can be dried on preweighed filters or dishes to obtain the dry weight. Due to their high water content, the dry weight can be as low as 1 to 2 per cent of the fresh weight. Cells should be dried at low temperatures (55 to 65°C), preferably either under vacuum or in a convection oven to avoid being caramelised. Filter-collected cells should be washed briefly with distilled water to remove medium solids before being dried. Cells which are drained but not filtered can also be dried for a relative comparison of biomass, but the solids in the medium retained by the cells will contribute greatly to the dry weight. If a more accurate estimate of the dry weight is needed, the percentage of the wet fresh weight contributed by the medium, multiplied by the solids content (largely the dissolved sugars), can be subtracted from the dry weight.

Increases in dry weight can be misleading, particularly soon after cells are subcultured. Cell cultures of many species will quickly absorb the sugars from the medium and convert them to starch granules inside the cells. This starch reserve will then be slowly utilised during the growth of the culture. If a growth curve of the culture is determined, starch formation will cause the dry weight to increase dramatically for 1 to 2 days and then to decrease slowly, while the fresh weight is still increasing. (The presence of starch granules can be confirmed by examining cells under a high-power microscope; starch granules will polarise light and will stain blue with 1 per cent iodine solution.) Thus, for initial studies, researchers will often monitor more than one parameter and then determine which is the best indictor of growth for the particular culture. Some cell lines will also concentrate and store substantial amounts of phosphate. Each new cell line must be evaluated for starch and phosphate accumulation, since some cell lines will accumulate neither, both, or only one of these substances.

Packed-cell volume

Cell suspensions can be transferred to calibrated conical screw-cap tubes and gently centrifuged at 200 to 500 × gram for 10 minute. The volume of cells and medium can be read directly from the tube.

Indirect measurements

In some instances, the entire cell mass cannot be harvested. For example, in long-term bioreactor experiments or immobilised-cell culture systems, only a small sample of the culture may be obtained. If the suspension is homogeneous and a representative sample can be obtained, any of the above methods can be used to estimate the entire contents of a bioreactor. At other times, the biomass can be estimated indirectly by monitoring the consumption of nutrients. A correlation coefficient will have to be established from each cell line, usually from small-scale shake flask experiments. If the culture does not accumulate starch, the growth of the culture can be correlated with the disappearance of sugars from the medium. If sucrose is the carbon source, many cultures will release invertase, which hydrolyses sucrose to fructose and glucose, and utilise the glucose before the fructose. In such cases, the combination of all three carbohydrates must be monitored by a method such as ion-exchange high-pressure liquid chromatography (HPLC) with refractive index detection. If glucose is used as the sole carbon source, a simple reducing sugar assay may be sufficient. If the culture does not accumulate phosphate reserves (immediately depleting the medium of phosphate), phosphate uptake can be inversely correlated with growth by using either colorimetric or chromatographic detection.

Nitrate is the major nitrogen source in most cell culture media. Very few cultures accumulate and store nitrate, and therefore depletion of nitrate from the medium is often will correlated with the growth of the culture. Several methods of nitrate quantitation have been developed, including colorimetric methods, HPLC, and capillary electrophoresis. Since nitrate (NO_3^-) is also the highest-molarity ion (40 mM in MS and 25 mM in B5), it is the major source of conductivity in the medium. It is often possible to correlate a decrease in conductivity with cell growth. For some experiments, the exact biomass is not required. For example, if the same inoculation and culture regimen is used each time, a certain conductivity value can be simply correlated with the correct time for harvest or elicitation.

Viability assays

At times, the percentage of live cells in a culture must be determined. This is often needed when cells have been subjected to harsh treatments as part of an experiment, such as heat shock, chemical or pathogen addition, or freezing during the development of cryopreservation conditions. It is also useful when cells have been unintentionally stressed such as during a culture room malfunction. Callus cultures can be dead but may not brown or collapse for months.

Several assays have been used, depending on the equipment available and the texture of the tissue. The best assay for viability is to subculture the cells and watch for a visible increase in cell mass. Unfortunately, this may take several days or weeks. Assays in which the stain (such as fluorescein diacetate or triphenyltetrazolium chloride) is meant to be taken up only by living cells are known to sometimes give false-positive results with cells which no longer divide.

Cytoplasmic streaming

A quick estimate of viability can be made by examining individual cells under a high-power microscope. The cytoplasm and some organelles can be seen circulating inside live cells. This procedure works well for fine cell suspensions but not for large aggregates or regenerating embryos and plantlets. If the researcher is unfamiliar with this phenomenon, both positive and negative controls (a sample of a known well-growing culture and a heat- or freezing-killed sample, respectively) should be used. False-positive results can come from mistaking Brownian motion for cytoplasmic streaming.

FDA staining

The diester fluorescein diacetate (FDA) will not fluoresce. However, it can pass through the plant cell membrane, where esterases will remove the acetate groups. The charged product molecule is unable to leave the intact cell, due to a pH gradient, and fluoresces strongly. Only live cells should emit fluorescence, since dead cells will leak the stain or will not have active esterases. A 0.5 per cent stock solution of FDA in anhydrous acetone can be stored at –20°C indefinitely. When needed, the stock is diluted 50 fold (final concentration, 0.01 per cent) in ice-cold culture medium. One drop of FDA solution is mixed with one drop of culture on a microscope slide, covered with a coverslip, and monitored for 5 to 30 minute. The slide is observed under a microscope with a UV/fluorescence attachment (excitation filter, 450 to 490 nm; barrier filter, 520 nm). If relevant, the total number of fluorescing cells can be counted and divided by the total number of cells (counted under visible light) to calculate the per cent viability.

Evan's blue staining

The blue stain is able to enter dead cells but cannot penetrate intact cell membranes. A solution of Evan's blue (5 mg/ml) in culture medium is prepared. The cells are mixed with stain solution and incubated for 5 minute; and then live (colourless) versus dead (blue) cells are counted under a microscope with bright-field illumination.

TTC reaction

Viable cells will reduce triphenyltetrazolium chloride (TTC) to a pink formazan product, which can diffuse out of the cells and into the reaction medium. Viability can be assessed visually or quantitatively in a spectrophotometer. This procedure can be used for both fine and coarse cells and callus, as well as whole embryos or leaf pieces. A 0.6 per cent solution of TTC in 0.5 M phosphate buffer (pH 7.0) is prepared. The tissue (100 mg of cells or plant tissue/3 ml of solution) is suspended and incubated overnight (15 hr.) at 30°C. The TTC solution is removed, the cells are washed with distilled water, and 7 ml of 95 per cent ethanol is added. The mixture is incubated at 80°C for 5 minute. The extract is cooled to room temperature, and the volume is adjusted to 10 ml with 95 per cent ethanol. The absorbance at 530 nm is measured.

Phenosafranin staining

Dead cells will be stained red with phenosafranin, but the dye cannot penetrate living cells, similar to Evan's blue. A 0.1 per cent solution dissolved in culture medium is used, and the cells are observed under bright-field microscopy.

Secondary-Metabolite Production

One application of plant cell cultures is the study of plant secondary-metabolic biosynthesis and the commercial production of such valuable natural products. Plant secondary products have been used for centuries in human medicine, such as codeine and morphine (painkillers) from opium poppy (*Papaver somniferum*) latex, atropine and scopolamine (neuroregulatory), quinine (antimalarial) from *Chinchona sp.* and diosgenin and digoxin (cardioactive) from the foxglove plant (*Digitalis* sp.). Modern bioactivity-guided screens have identified a range of new important compounds, especially the anticancer compounds vincristine and vinblastine from periwinkle (*Catharanthus roseus*), taxol from the yew tree (*Taxus* sp.), and camptothecin (*Camptotheca acuminata*). Some medicinal plant parts are consumed whole or as a complex extract, such as ginseng root. There are also valuable food ingredients including colouring agents such as anthocyanins from many species and flavour extracts such as vanilla from the vanilla bean (a species of orchid). Fragrant oils such as mint, rose, and jasmine have always commanded high prices in the perfume industry. Some plant extracts are used as pesticides in organic gardening, including rotenone from tropical legumes, pyrethrins from the flowers of the pyrethrum daisy (*Chrysanthemum coccineum* and other species), and azadirachtin from the neem tree (*Azadirachta indica*).

Many studies have investigated the production of these and hundreds of other secondary metabolites in plant tissue culture. In some cases, cell cultures provide a source of homogeneous, highly active cells, ideal for studying the biosynthetic pathways by labeled-precursor feeding, enzyme purification, and cloning of the genes encoding pathway enzymes. Other studies have focused on increasing the productivity of cultures, with a goal of profitably producing these metabolites. Many of the above products are currently produced by growing the plants, often in tropical regions, and harvesting the processing the appropriate parts. This means that production is subject to seasonal and environmental variation, disease, competition with other crops and land uses and political pressures such as environmental concerns or international embargoes. The growth of the source plant may be slow, such as yew trees (for taxol) or ginseng roots. Occasionally, cell cultures make high levels of one or a few compounds, which would not normally accumulate in the intact plant, decreasing downstream purification costs. For example, elicited *Papaver somniferum* cultures produce very high levels of sanguinarine, while the most competitive whole-plant system (*Macleaya cordata*) accumulates more sanguinarine but only as a complex mixture with other undesirable alkaloids, adding several expensive chromatographic steps to the purification.

In some cases, plant tissue cultures might appear to be an economic source of metabolites. Unfortunately, due to their slow growth and often low volumetric productivity, only extremely valuable compounds may be profitably produced. Cell cultures are also often capable of degrading the desired products, and cell lines can suddenly lose their productivity, making industrial production difficult. Many strategies have been used to improve production, including selecting for higher-producing cell lines, developing production media, and inducing production by elicitation.

Constitutive production and strain improvement

In certain species, media which allow good growth of the cultures will also allow good accumulation of secondary metabolites. In general, compounds which accumulate in root tissues or throughout all organs of a plant have been most easily produced at high levels in cell cultures. This may in part be due to the lack of light or the hormone regimes; it is often said that cell culture metabolism most closely resembles that of plant roots.

Production can vary greatly among cultures of the same species. When initiating cell cultures for secondary-metabolite production, many separate cultures should be initiated from individual seedlings, and plants or seedlings should be obtained from as many sources (companies, botanical gardens, and research laboratories) as possible. These should be evaluated separately to determine if there is a variation in production between individuals, cultivars, varieties, or other genotype designations.

In some cases, product accumulation decreases either rapidly or gradually as the culture is subcultured, possibly due to natural selection for faster-growing cells, especially in suspension. This natural variation can also be used advantageously, in that occasionally a subpopulation of high producing cells can be selected. Visual selection was successfully applied to the production of shikonin, a red alkaloid pigment with antibacterial activity. For initial rounds of selection, portions of callus which were darker red were picked by hand and subcultured separately; for later rounds of selection, quantitative chemical analysis was used. Fluorescence microscopy was used to select cell lines with high berberine production. The berberine content was increased from approximately 5 per cent unselected cells to 13.2 per cent in a selected line.

Production media

Many cell lines will accumulate only low level of secondary metabolites in standard growth media, but production can be increased severalfold by altering one or more medium components. A wide range of production media have been developed, and a few trends have been established. In general, the growth of the culture may slow or stop completely. Production media often contain lower levels of growth regulators or none at all, or they contain weaker auxins such as NAA or IBA instead of 2,4-D. For anthocyanins (coloured phenolic compounds), limiting the growth of the culture by restricting either the nitrate or phosphate supply while simultaneously providing very high concentrations of sugars has greatly improved production. Increasing the sucrose concentration from 20 to 80 g/litre and decreasing the nitrate concentration from 25 to 2.5 mM in B5 increases anthocyanin accumulation 10-fold, while phosphate limitation together with a high sucrose concentration is also successful. In the production of alkaloids (nitrogen-containing compounds), low phosphate levels are beneficial, but high nitrate levels increase accumulation in some cases and are inhibitory in others.

In many cases, fully dedifferentiated tissue cultures will not accumulate the same valuable product as the parent plant or will accumulate only small amounts, sometimes only when cells are allowed to differentiate back into shoots or roots. For example, thiophenes are light-activated insecticidal and antimicrobial compounds that are highly toxic to both plant pests and host plant cells; they are accumulated only in specialised veins running longitudinally inside the roots. For unknown reasons, opium alkaloids accumulate only in specialised lacticifers in the above-ground portions of plants, and production in tissue culture has been associated only with redifferentiation into shoots. Similar observations have been made for the famous anticancer compounds vincristine and vinblastine from *Cantharanthus roseus*. These molecules are heterodimers derived from the indole alkaloids catharanthine and vindoline. While high levels of catharanthine (5 to 20 times that in the plant) accumulate, no significant levels of vindoline, vincristine, or vinblastine have been found in homogeneous cell cultures, even though this has been one of the most thoroughly investigated systems. One of the enzymes in vindoline biosynthesis is light activated, while another may require the differentiation of functional chloroplasts, which are absent under standard cell culture conditions.

Elicitation

Many economically valuable plant secondary metabolites serve as defense compounds for the producing species, protecting them against attacks by insect herbivores and microbial pathogens. Some of these compounds are absent or present only at very low levels in healthy tissues, but they accumulate to very high levels in a narrow zone surrounding the site of microbial infections; these are referred to as phytoalexins (from the Greek for 'plant warding-off agents'). It was observed that contamination of a plant callus culture by a fungus resulted in the accumulation of a pigmented antifungal alkaloid. Later, it was determined that live fungus was not required to cause the response but that either autoclaved fungal culture filtrates or a number of purified compounds could 'elicit' phytoalexin production and other defense responses, in both callus and suspension cultures. The process was referred to as 'elicitation,' because the regulatory mechanisms involved were not known at the time; it has subsequently been shown that in most cases, the increase in the production of secondary metabolites is the result of an increase in the transcription of the genes encoding the corresponding biosynthetic pathway enzymes.

Elicitors are classed as biotic or abiotic. Abiotic elicitors include heavy metals such as copper or mercury salts and UV light. Biotic elicitors include many components of fungal cell walls (chitin, chitosan, β-glucan, peptidoglycans and arachidonic acid), compounds excreted by plant pathogens (cellulase, ribonuclease, pectinase, antibiotics, various sugar polymers and proteins) and compounds released from plant cell walls during degradation by plant pathogens (pectin fragments). The type of elicitor that is most effective in inducing secondary metabolism in a particular plant cell culture varies widely. The elicitor does not have to be derived from a pathogen of the source plant. In general, members of the same plant genus will respond well to the same elicitors but members of the same family may not. For example, cultures of several legume species (alfalfa, soyabean, and green bean) are highly elicited by a crude β-glucan derived from *Saccharomyces cerevisiae* but are unresponsive to chitin. Solanaceous species (tobacco, potato, and tomato) are all elicited by cellulases, arachidonic acid, and β-glucans but while two members of the poppy family, *Papaver somniferum* and *Eschscholtzia californica*, are both elicited by β-glucans, only the former is responsive to chitin hydrolysates.

In the *Eschscholtzia* system, in which elicitation was first studied, the control cultures accumulate a low level of one alkaloid. Following addition of the elicitor, the level of this alkaloid increases and five related alkaloids appear; the total benzophenanthridine alkaloid content of suspension cultures increases more than 20-fold compared to that in unelicited cultures. In some cultures, such as soyabean and green bean, no pterocarpan phytoalexins are present before elicitor addition but reach maximum levels within 24 to 48 hr. after elicitation. In an attempt to use elicitation to stimulate the accumulation of opium (morphinan) alkaloids in *Papaver somniferum* cultures, the accumulation of sanguinarine, a member of a different (benzophenanthridine) class of alkaloids but one which is also inducible by pathogen attack on poppy plants.

The optimal time for addition of the elicitor is usually in the middle to late growth phase of the culture, often 3 to 7 days after subculture. Elicitor addition before or after this time results in little or no response from the plant cells. The optimal time may shift; for example, as a cell suspension is repeatedly subcultured and the net growth rate increases, the optimal time will become earlier by as much as 2 or 3 days. Elicitors are often effective when added to standard growth media, although combining a production medium with elicitation may result in even higher production levels. Many elicitors cause a browning of the cell cultures and may result in the death of a high percentage of the cells.

The first secondary metabolite commercially produced by plant cell cultures was shikonin, a pigmented antibacterial alkaloid which was normally extracted from the older (3 to 7 year old) roots of *Lithospermum erythrorhizon*. By using a combination of strain selection and production medium, the investigators found that the cultures could accumulate 12 per cent dry weight of shikonin, versus 1 to 2 per cent in older roots. While other products have come close to commercialisation, in general, producers have defaulted to using field-grown plants as the source. The most recent promising candidate for tissue culture production is taxol, due to its high value and environmental concerns regarding the harvesting of wild trees. While plantation production and semisynthetic methods are currently the favoured production routes, recent manipulations in explant sources and culture conditions have resulted in the accumulation of taxol in plant cell cultures at levels several times higher than that found in the tree bark.

Regeneration

Even individual plant cells or protoplasts, if cultured properly, can regenerate into fully developed, fertile plants. This property is referred to as 'totipotency'. In current applications, small explants (rather than individual cells) are first induced to form calli, which are then induced to regenerate into a number of plants. There are two modes of regeneration, organogenesis and somatic embryogenesis.

In organogenesis, leaf explants, callus cultures, or suspension cultures differentiate directly to form shoots, which can then be induced to form roots. Generally, the procedures involve induction of rapid cell division and/or callus formation on media containing high auxin levels, followed by a medium containing low auxin and high cytokinin levels to induce shoot formation. For example, for tobacco, young leaves are surface sterilised, cut into 1-cm squares, and placed onto MS agar containing 4.5 μM 2,4-D and 30 gram of sucrose per litre for 2 to 4 weeks to induce callus formation. The explants are then transferred to MS agar supplemented with 5 μM BAP and placed under lights; shoots appear within a few weeks. Young shoots are cut off with a scalpel and rooted in MS with 20 g of sucrose per litre without hormones. Organogenesis is easily accomplished for most solanaceous species (tobacco, tomato, potato, petunia, etc.) and has been reported for many dicot species but is rarer among monocots.

In somatic embryogenesis, embryos or embryo-like structures are produced from somatic cells, as opposed to the maturation of zygotic cells in plant flowers. Somatic embryogenesis was first demonstrated for carrot cultures in 1964 but has since been demonstrated for many species, both dicots and monocots. For carrot, callus cultures are simply initiated on MS agar supplemented with 4.5 μM 2,4-D and 30 g of sucrose per litre for 1 month; then a fine, rapidly growing suspension is initiated in the same medium without agar. After 2 to 4 weeks, the cells are washed with the same medium without 2,4-D and incubated in the dark to induce embryo formation. Embryos appear after 8 to 10 days and continue to increase in size for 2 weeks. The embryos are then transferred to MS agar medium (with sucrose and no hormones) for plantlet formation; this last step is referred to as conversion or germination of the embryos. In some systems, many embryos are formed but not all of them will germinate. The addition of abscisic acid may aid in the development of normal embryos, and supplementation of the medium with proline and glutamine can greatly increase embryogenesis.

Micropropagation

In both organogenesis and somatic embryogenesis systems, hundreds of plantlets can be generated from a single explant. This is one method by which plants can be multiplied for commercial purposes. Another method is simply establishing sterile shoot cultures of plants, from which axenic cuttings are made and rooted in culture; cuttings can be made at any time of the year and will usually grow more quickly under plant tissue conditions than in the field. Either method is referred to as micropropagation. Many houseplants and ornamentals are propagated in this manner, allowing producers to introduce new varieties in a short time. Crop species such as blueberries, strawberries, garlic, and potatoes are also propagated by micropropagation to produce virus-free plants. The regeneration step can also be used to induce secondary-metabolite accumulation.

Transformation

Genetic transformation of plant cells, and their subsequent regeneration into transgenic plants was first accomplished in the early 1980s for tobacco and is now routine for many species. Important crop species such as corn, wheat, rice and soyabean, which were deemed impossible to transform or regenerate only a few years ago, are now routinely manipulated. Usually, breakthroughs in transformation protocols of so-called recalcitrant species are first made with a highly regenerable cultivar. This transformed cultivar can then be crossed with elite high-producing cultivars to introduce the transgene into a number of plant lines.

In very early versions of plant transformation protocols, foreign DNA was introduced into plant protoplasts (plant cells treated with cellulase and other digestive enzymes to remove the cellulose cell walls) by using electroporation or polyethylene glycol. Since many plants are regenerated from protoplasts only with great difficulty, this method has not been widely applied. Most plant transformation has been carried out by *Agrobacterium tumefaciens*-mediated DNA delivery. Wild-type A.*tumefaciens* causes crown gall disease by transferring a segment of plasmid DNA (called the T-DNA, for transfer DNA) containing bacterial genes encoding auxin biosynthesis and other pathways into wounded plants, which in turn cause dedifferentiated callus tumours to form, releasing nutrients to the surrounding A.*tumefaciens* organisms in the soil. A.*tumefaciens* strains have been modified (disarmed) such that the auxin biosynthetic genes have been deleted so they can no longer cause disease but the strains still contain the *vir* genes required for T-DNA transfer. Foreign genes in interest, including a selectable marker, are inserted into binary vectors, such that the foreign genes are between the ends of the T-DNA region. When in contact with cultivated plant cells, usually at the beginning of the callus induction phase of either organogenesis or somatic embryogenesis, the disarmed A.*tumefaciens* strain transfers the foreign gene construct. Selection pressure is applied (usually an antibiotic or herbicide in the medium), so that the only cells which are able to grow are those which received the foreign T-DNA (including the selectable marker, usually a drug resistance or herbicide resistance gene). The A.*tumefaciens* strain is then killed by antibiotics (such as penicillin derivatives) which do not act against plant cells. The surviving transformed plant cells are then induced to regenerate into transgenic plants.

A.*tumefaciens*-mediated transformation has been used for many dicot species but has been less effective with monocots. A more recently developed DNA delivery system is called particle bombardment or the biolistic approach. Early versions of biolistic equipment used the explosive force of gunpowder to propel the 1-μm tungsten or gold particles, coated with the desired foreign gene construct. The current version of the equipment (patented by Dupont but marketed through Bio-Rad) uses high-pressure helium gas and calibrated membranes (rupture disks) as a safer and more reproducible driving force. Transgenic plants have been produced by 'shooting' leaf explants, embryogenic suspensions, or young embryos followed by applying drug or herbicide selection pressures, as described for the A.*tumefaciens* system, and then by regeneration. The biolistic approach was the first system to allow routine transformation of many monocot species such as corn, wheat, and rice, but it is also popular for dicot transformation.

Plant transformation and regeneration has resulted in the release and marketing of many genetically engineered plants. Current crops in the field include insect-resistant corn and cotton expressing the insecticidal *Bacillus thuringensis* toxin protein, herbicide-resistant soyabeans that are resistant to glyphosate and bialophos, and ripening-delayed tomatoes. Many more products are under development, including plants with altered primary-metabolite (fatty acids, oil) and secondary-metabolite accumulation.

PLANT BIOTECHNOLOGY

Plants are the primary source of food for the human race and only by correct management of plant agriculture can the present human populations continue to be fed. The flow of energy from sunlight through plant photosynthesis is at the heart of the importance of plants in the world economy. In the highly technology-driven world in which we now live the links between plants and people are easily overlooked (Fig. 10.1).

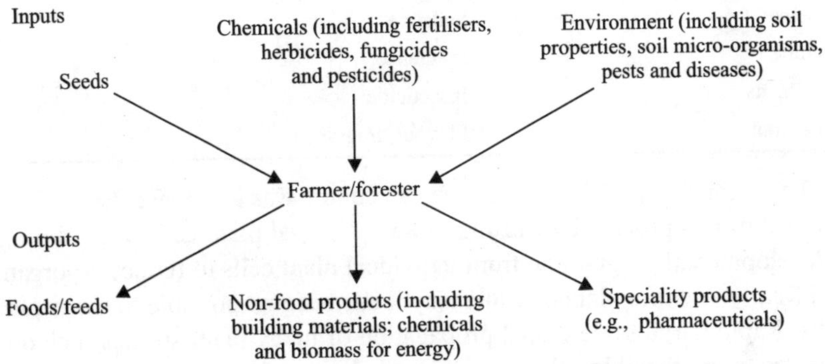

Fig. 10.1. A scheme illustrating the production and uses of plants.

Since early times human beings have sought to improve the quality and productivity of agriculturally important plants. This was done by selection and traditional breeding procedures – a painstakingly slow and difficult process. But it has been a remarkably successful commitment as witnessed by the high quality of the present food plants such as corn, rice, wheat, potato, etc., – all far removed from their ancestral beginnings. Traditional breeding programmes involving sexual crosses will continue to dominate the approach to improve the agronomic characteristics of food crops but will be increasingly augmented by new techniques incorporating micropropagation, protoplast fusion and genetic engineering.

As early as 1939 it became possible to isolate small numbers of cells from certain plants and to keep them alive indefinitely in artificial cultivation.

The cultivation of these tissues required the presence of a plant hormone that allowed the cells to propagate in an unorganised manner resulting in an amorphous mass of cells. Advances were soon made in cultivation techniques, achieving rapid growth rates in chemically defined media. These individuals or groups of cells were treated like microbial suspensions and were able to grow under aerated and shaken conditions, initially in flasks but subsequently in large traditional bioreactors. By this method it is possible to produce important secondary metabolites with high commercial value (Table 10.4).

Table 10.4. Potential uses for plant secondary products.

Compound	*Uses*
Vinblastine/vicristine	Leukaemia
Ajmalicine	Circulatory problems
Digitalis	Heart disorders
Quinine	Malaria, flavour
Codeine	Sedative
Jasmine	Fragrance
Pyrethrins	Insecticide
Spearmint	Flavour, fragrance

The second major advance in plant cell culture was to achieve the complete reversal of this process by causing these individual plant cells to go through a developmental programme from individual plant cells to tissues, to organs, and finally to entire plants. In this way it has become possible to clone plant cells. Rapid, large-scale clonal propagation of many plant species including trees is now feasible. Small tissue explants of many species can be aseptically removed from the parent plant and artificially maintained and increased in number by suitable control of the medium. The process can be rapid and produces high quality, uniform plants. Outstanding examples of this technology have been the recent repetitive cloning of oil palms and coffee plants from callus tissues, producing unlimited numbers of stable types. This area of micropropagation not only allows rapid propagation or mass production of identical clones of plant species but also has the following uses:

1. Elimination of viruses and other pathogens.
2. Storage of essential germ plasm instead of conventional seeds.
3. Embryo rescue.
4. Production of haploids by another and ovary culture (gametoclonal variants), useful for cereals.

It is now possible to take plant cells and subject them to the battery of manipulative techniques long practised in industrial microbiology, e.g., mutation, strain selection and process development. Thus, the genetic diversity of plants may be altered without the normal sexual process of fertilisation: by production of haploid, triploid and tetraploid cells; by the use of protoplast fusion between different species and even genera; and by transformation, i.e., transferring DNA from one plant cell (or even another type of organism) into the cells of another. The techniques of recombinant DNA technology are now available in practical form to the plant technologist (Fig. 10.2).

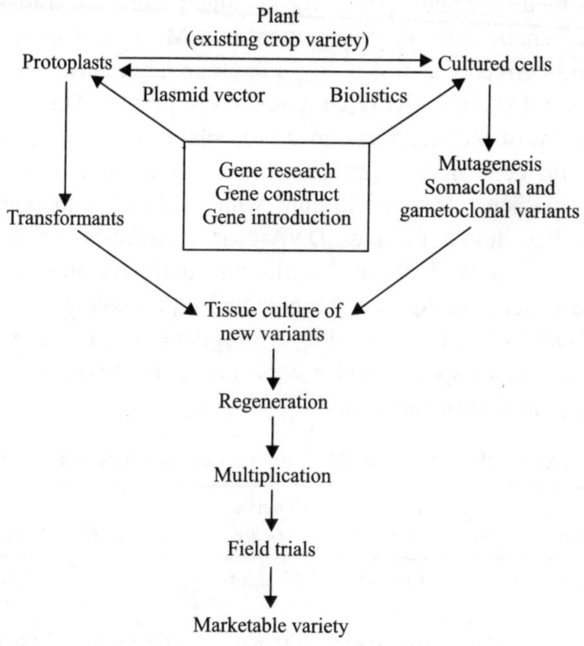

Fig. 10.2. Experimental approaches used to create new plant varieties by biotechnology.

Protoplasts can be produced from most plant cells by digesting away the cell wall and maintaining the protoplasts in a suitable osmotic medium. Many types of protoplast can be induced to reform cell walls and to divide to from cell colonies. Some plants, including potato, pepper, tobacco and tomato, can be fully regenerated form protoplasts. Full regeneration for important cereal crops is not yet possible. Some protoplasts may be fused with other species, allowing a novel mixing of genetic traits.

With tissue culture of callus or large structures, regeneration leads mainly to uniformity of plants. In contrast, regeneration from single plant cells or protoplasts can often be accompanied by minor extensive changes in the final plant phenotype. This has been termed *somaclonal variation* and is widely used as a means of crop improvement, in particular with respect to yield and disease resistance (Fig. 10.2).

The first practical system for genetically manipulating plants occurred in 1983 with the discovery of the ability of the bacterium *Agrobacterium tumifaciens* to transfer part of its Ti plasmid (T-DNA) into the host plant genome. In this way foreign genes can be inserted into the plasmid DNA and then integrated into the plant genome. Transformants can be identified by selection methods, adult plants reconstituted from the transfected cells and the new genetic material transmitted as a Mendelian trait. Nowadays, the Ti plasmid-derived vector is used routinely in laboratories throughout the world for transforming dicotyledonous crop plants. This method was unsuitable for most monocotyledonous crop plants such as maize, wheat or rice. Successful gene transfer into the monocotyledonous crop plants was achieved by biolistics or 'particle gun' which was used to bombard plant cells with genes in the form of DNA-coated particles. In this way the microprojectile penetrated the cell walls and delivered the DNA into the nucleus. The intact cells could be regenerated into whole plants (Fig. 10.3). By these methods it is now possible to introduce advantageous new traits into almost any plant species and a wide range of agricultural crop plants have been manipulated (Table 10.5).

Table 10.5. Crops into which single purified genes have been introduced.

Cereals	*Beverage plants*	*Fibre*	*Legumes/ oil seeds*	*Vegetables, fruits*	*Flowers*	*Pasture*	*Trees*
Rice	Coffee	Cotton	Linseed	Carrot	Rose	Lucerne	White spruce
Maize		Flax	Soyabean	Tomato	Chrysan-themum	Orchid	Walnut
Wheat			Rape	Strawberry	Carnation	Grass	Apple
Rye			Sunflower	Celery	Geranium		
			Canola	Cauliflower			
			Palm	Tobacco			
				Sugar beet			
				Cucumber			
				Potato			

Ti plasmids have also been used to insert 'antisense genes' in order to negate the functions of specific plant genes concerned with an undesirable phenotype. Such genes are produced by reversing the orientation of a gene in relation to its promoter. The transcribed product cannot be translated into a normal protein thus preventing its function within the plant.

Application of Genetic Engineering to Plant Agriculture

What are the current main goals for the application of genetic engineering to plant agriculture? In essence genetic engineering releases the plant breeder from two major constraints:

1. Limits imposed by interspecific barriers.
2. Lack of precision inherent in traditional breeding methods.

Presently genetic engineering methods allow the introduction of single-gene characteristics into plant cells and this should extend later to multiple genes and indeed to whole biochemical pathways. The main introduced improvements will initially include:

1. Improved resistance to specific herbicides.
2. Improved resistance to insect pests and microbial diseases.
3. Improved post-harvest characteristics.

Improved resistance to specific herbicides

The killing of plant weed species by the application of selective herbicides gives a growth advantage to commercial crop plants. Such compounds have been designed to disrupt the growth of certain weed species without affecting the particular crop plant. However, there is increasing opposition to the continued use of such chemical compounds on the grounds of environmental and human health considerations. Herbicide-tolerant crop plants have now been produced by genetically engineering plant genomes resistant to specific herbicides. This has been seen as a way of producing more effective, less costly and more environmentally compatible weed control. This will permit reduced overall herbicide use, and a return to broad spectrum herbicides which have low mammalian toxicity and are rapidly degraded by soil micro-organisms. Successful crops so far in use include the potato and oil seed rape. Weed control is an indispensable aspect of modern agriculture.

Improved resistance to insect pests and microbial diseases

Gene transfer into crop plants to impart insect or microbial resistance is major new area of research into plant protection. Most efforts have been directed towards protecting standing crops in the field, whereas little attention has yet been given to insect pests in post-harvest storage, where huge losses can occur particularly in developing countries. Genes from *Bacillus thuringiensis* (Bt) have been introduced into several crops, including tomato and cotton and field-testing has demonstrated impressive results against many pests.

Microbial diseases, in particular fungal and viral, remain one of the major factors limiting crop productivity world wide, with continuing huge losses set against large cash inputs for pesticide treatment. Much improved resistance to viruses by the integration of genes for viral coat proteins has now been achieved in several crop plants, particularly rice. These approaches could well be the most advantageous aspects of all plant biotechnology. The necessary widespread use of insecticides, fungicides and pesticides for crop protection undoubtedly has damaging effects on the environment and consequently it is imperative to improve the control of pests and diseases by genetic means.

Improved post-harvest characteristics

Losses during storage and transport of some crops can be as high as 40 per cent in USA and Europe and as high as 80 per cent elsewhere. While a great deal of this loss will be due to diseases and pests, with soft fruits and vegetables there can be bruising, heat and cold damage, over-ripeness, off flavours and odours, etc. Most of these physiological changes result from endogenous enzyme activity. Can such activity be genetically stopped or slowed down? In the tomato the enzyme polygalacturonase breaks down cell wall constituents, thus leading to softening of the fruit during ripening. This process is independent of colour development. In normal conditions, if the tomato is left to ripen on the vine and develop full colour the softening process is also occurring, so creating an easily bruised and damaged fruit on shipment. By inhibiting the polygalacturonase by antisense genes the tomato can remain on the vine until mature and be transported in a firm solid state. The Flavr Savr tomato has been engineered by Calgene in the USA with improved flavour and shelf-life and is now marketed in the USA. Such techniques will now be used in a wide variety of soft fruits. Other studies are considering ethylene synthesis/inhibition as a means of controlling fruit and flavour maturation.

Genetic manipulation of flower and leaf colour, abundance of flowers, perfume, and shape are now major targets for the decorative plant industries.

Production of high value oil products

There is increased interest in the genetic modification of oil-producing plants such as soya bean, rape and canola to produce a wide range of industrial lubricants, cosmetic compounds, detergents that are biodegradable, etc. The proportion of plant agricultural production directed towards non-food uses has now passed 20 per cent in many developed countries, e.g., cotton, tobacco, natural rubber and, of course, wood. The discovery that plants can synthesise functional human antibody fragments opens up a whole new area for plant biotechnology. The use of plants as antibody factories will avoid the often controversial industrial use of animals.

The first stages of the plant biotechnology revolution have now arrived and are being assessed by a somewhat sceptical and unprepared public. Successful commercial products will need to meet a genuine market demand.

Forestry

Throughout the world there is an escalation in demand for wood-derived products and many major countries such as those of Europe, India and Japan are now in deficit. This will be further compounded by increased pollution such as 'acid rain' and the huge losses now occurring in the rain forests of South America and Asia by indiscriminate felling. Forests have not had the extensive research and management bestowed on other cultivated tree crops such as rubber, coffee and citrus. The long generation times of most trees (5–20 years) when compared to herbaceous crop plants can well explain the lack of comprehension of their life cycles.

However, biotechnology will play an important role in achieving an increase in production as well as bringing major improvements to the quality of the trees. Tissue culture technology such as micropropagation, somatic embryogenesis (induction of single cells or cell aggregates to develop into embryo-like structures from which a shoot and root develop), selection of somaclonal and gametoclonal variants and gene transfer are being developed to improve forests (Fig. 10.3). Targets for genetic improvements include increased productivity and biosynthesis, increased photosynthetic ability, disease, insect resistance, etc.

Trees have long generation times and, as a consequence, genetic improvements will be slow. Loblolly pine has been transformed by *Agrobacterium tumefaciens* and this may allow gene transfer techniques similar to those used with other plant species.

Biological Control

The use of chemical pesticides has led to dramatic improvements in the production levels of agriculture and forestry. Their widespread use is the dominant reason that fewer people can now produce more food at less cost than ever previously. The pesticide market is dominated by synthetic chemicals and will undoubtedly remain so. However, consumers are becoming increasingly concerned about food quality and possible carry-over of pesticide residues into food products. Biotechnologists are now actively examining alternatives to chemical pesticides as a means of controlling agricultural pests and diseases. An obvious approach would be to use naturally occurring biological means of controlling these problems. It is a fact of life that all organisms have their own specific diseases and predators. In the present context, biological control will refer to the use of micro-organisms applied in the field to control pests and diseases. In a similar way insect predators can also be used for control purposes.

The most successful biocontrol agent is *Bacillus thuringiensis*, a sporeforming bacterium containing crystalline protein inclusions. The proteins are highly toxic to insect pests but specific in their activity. They have been widely used for over 30 years against lepidopteran (caterpillar) pests. The commercial application of these biopesticides is usually as formulations of spores and crystalline inclusions of disrupted *B.thuringiensis* and are applied at 4–6 gram ha^{-1} or about 4^{20} molecules ha^{-1}.

In recent years the toxin genes have been isolated and sequenced and recombinant DNA-based products produced and approved. These genes have also been incorporated into various plant species and expressed in the plant tissue. This now offers the farmer either direct application of the microbial biopesticides or protection of the plant directly by toxic gene expression in the plant tissue. A large number of companies are now involved in the research, development and production of *B.thuringiensis* toxins. In most cases the companies are relatively small, innovative and research driven, but tend to form associations with well-established agrochemical companies. While the *B.thuringiensis* toxins dominate the market there are many examples of viral and fungal biopesticides that are achieving an increased market share.

The commercial potential for these products will be determined by their efficacy, cost–benefit ratios when compared to synthetic chemicals, ease of use and spectrum of activity (Table 10.6). The comparisons between chemical and microbial pesticides is shown in Table 10.7.

It is anticipated that biocontrol agents may ultimately increase their market share of global pesticide applications to about 5 per cent. However, should consumer opinion move strongly against chemical pesticides then the BCA option will become a meaningful alternative.

Table 10.6. Requirements for a micro-organism to be a successful biological control agent (BCA).

It must have a substantial market-size and consumer demand
It must be equal to or better than chemical pesticides in performance and persistence
The product must be safe, with low mammalian toxicity and little effect on non-target species
It should remain stable on storage
Mass production by bioprocess technology should be cheap
BCAs should be applied using conventional technology without recourse to major changes in current agricultural practices

Table 10.7. Chemical versus microbial pesticides.

Description	*Chemical pesticide*	*Microbial pesticide*
Product use		
Speed of action	Usually rapid	Can be slow
Killing efficacy	Often 100%	Usually 90–95%
Spectrum of effect	Generally broad	Generally narrow
Resistance of target organism	Often developed	Not yet seen
Product safety		
Toxicological testing	Lengthy and costly	Low cost
Environment hazards	Many well-known examples	None yet shown
Residues	Interval to harvest usually required	Crop may be harvested immediately

11

Biotechnology of Animals

INTRODUCTION

Animal agriculture in the form of cattle, pigs, sheep, poultry and fish represents a major aspect of food production world wide. While many of these animals are produced for their meat alone, others contribute to human nutrition by way of milk and egg production. In the developed world animal production is highly intensified and technologically driven. Animal production will reflect on quality of feed, availability and need of growth hormones, pesticides, antibiotics and vaccines, good animal husbandry and welfare, and increasingly selective breeding, molecular biology, embryo manipulation and gene transfer.

As with plant agriculture early animal breeders sought to identify worthwhile properties in animals and to perpetuate them into future offspring. Selective breeding aims to increase the frequency of a large number of genes that work together with the remainder of the animals's genome to produce the desired phenotype. For example, between 1945 and 1993, selective breeding increased milk production of the average dairy cow by a factor of 3. Further, when we consider the huge variety of dogs it should be remembered that they are all one species, capable of easily interbreeding and the present pedigree varieties have arisen by carefully controlled selective breeding programmes. For many farm animals conventional breeding has already achieved high producing animals but it is increasingly apparent that the increases in productivity possible by this means would seem to be approaching a plateau. To sustain an ever-increasing world population ways must be achieved to meet this increasing demand for animal products.

The term *tissue culture* refers to the culture of whole organs, tissue fragments as well as dispersed cells on a suitable nutrient medium. It can be divided into (i) organ culture; and (ii) cell culture mainly on the basis of whether the tissue organisation is retained or not. In *organ cultures*, whole embryonic organs or small tissue fragments are cultured *in vitro* in such a manner that they retain their tissue architecture. In contrast, *cell cultures* are obtained either by enzymatic or mechanical dispersal of tissues into individual cells or by spontaneous migration of cells from explants; they are maintained as attached monolayers or as cell suspensions.

Freshly isolated cell cultures are called *primary cultures*; they are usually heterogeneous and slow growing, but are more representative of the tissue of their origin both in cell type and properties. Once a primary culture is subcultured, it gives rise to *cell lines*, which may either die after several subcultures (such cell lines are known as *finite cell lines*) or may continue to grow indefinitely (these are called *continuous cell lines*).

Usually, normal tissues give rise to finite cell lines, while tumours give rise to continuous cell lines. But there are several examples of continuous cell lines, which were derived from normal tissues and are themselves nontumorigenic. The evolution of continuous cell lines from primary cultures is supposed to involve a mutation, which alters their properties as compared to those of finite lines.

A variety of areas of research in biotechnology depend largely on tissue culture and, therefore, these areas have given the desired impetus to the development of tissue culture methods in animals. These areas include the following: (i) production of antiviral vaccines, which required the standardisation of cell lines for the multiplication and assay of viruses; (ii) cancer research, which required the study of uncontrolled cell division in cultures; (iii) cell fusion techniques; (iv) genetic manipulation, which is possible in cells or organs in culture; (v) monoclonal antibodies whose production, requires cell lines in culture; (vi) production of pharmaceutical drugs using cell lines; (vii) chromosome analysis of cells derived from womb; (viii) study of the effects of toxins and pollutants using cell lines; (ix) use of artificial skin; (x) study of the function of nerve cells (even though neurons can not be propagated *in vitro*, without resorting to the use of transformed cells.

ADVANTAGES AND DISADVANTAGES OF TISSUE CULTURE

Tissue culture offers a number of advantages for a variety of studies and applications, but there are also disadvantages, so that tissue culture should not be used indiscriminately. These advantages and disadvantages are listed in Table 11.1.

Table 11.1. Advantages and disadvantages of the use of animal tissue culture.

Advantages	*Disadvantages*
Controlled physiochemical environment (pH, temperature, osmotic pressure, O_2, CO_2, etc.	Expertise is needed, so that behaviour of cells in culture can be interpreted and regulated.
Controlled and defined physiological conditions (constitution of medium, etc.)	Ten times more expensive for same quantity of animal tissue; therefore reasons for its use should be compelling
Homogeneity of cell types (achieved through serial passages).	Unstable aneuploid chromosome constitution.
Economical, since smaller quantities of reagents are needed than *in vivo*.	
Legal, moral and ethical questions of animal experimentation are avoided.	

LABORATORY FACILITIES FOR TISSUE CULTURE

The laboratory facilities for animal culture (as also for plant tissue culture) should allow for the following: (i) sterile handling; (ii) incubation (iii) preparation of glassware, media and tissue; (iv) wash-up; (v) sterilisation and (iv) storage. The clean area for handling should be available at one end of the room and the wash up and sterilisation at the other end, with preparation, incubation and storage in between.

ORGAN CULTURE

In vitro culture and growth of organs or parts thereof in which their various tissue components, e.g., parenchyma and stroma, are preserved both in terms of their structure and function so that the cultured organs resemble closely the concerned organs *in vivo* is called *organ culture*. In such cultures, new growth is in the form of differentiated structures, e.g., glandular structures in case of glands, small bronchi in case of lung tissues, etc., in tissues lined with one or the other type of epithelium, the epithelium differentiates in a pattern similar to that in the concerned organs *in vivo*.

The cultured organs retain their physiological features, e.g., hormone-dependent organs remain hormone-dependent and endocrine organ go on secreting the specific hormones. In addition, the morphogenesis in cultured foetal tissues is more or less comparable to that *in vivo*. In case of organ cultures, outgrowth of isolated cells from the periphery of explants is minimised by manipulating the culture conditions.

Technique

The technique of organ culture may utilise one of the following five approaches.

Plasma clot

In this approach, the explant is cultured on the surface of a clot consisting of chick (or other) plasma and chick embryo extract contained in a watchglass; therefore, it is also called *watchglass technique*. The watchglass may or may not be closed with a glass lid sealed with paraffin wax. This has been the classical standard technique for studying morphogenesis in embryonic organ rudiments. It has been also modified to study the action of hormones, vitamins, carcinogens, etc. on adult mammalian tissues.

One of the chief disadvantages of all plasma clot methods is that the clot liquefies in the vicinity of explants so that they become partly or fully immersed in the medium. The duration of culture is rather short (less than 4 weeks) and biochemical analysis is not possible due to the complexity of the medium.

Raft methods

In this approach the explant is placed onto a raft of lens paper or rayon acetate, which is floated on serum in a watch-glass. Rayon acetate rafts are made to float on the serum by treating their four corners with silicone. Similarly, floatability of lens paper is enhanced by treating it with silicone. On each raft, four or more explants are usually placed. In a combination of raft and clot techniques, the explants are first placed on a suitable raft, which is then kept on a plasma clot. This modification makes media changes easy, and prevents the sinking of explants into liquified plasma.

Agar gel

In this approach, the medium (consisting of a suitable salt solution, serum, chick-embryo extract or a mixture of certain amino acids and vitamins) is gelled with 1 per cent agar. This method avoids immersion of explants into the medium and permits the use of defined media. Generally, explants need to be subcultured on fresh agar gels every 5–7 days. This method has been used to study many developmental aspects of normal organs as well as of tumours.

Grid method

The *grid method* utilises 25 mm × 25 mm pieces of a suitable wire mesh or perforated stainless steel sheet whose edges are bent to form 4 legs of about 4 mm height. Skeletal tissues are generally placed directly on the grid but softer tissues like glands or skin are first placed on rafts, which are then kept on the grids. The grids themselves are placed in a culture chamber filled with fluid medium up to the grid; the chamber is supplied with a mixture of O_2 and CO_2 to meet the high O_2 requirements of adult mammalian organs.

Cyclic exposure to medium and gas phase

This technique has been successful in long-term (up to 4–5 months) culture of human adult tissues like oesophagus, mammary epithelium, uterine endocervix, etc. The explants are intermittently exposed to the fluid medium and the gas phase. The number of explants per dish varies from 2–18 depending on the organ cultured. The explants are attached to the bottom of a plastic culture dish and are covered with fluid medium.

Advantages

1. The explants remain comparable to the *in vivo* organs both in structure and function, which makes them more suitable than cell cultures for physiological studies.
2. The development of foetal organs *in vitro* is comparable to that *in vivo*. Hormone-dependent organs remain so, while endocrine organs secrete the specific hormones.
3. Therefore, organ cultures provide information on the patterns of growth, differentiation and development, and on the influences of various factors on these features.
4. In some cases, organ cultures may replace whole animals in experimentation as the results from them are easier to interpret. The results obtained with organ cultures usually give an idea of the *in vivo* events; this often reduces considerably the number of experiments necessary with whole animals to investigate a given problem.

Disadvantages

1. Results from organ cultures are often not comparable to those from whole animal studies, e.g., in studies on drug action, since the drugs are metabolised *in vivo* but not *in vitro*.
2. Organ cultures can be maintained only for a few months. But it may be desirable to study the effects of certain factors for several months. In such cases, the organs treated *in vitro* may be transplanted into suitable host animals, e.g., nude mice.

Applications

1. Studies on the patterns of growth, differentiation and development of organ rudiments, i.e., foetal organs, and the influences of various factors like hormones, vitamins, etc., on these parameters.
2. The action of drugs, carcinogenic agents, etc. on the animal organs is studied *in vitro* to at least serve as a guide for the events in whole animals.
3. Perhaps the most dramatic application of organ cultures is to produce tissues for implantation in patients; this is often called *tissue engineering*. Human skin has been successfully produced *in vitro* and used for transplantation in more than 500 cases of serious burns, ulcers, etc. The ultimate objectives of tissue engineering are to reconstitute body parts *in vitro* (i) for use as grafts or transplant; and (ii) to be used as models for studies on drug delivery and action. reagents. It is expected that cartilage tissue developed *in vitro* (*artificial cartilage*) will be available for human implantation in cases of injuries, arthritis, etc.; results with rabbits have been encouraging. It is hoped that studies will permit the culturing and constitution of bones, liver, pancreas, etc.

CELL CULTURES

Cell cultures may contain the following three types of cells: (i) stem cells; (ii) precursor cells; and (iii) differentiated cells. *Stem cells* are undifferentiated cells, which have unlimited capacity for poliferation, and they can differentiate under correct inducing conditions into one of several kinds of cells; different kinds of stem cells differ markedly in terms of the kinds of cells they will differentiate into. *Precursor cells* are derived from stem cells, are committed to differentiation, but are not yet differentiated; these cells retain the capacity for proliferation. In contrast, *differentiated cells*, usually, do not have the capacity to divide. Some cell cultures, e.g., epidermal keratinocyte cultures, contain all the three types of cells. In such cell cultures, stem cells constantly provide new cells which develop into precursors; the precursor cells proliferate and mature into the differentiated cell types. Thus stem cells are necessary for the maintenance of such cultures, which by their nature are heterogenous. On the other hand, other cell cultures, e.g., fibroblast cultures, contain a more or less uniform population of dividing cells at low cell densities ($< 10^4$ cells/cm^2), but at high cell densities (10^5 cells/cm^2) are uniformly composed of nonproliferating differentiated cells. The cells begin to proliferate once the cell density is appropriately reduced.

Differentiation and cell proliferation are affected by, in addition to cell density, factors like serum, Ca^{2+} ions, hormones, cell to cell and cell to matrix interactions, etc. Generally, cell proliferation is promoted by low cell density, low Ca^{2+} (100–600 µM), and high growth factor levels, while differentiation is promoted by the exact opposite conditions and by the presence of differentiation inducing factors, e.g., cortisone, nerve growth factor, etc. The proportions of stem, precursor and differentiated cells are markedly affected by the source tissue used for obtaining the cultures. For example, cultures derived from embryos and those derived from even adult tissues where continuous cell renewal occurs naturally, e.g., intestinal epithelium, haematopoietic cells (cells producing blood cells), etc., stem cells are likely to be more frequent than in other cell cultures. In contrast, cell cultures from tissues where renewal occurs only under stress, e.g., fibroblasts, muscle, etc., may contain only precursor cells.

Cell cultures can be grown as (i) monolayers or as (ii) suspension cultures. Propagation in suspension cultures is limited to haematopoietic cell lines, ascites tumours and transformed cells (those cells that have become phenotypically modified during *in vitro* culture to become anchorage-independent and are able to grow in layers of several cells thick, as against monolayer growth of nontransformed cells). Therefore, cells in culture need a surface or substrate to adhere to so that they are able to proliferate. Cells that are unable to adhere to a substrate are unable to divide, i.e., their growth is *anchorage-dependent*.

Substrate on Which Cells Grow

The surface available for attachment of cultured cell is called substrate. The majority of vertebrate cells cultured *in vitro* have been grown as monolayers on an artificial substrate. Spontaneous growth in suspension is restricted to hemopoietic cell lines, rodent ascites tumour cell lines and a few other selected cell lines. With these exceptions, most cells need to spread out on a substrate in order to proliferate. Inadequate spreading will inhibit cell proliferation. Cells shown to require attachment for growth are said to be anchorage dependent.

However, cells which have undergone transformation, frequently become anchorage independent and can grow in suspension. (Note: In microbiology and plant cell culture, transformation means uptake of new genetic material but in animal cell culture, particularly mammalian cell culture, transformation means phenotypic modification of cells in culture, without involving the uptake of new genetic material, which is described as transfection to distinguish it from transformation). Whenever, a substrate is needed, it may be adhesive (glass, plastic, palladium, etc. or a metallic surface; these surfaces may sometimes be vigorously modified) or nonadhesive (agar, agarose, methocel) and will be briefly discussed in this section.

Glass as a substrate

Glass is the most commonly used substrate, in the form of slides, coverslips, flasks and test tubes, which may or may not be vigorously modified. It is cheap, easily washed without losing its growth supporting properties, can be readily sterilised by dry heat (oven) or moist heat (autoclave) and is optically clear for microscopic examination of cultures. Treatment with strong alkali renders glass unsatisfactory for culture unless it is neutralised by acid wash.

Disposable plastic as a substrate

Polystyrene is often used as the most common and cheapest substrate, to be discarded after single use. This plasticware is supplied sterile and is not autoclavable. However, it is hydrophobic and needs to be treated by gamma irradiation or some chemical or with electric arc to produce a charged surface, which thus becomes wettable. One may, however, have to determine the plating efficiency and growth rate of cells, since great variation occurs in the plasticware manufactured by different commercial companies.

Other plastics used for tissue culture include the following: polyvinylchloride (PVC), polycarbonate, polytetrafluoroethylene or teflon (PTFE), thermanox (TPX), etc. In several of these other plastics, one should try to grow a regular monolayer and then attempt to culture or clone cells on it with and without pretreatment. PTFE films are available as displosble Petri dishes or as membranes to be incorporated in an autoclavable reusable culture vessel. These dishes have two other advantages: (i) substrate is permeable to O_2 and CO_2 and (ii) plastic is thin, and is suited to sectioning for microscopy.

Substrates involving polystyrene, sephadex and polyacrylamide are also available in the form of beads for propagation of anchorage dependent cells in suspension.

Palladium and metallic surface as substrates

For 90 per cent of animal cell and tissue cultures, glass or plastic is employed as a substrate. However, other substrates have been used for specialised applications. For instance, palladium deposited on agarose (using electron microscopy shadowing equipment) was used as a substrate for growth of fibroblasts and glia. Stainless steel discs and other metallic surfaces were also used.

Treatment of substrate surfaces

The substrate surfaces described above can be treated to improve cell attachment. The treatments with culture medium meant for another culture, or with purified fibronectin (1ng/ml) added to the medium or with collagen are commonly used. Coating with collagen can be achieved, using rat tail collagen or other commercially available forms, and pouring the collagen solution over the surface of the dish, draining the excess, and permitting the residue to dry. After coating with collagen, the substrate may be sterilised under UV and used either immediately or may also be stored. Sometimes on the inert coating of collagen (or gelatin or poly-D-lysine), a monolayer of an appropriate cell type may be used to provide a matrix for the maintenance of some specialised cells.

Feeder layers on substrate

Sometimes, on the substrate surface a monolayer of special kind of cells is used, partly to supplement the culture medium for growth and partly for conditioning of the substrate by cell products. These monolayers are called feeder layers, since they also feed the growing cultures. The feeder layer may consist of mouse embryo fibroblasts, normal fetal intestine, glial cells, etc. and has been effectively used for selective growth of breast and coelomic epithelium, central and peripheral neurons, and also for growing cells used for the production of transgenic animals through transfection.

Gas phase for tissue culture

The animal cell and tissue cultures also require for their growth a gas phase consisting of O_2 and CO_2, even though cultures (cell culture vs organ culture) may vary in their oxygen requirement. While cell cultures may grow on an atmospheric oxygen concentration or even lower O_2 concentration, some organ cultures (e.g., late embryo, newborn or adult) may require upto 95 per cent O_2. While O_2 is directly available to cells in cell cultures, in organ cultures there may be a problem of diffusion. Since there is a trend towards the use of serum free media, O_2 concentration may become an important factor in tissue cultures in future. CO_2 concentration, which is a complex factor, also needs to be manipulated, since it also influences the pH and the HCO_3^- ion concentration of the culture. At very low concentration of cells, it is necessary to add CO_2 in gas phase of sealed flasks of most cultures. However, in high concentration of cells, CO_2 is liberated from the cultured tissue or cells. Therefore, it is not necessary to add CO_2 in gas phase of such sealed flasks with high concentration of cells; but instead sometimes we may need to slacken the cap of a culture flask and allow excess CO_2 to escape.

Modification of Substrate Surface

The most important factor in cell attachment to a substrate is the net negative charge of its surface and the high electrostatic energies present. Since animal cell surface also has a net negative charge, cross-linking with glycoproteins and/or divalent cations (Ca^{2+}, Mg^{2+}) is required for cell adhesion. A glycoprotein involved in cell attachment is fibronectin which is present in the serum used in many culture media, and is synthesised by many cells. In view of this, the substrate surface may need modifications, some of which are summarised below.

1. Glass and metal surfaces are naturally negatively charged, and no treatments are necessary. Repeatedly washed glassware can be treated with magnesium acetate to regain cell adhesion properties.
2. Organic surfaces, e.g., plastics, must be made wettable and negatively charged. This is readily achieved by certain chemical (e.g., oxidising agents, strong acids) or physical (e.g., UV irradiation, high voltage discharge, etc.) treatments. Plasticware are supplied after one such treatment.
3. In some special cases, e.g., for culture of neurons, muscle cells, etc., the substrate surface is made positively charged by coating it with gelatin, collagen, or polylysine. A tissue culture grade collagen is commercially available; alternatively, it is prepared from rat tails.

Serum

The serum provides various plasma proteins, peptides, lipids, carbohydrates, minerals, and some enzymes. Serum serves the following major functions.

1. It provides the basic nutrients for cells; the nutrients are present both in the solution as well as are bound to the proteins.
2. It provides several hormones, e.g., insulin, which is essential for growth of nearly all cells in culture, cortisone, testosterone, prostaglandin, etc.
3. It contains several growth factors, e.g., platelet derived growth factor (PDGF), transforming growth factor β (TGF-β), epidermal growth factor, etc., these are present in concentrations of μg/l. Both

hormones and growth factors are involved in growth promotion and specialised cell function. A given hormone or growth factor may stimulate growth of one cell type, may have no effect on another and may even be inhibitory to some others. For example, PDGF induces proliferation in fibroblasts, but induces differentiation of some types of epithelia. Further, proliferation of a single cell type may be induced by more than one growth factor, e.g., fibroblasts respond to PDGF, epidermal growth factor, fibroblast growth factor and somatomidins.

4. A major role of serum is to supply proteins, e.g., fibrobnectin, which promote attachment of cells to the substrate. It also provides spreading factors that help the cells to spread out before they can begin to divide. Although cells do produce these factors, but trypsinised cells are usually unable to attach to the substrate.
5. It provides several binding proteins, e.g., albumin, transferrin, etc., which carry other molecules into the cell. For example, albumin carries into cells lipids, vitamins, hormones, etc. Transferrin usually carries Fe in a nonbasic form, but binding of transferrin to its receptor in cell membrane is believed to be mitogenic.
6. It increases the viscosity of medium and, thereby, protects cells from mechanical damages, e.g., shear forces during agitation of suspension cultures.
7. Protease inhibitors present in the serum protect cells, especially trypsinised cells, from proteolysis.
8. The serum also provides minerals, like Na^+, K^+, Fe^{2+}, Zn^{2+}, etc.
9. It also acts as a buffer.

However, there are several disadvantages of using serum in the culture medium; these are summarised below.

1. Serum may inhibit growth of some cell types, e.g., epidermal keratinocytes.
2. Serum may contain some cytotoxic or potentially cytotoxic constituents. For example, foetal calf serum contains the enzyme polyamine oxidase, which converts polyamines like spermidine and spermine (secreted by fast growing cells) into cytotoxic polyamino-aldehydes.
3. There is a large variation in serum quality from one batch to another; this requires costly and time consuming testing every time a new batch has to be used.

4. Some growth factors may be inadequate for specific cell types and may need supplementation.
5. It interferes with downstream processing when cell cultures are used for production of biochemicals.
6. The supply of serum is always lower than its demand.

Media with Serum

Although natural media are very useful and convenient for a wide range of uses, they suffer from disadvantage of poor reproducibility due to lack of knowledge of the exact constitution. For this reason, synthetic media have been designed for a variety of tissues and purposes, like the following: (i) media for immediate survival, (ii) media for prolonged survival.

Serum free Media

The use of serum in culture media has several disadvantages, which led to the development of many serum free media. Some of the disadvantages include the following: (i) serum varies from batch to batch and deteriorates within one year; (ii) changing serum batch requires fresh testing, so that the replacement is as close to the previous batch as possible; (iii) if more than one cell types are used, each may require different serum batch, so that a number of batches may need to be maintained and co-culturing may be difficult; (iv) the demand of serum usually exceeds the supply for a variety of reasons; (v) when cell culture is used for downstream processing to recover cell products, the presence of serum is an obstacle to purification; (vi) serum increases the cost of the medium, since its cost is ten times the cost of its constituents if used as a substitute; (vii) serum may stimulate undesirable growth and may even inhibit growth in some cases.

The major advantage of serum free media is the ability to make a medium selective for a particular cell type, since each cell type appears to require a different recipe. However, the removal of serum also requires (i) high degree of purity of reagents and water and (ii) high degree of cleanliness of all apparatus. This is necessary, since removal of serum removes the protective and detoxifying action, which some of the serum proteins have.

The serum free media may be developed using any of the following two approaches: (i) A known recipe for a related cell type may be used, and individual constituents may be altered, until the medium is optimised, as has been done by *Ham*. This is time consuming, and sometimes takes as many as three years for developing a new medium. (ii) In the second approach, an existing medium like *RPMI 1640* or *Ham's F12* or *DMEM (Dulbecco's modified minimal essential medium)* is taken and a shorter list of constituents like selenium, transferrin, albumin, insulin, androgen, hydrocortisone, estrogen, etc. is used for manipulation. Using these approaches, many recipes for serum free media have been prepared and are available for commercial use or can be made up in the laboratory. However, despite several advantages associated with the use of serum free media, the serum free media will not be put to very common use, till better, convenient, cheap and less temperamental serum free media are developed. Following factors discourage the use of these serum free media: (i) simplicity of media with serum; (ii) lack of reliable source of most serum free media and (iii) the considerable investment in time, effort and resources needed for developing serum free media. In view of this, one would have to make a judicious choice of a medium with or without serum depending upon the type of cells or a cell line.

INITIATION OF CELL CULTURES

The initiation of cell cultures may be conveniently dealt with under the following heads: (i) preparation and sterilisation of the substrate (culture vessels); (ii) preparation and sterilisation of the medium; (iii) isolation of explant; (iv) disaggregation of the explant; and (v) subculture and cloning.

Preparation and Sterilisation of Substrate

Plasticware are supplied sterilised and ready for use; treatments are usually not necessary. Since they are disposable and not reused, they are costlier than glassware which are adequate for most purposes. However, glassware must be carefully washed with nontoxic detergent following, preferably, an overnight soak. They should be thoroughly rinsed in tap water and finally in deionised or distilled water. Glassware are kept in containers or wrapped in aluminium foil and usually sterilised in dry heat (160°C for 1 hour), while screw caps are autoclaved separately.

Preparation and Sterilisation of Medium

The various media constituents and other reagents used in cell cultures must be carefully sterilised either by autoclaving or by filtration. Heat stable constituents like water, salts, supplements like peptone or tryptose, etc. are autoclaved at 121°C for 20 minute. But heat labile constituents like serum, trypsin, proteins, growth factors, etc. must be sterilised by filtration through a 0.2 mm porosity membrane filter. Each filtrate should be tested for sterility to avoid failure due to contamination. In case of soda glass, caps should be left slack to avoid breaking during autoclaving.

Autoclaving is preferred to filtration since it is cheaper, needs less labour and is uniformly effective. Therefore wherever possible, autoclaving should be resorted to and autoclavable versions of the media should be used.

Isolation of Explants

Explants should be dissected out following an appropriate protocol developed to avoid contamination and retain cell survival. In general, embryos and other contents of uterus are free from contamination, while in adult tissues bacterial contamination may be common. In simple terms, the site of opening is sterilised with 70 per cent alcohol, the tissues are removed aspectically, and usually placed in a balanced salt solution, e.g., EBSS or HBSS, supplemented with antibiotics.

Disaggregation of Explants

Cell cultures are generally started from disaggregated explants. Occasionally, cells do migrate out of the cultured whole tissues; such cells are then separated by trypsin or other enzymatic treatment to establish cell cultures. But in general, the tissues are first disaggregated in one of the following ways: (i) mechanical disaggregation; (ii) enzymatic disaggregation; and (iii) EDTA treatment.

Mechanical disaggregation

The tissue is chopped into fairly small pieces, say, of 1 mm^3 in size. These are then cultured into a suitable vessel; the pieces attach to the substrate either due to their own adhesiveness, the dish may be scratched to facilitate attachment, or clotted plasma may be used.

Enzymatic disaggregation

Tryspin (0.25 per cent crude or 0.01–0.05 per cent pure) and collagenase (200–2000 units/ml, crude) are the most commonly used enzymes, but other enzymes like elastase, mucase, pappain, etc. have also been used. Trypsin is preferred because it is effective for many tissues, is tolerated by most cells, and its residual activity is neutralised by serum (or by a tryspin inhibitor in case of serum-free media). The dissociated cells are collected, washed free of the enzyme and cultured.

EDTA treatment

Some tissues like epithelium require Ca^{2+} and Mg^{2+} ions for their integrity. Disaggregation of such tissues is readily achieved by a treatment with EDTA (ethylene-diamine-tetra-acetic acid) prepared in a balanced salt solution.

This treatment is generally used for separation of cells from established cultures of epithelial tissues.

Subculture

When mechanically or enzymatically disaggregated tissue pieces or cells are cultured, they attach to the substrate, and a monolayer of cells develops from them. These cultures are called *primary cultures*. Primary cultures do not contain many types of cells present in the tissue because they are unable to attach to the substrate and survive; the dead cells are easily eliminated when the medium is poured out during subculture. Further selection will occur during growth of primary cultures and of subsequent cell lines derived from them as faster growing cell types will gradually increase and the slower growing types will go on declining with time.

EVOLUTION OF CONTINUOUS CELL LINES

Most of the cell lines will grow for a limited number of subcultures (usually, 8–10 subcultures; total culture period about 12 weeks), after which they will either cease to proliferate and die out or will give rise to continuous cell lines that can be grown indefinitely. The phenomenon that generates a continuous cell lines is called *in vitro* transformation, which may either occur spontaneously, or may be chemically or virally induced. The word transformation is used because continuous cell lines have marked similarities with malignant cells, and are often tumorogenic.

The common features of continuous cell lines are summarised below.

1. Smaller, more rounded, less adherent cells with a higher nucleus/cytoplasm ratio.
2. Increased growth rate; cell doubling time reduced to 12–36 hour from 36–48 hours for finite cell lines.
3. Aneuploid (between 2n and 4n) chromosome number showing considerable variation among cells of a cell line.
4. Reduced serum dependence.
5. Increased cloning efficiency.
6. Reduced dependence on substrate adhesion (anchorage) and increased tendency to grow in suspension.
7. Many of the cell lines become malignant, but some cell lines are normal and nonmalignant. The changes in growth control are under distinct positive (oncogene) and negative (tumour suppessor genes) control genes.

8. Ability to grow upto higher cell densities, i.e., growth is less dependent on cell density.

The faster growth rate, lower serum requirement, ability to grow in suspension, independence from cell density are distinct advantages of continuous cell lines. But their limitations relate to greater chromosome instability, loss of tissue-specific markers and differences from the phenotypes of donor tissues.

Maintenance of Cell Lines

Thousands of cell lines have been developed from human and other animal tissues; many of them are finite, while others are continuous cell lines. The continuous cell lines themselves may have originated from normal (through transformation) or tumour tissues. A cell line is multiplied and divided into a (i) seed stock and (ii) a working or distribution stock. The seed stock is characterised and then maintained in frozen state, preferably in liquid nitrogen at –196°C. Seed stock ampoules are used to produce distribution stocks so that the cell line does not change with time due to genetic instability, selection, senescence or transformation. In contrast, working stock or distribution stock consists of material multiplied from the seed stock and used for further studies or for distribution.

LARGE SCALE CULTURE OF CELL LINES

Cell cultures are used for obtaining useful products like biochemicals (interferon, interleukins, hormones, enzymes, antibodies, etc.) and virus vaccines (polio, mumps, measles, rabies, foot and mouth, rinderpest, etc.). For these objectives, large scale cell cultures are essential; fermenters of upto 10,000 L are used for this purpose. The scaling up of cell cultures may be done as (i) monolayer cultures; (ii) suspension cultures; or (iii) immobilised cell systems. For obvious reasons, scaling up of monolayer systems is more difficult than of others.

Monolayer Cultures

Monolayer cultures are essential for anchorage-dependent cells. Scaling up of such cultures is based on increasing the available surface area by using plates, spirals, ceramics and microcarriers (most effective).

The various culture vessels used are given below:

1. Roux bottle.
2. Roller bottle.

3. Multitray unit.
4. Synthetic hollow fibre cartridge.
5. Opticell culture system.
6. Plastic film.
7. Heli-cell vessels
8. Bead bed reactors.
9. Heterogeneous reactors.
10. Microcarrier cultures.

Suspension Cultures

In scaling up, both chemical (O_2, pH, medium constituents and removal of waste) and physical (the configuration of bioreactor and power supplied to the reactor) factors have to be optimised for good results. The medium must be suitably stirred to keep the cells in suspension and to make the culture homogeneous; it becomes increasingly difficult with the scaling up. Various types of stirrers range from simple magnetic stirrers, flat blade turbine impellers, to marine impellers, to those using pneumatic energy, e.g., airlift fermenter, and those using hydraulic energy, e.g., medium perfusion. Improved mixing can be obtained by changing the design of stirrer paddle or by using multiple impellers. The objective of stirring is to achieve good mixing without causing damage to the cells. Vibro-mixer achieves stirring by vertical reciprocating motion of 0.1–3 mm at a frequency of 50 cycles/second of a mixing disc fixed horizontally to the agitator shaft. These stirrers cause random mixing, less foaming and lower shear forces.

It is important to supply sufficient O_2 without damaging the cells. Mean O_2 utilisation rate by cells is about 6 mg $O_2/10^6$ cells/hr. But O_2 is only sparingly soluble in culture medium; the oxygen transfer rate (OTR) from gas phase into medium is about 17 μg/cm^2/hr. Therefore, surface aeration can support about 50×10^6 cells in a 1l culture vessel.

Aeration may be achieved by *medium perfusion*, in which medium is continuously taken from culture vessel, passed through an oxygenation chamber and returned to the culture. The cells are removed from the medium taken for perfusion so that the medium can be suitably altered, e.g., for pH control. Perfusion is used with glass bead and, particularly with microcarrier systems. Where considered safe and desirable, O_2 supply in the culture vessel can be enhanced from the normal 21 per cent to a higher value and the air pressure can be increased by atmosphere. This increases the O_2 solubility and diffusion rates in the medium, but there is a risk of O_2 toxicity.

The reactors used for large scale suspension cultures are of 3 main types:

1. Stirred bioreactors.
2. Continuous-flow cultures.
3. Airlift fermenters.

Immobilised Cultures

Cultures based on immobilised cells offer several advantages: (i) higher cell densities (50–200 × 10^{-6} cells/ml); (ii) stability and longevity of cultures; (iii) applicability to both suspension and monolayer cultures; (iv) protection of the cells from shear forces due to medium flow (in case of many systems); and (v) less dependence of cells at higher densities on external supply of growth factors, which saves culture cost.

There are two basic approaches to cell immbolisation: (i) immurement; and (ii) entrapment.

Immurement cultures

In such cultures, cells are confined within a medium permeable barrier. Hollow fibres packed in a cartridge are one such system. The medium is circulated through the fibre, while cells in suspension are present in the cartridge outside the fibre.

Entrapment cultures

In this approach, cells are held within an open matrix through which the medium flows freely. An example is the *Opticell* in which the cells are entrapped within the porous ceramic walls of the unit.

GENETIC ENGINEERING FOR TRANSGENIC ANIMALS

Selective breeding is a painfully slow process and, especially with larger animals with long gestation periods, can take many years to establish desired phenotypic changes. However, the advent of recombinant DNA technology and its application to animal breeding programmes could greatly increase the speed and range of selective breeding. The first recorded examples of the transfer of a foreign gene into an animal by recombinant DNA technology was the insertion and expression, into the mouse genome, of a rat gene for growth hormone. The subsequent progeny were all much larger than the parents. This 'Super Mouse' gained much public attention as it was the first example of a transgenic animal, i.e., an animal that has acquired novel genetic material by artificial means rather than by the normal route of sexual reproduction.

Subsequently, there has been external speculation on the economic potential of transgenic farm animals and there can be little doubt that this should become a highly lucrative world-wide industry with great benefits to humankind. It is perhaps the most controversial of all areas of modern biotechnology. Notwithstanding, some of the main opportunities where this new technology can be envisaged with animal breeding programmes are listed in Table 11.2. Of particular relevance will be improvements in meat production from a wide range of farmed animals including fish, improved milk yields and quality, and disease-free animals. Undoubtedly the most unusual but commercially feasible project is the use of certain lactating animals such as sheep, pigs, rabbits and cows to produce novel secretions of human proteins in their milk, which can then be extracted and used pharmaceutically.

Table 11.2. Anticipated changes involving transgenic animals.

Efficiency of meat production	Wool quality and quantity
Improved quality of meat	Disease resistance in animals
Milk quality and quantity	Production of low cost pharmaceuticals and
Egg production	biologicals

How can novel DNA be incorporated into animal genomes and then stably inherited into the offspring? At the present time the most successful method for gene transfer into livestock is by microinjection into the pronucleus of fertilised eggs. Microinjection techniques make use of finely constructed glass needles that allow the injection of purified DNA into the fertilised eggs of the chosen species. The eggs are then surgically transferred into hormonally synchronised surrogate mothers (Table 11.3). Unlike mice and pigs, in sheep and cattle litter size is limited to one or two and therefore, large numbers of the latter animals have to be employed as recipients for the microinjected eggs.

Table 11.3. Sequences necessary to establish transgenic animals.

Identification and construction of foreign gene (genetic engineering)
Microinjection of DNA directly into pronucleus of a single fertilised egg
Implantation of these engineered cells into surrogate mothers
Bringing the developing embryo to term
Proving that the foreign DNA has been stably and heritably incorporated into the DNA of at least some of the newborn offspring
Demonstrating that the gene is regulated will enough to function in its new environment

Transgenic pigs, sheep and cattle have now been obtained, although the frequency of success is only about 1 per cent compared with 2–5 per cent with mice. This low efficiency of the technology will continue to exert some limitation to wider acceptance. However, with fish the eggs are fertilised externally, thus eliminating many of the complicated techniques required in mammals to harvest ova, fertilise them and then introduce the embryos into foster mothers. Successful fish transgenics can be as high as 70 per cent.

The ultimate goal of animal breeders will be to introduce specific economically important traits into commercial livestock (Table 11.2). However, the knowledge of the mechanisms regulating gene expression in higher animals is limited and this restricts the ability to construct transgenic animals.

A novel and commercially realistic use of transgenic animals is the production of human proteins/pharmaceuticals in transgenic lactating animals. Transgenic constructs that allow the mammary glands of lactating animals to secrete high value human proteins are now possible and will undoubtedly be the first truly commercial use of transgenic animals for product formation. The animals will in fact become bioreactors producing pharmaceutical products previously only produced in culture by transgenic micro-organisms. Gene constructs for human coagulation factor IX (some haemophiliacs lack a blood-clotting agent called factor IX) have been successfully inserted into the sheep genome and while expression levels are still low, factor IX is present and the trait is heritable. The potential of transgenic animals to secrete a wide range of commercially valuable healthcare products is almost unlimited and should be realised in the future. Pharmaceutically used transgenic sheep will not be allowed to enter the human food chain, thus removing possible public outcry concerning the consumption (cannibalism) of the human gene inserts. Contrary to some popular viewpoints, transgenic animal studies are not about producing animal monsters and freaks but, rather, introducing specific, economically significant traits into livestock that will have benefits to humankind. Transgenic animals have also contributed hugely to the under-standing of gene function in animal species.

Genetically Engineered Hormones and Vaccines

The pituitary gland of animals secretes growth hormones that can have major influence on how the animal grows and, in lactating animals, on milk production. In the 1980s the gene responsible for bovine growth hormone (somatotropin or BST) production was successfully isolated and transferred into bacterial cells to produce large quantities of BST. When cows were injected with about 30 mg of BST there was significant increase in milk production (10–30 per cent) but continued increased yields depended on regular injections. Such improvements in milk production could lead to fewer animals producing the same volume of milk. There is no evidence of increased concentrations of BST in the milk nor that the constituents of milk are in any way altered. Comparable studies with genetically engineered pig somatotropin (PST) has shown that body fat can be reduced by up to 80 per cent and feed efficiency increased by 20 per cent. The approval of BST by the US Food and Drug Administration in November 1993 represented the first biotechnology-derived pharmaceutical to be commercialised for animal agriculture in the USA.

BST is the first genetically engineered product in agriculture that has been intensively examined for its economic impact. This is due mainly to the vast importance of milk to most Western economies and the positive product image of milk for human health. Animal welfare is also of major current concern.

From a scientific point of view BST has been clearly demonstrated to be a safe product and is now permitted in many countries, in particular the USA, where it is marketed by Monsanto under the tradename Posilac. But many consumer organisations continue to oppose the use of this method. In Europe the European Commissioners have recommended a continuation of the earlier ban for a further five years on grounds that are more politically based than for health and safety reasons. In a Europe struggling to control over-production of milk is a milk-boosting hormone really needed? Furthermore, the large number of small dairy farmers who might go out of business has considerable political impact.

From an animal welfare consideration there is now evidence that pigs transgenic for PST suffered skeletal and joint problems, and cows have increased mastitis. Some would consider that BST and PST were the wrong biotechnology-derived products to lead the way for public acceptance of new biotechnology. The public viewpoint of these biotechnology products is that they have been profit-driven and not need-driven. Much has been learned from these products that may be applicable to the introduction of other new biotechnology products in agriculture.

Animal vaccine production has been developed against many microbial diseases. Animal diseases are a major cause of suffering in all forms of animal farming, and if disease could be reduced or eliminated then animal welfare as well as yields would improve. By improving the health of the animals (e.g., in intensive farming) the number of animals required to give the same final yield-could be reduced.

The production of genetically engineered animal vaccines has been a major, if somewhat unheralded, success story in biotechnology. Numerous vaccines have been developed for specific cattle, pig, poultry, sheep and fish diseases. The commercial success is considerable, though not dramatic, and has led to much reduction in animal hardship. A new genetically engineered vaccine against rabies is now being scattered via dead chicken heads in many parts of Europe where rabies is endemic in the wildlife. The future for this type of diseasc prevention is immense.

DIAGNOSTICS IN AGRICULTURE

In traditional analytical methods applied to plant and animal agriculture, be it chemical or microbiological, the primary aim has been to isolate or separate out the analyte from the complex chemical milieu of the sample. Such methods normally require an operator of considerable chemical analytical experience and are time consuming and expensive. However, new methods based on biotechnology-derived techniques are revolutionising many aspects of agricultural analysis, not only by being able to equal the sensitivity of the established methods but also by being able to carry out many determinations *in situ* without the need for complex isolation procedures. Furthermore, these methods are usually cheaper, more easily adapted to automation and more rapid. Often, too, they can be performed by relatively unskilled operators (Table 11.4).

Table 11.4. Criteria required for successful rapid methods.

They should:
Be fast, accurate and reliable
Be simple to operate and have low running costs
Have readily available and stable reagents
Have minimal labour requirements
Have a high degree of sensitivity and specificity

Here, the impact of three new biotechnology-derived rapid methods, namely *immunoassays*, *DNA probes* and *biosensors*, will be examined in the context of plant and veterinary diseases and physiological monitoring of animals. In almost all cases these new and rapid methods were first developed were first developed with the massive human healthcare market in mind, where economic rewards were obvious, with huge sales anticipated in hospitals, surgeries, and for home care and monitoring. The veterinary market has also been a highly profitable activity, while in plant agriculture the new methods are increasingly being used.

Immunoassays in general, but specifically those using monoclonal antibodies, are widely recognised for their commercial success in clinical and veterinary diagnostics. Indeed, these diagnostic tests make up a major part of new biotechnological products presently on the market. Immunoassays differ from other analytical methods in that the high technology resides in the molecules rather than the apparatus. Nucleic acid probe technology is based on the principle of hybridisation of complementary sequences of DNA or of DNA and RNA. The respective nucleotide strands must have exact, corresponding sequences of nucleotides for exact hybridisation or alignment to occur, thus a given strand can hybridise only with its complementary strand. This high level of specificity has now been directed to identify micro-organisms in complex mixtures – the DNA probe or hybridisation assay.

Using these diagnostic methods it is now possible to detect microbial diseases in animals at very low levels of infection in body fluids or tissues and to be able to isolate an animal before it becomes infectious. A wide range of animal diseases can now be more easily monitored by 'user friendly' diagnostic kits applied by the veterinarian or the farmer. Early diagnosis can be an essential prerequisite for containment and elimination of infectious diseases (Table 11.5). One of the largest areas of application for diagnostic kits is in measuring fertility hormones in animal blood or milk, e.g., progesterone, oestrogen sulphate and equine gonadotrophin. Illegal use of growth hormones and antibiotics can also be monitored. Several hundred diagnostic kits are now available world wide.

Table 11.5. Successful and anticipated uses of rapid diagnostic methods with animal diseases.

Companion animals
Feline leukaemia, canine heartworm, rheumatoid arthritis
Poultry
Screening of avian reovirus; coccidiosin, salmonellosis, respiratory infections
Large animals
Trichinosis, mastitis, leukaemia, brucellosis, rinderpest, trypanosomiasin, swine fever, foot-and-mouth virus

Immunochemical technology using monoclonal antibodies is now widely used for the analysis of pesticide residues in foods, together with toxic microbial products such as mycotoxins. Specific plant diseases can now be detected in a crop at very early stages and appropriate treatment applied earlier. Such methods can also be used to analyse the phytosanitary quality of seed products for sale and to certify that potatoes, bulbs, fruit trees and ornamental plants are disease free. These methods now allow more efficient crop breeding and international trade. Genetic finger printing allows easy differentiation between plant varieties.

It is now not difficult to develop antibodies specific to known fungal, bacterial and viral diseases. A wide range of rapid diagnostic kits are now commercially available, especially for the high value horticultural markets.

12

Food and Beverage Biotechnology

INTRODUCTION

Food production is the largest world-wide industry and in industrialised nations the expenditure on food can account for at least 20–30 per cent of house-hold budgets. The food industry has evolved through specialist trades or occupations, e.g. butchering, bakering and confectionery, to national and multinational organisations involved in the manufacture and distribution of food on a world-wide scale. With the improvement in means of transportation, foods are available on a global basis and developments in food preservation methods give independence for seasonal availability. In essence, the food industry now serves the function of supplying society with high quality, wholesome foods, all the year round, and at a distance in time and location from the place of primary production.

The food chain has its origins in production agriculture, with the planting of the seed or the rearing of animals, and concludes with the utilisation of the food products by the consumer. Apart from fruits and vegetables, most food raw materials, e.g. cereals and meats, will require some degree of processing. The link between the products of the farm and the consumer is the food processing industry, whereby relatively bulky, perishable, raw agricultural products are transformed into shelf-stable, convenient and palatable foods and beverages. Food biotechnology is concerned with the integration of both modern biological knowledge and techniques and current bioengineering principles in food processing and preservation. This will achieve an elevation of the scientific and technological basis of industrial food processing and preservation to that presently achieved in other advanced biotechnological industries, e.g. antibiotic production. In the next decades, we will witness increased efforts to integrate the production of agricultural materials, their processing into foods and subsequent utilisation.

The challenge is to recognise the potential of biotechnological techniques to fulfil the food requirements of today's society both for developed and developing nations. Food biotechnology encompasses a wide range of options for improved quality, nutrition, safety and preservation of foods. Clearly, no single biotechnological advance will revolutionise the food industry, whereas economics, customer acceptance and regulatory rather than scientific hurdles will have a major influence on the range and spread of food biotechnology applications. Modern biotechnological techniques will have considerable importance in influencing trends in the food market, namely cost, preservation, taste, consistency, colour, safety and, above all, health aspects.

The food and beverage industries are very different from the pharmaceutical industry; their products are cost and marketing driven rather than technology driven. Research and development in most of the food and beverage industries is usually less than 1 per cent of sales, is very process oriented and enjoys little patent protection. Since most food and drink products are high volume, low cost items, it is inevitable that market research has become more significant than basic research. Some products such as organic acids, amino acids and gums now increasingly used by the food and drinks industry are in the middle price range, while only a few really high priced products will have a viable future (e.g. sweeteners and flavourings).

The food and beverage industries are high in terms of turnover and labour employed and are very diverse, ranging from small individual producers to giant multinationals. The impact of biotechnology on the food and beverage industries can be anticipated in two directions:

1. *Agronomic*: Increased plant and animal yields, extended growth range and environments from which the farmers will mainly benefit.
2. *Non-agronomic*: Improving plants and micro-organisms to provide benefits to the food producer, retailer or consumer (Table 12.1).

New developments in biochemical engineering could also be of advantage to those industries using mechanical (e.g. grinding), physical (e.g. membrane separation, cooking) and chemical (e.g. hydrolysis, salting) methods.

FOOD AND BEVERAGE FERMENTATIONS

Fermented foods and beverages have a significant role in all societies and result from the action of micro-organisms or enzymes on a wide range of agricultural materials with associated desirable biochemical changes giving significant organoleptic improvements to the final product. As a result of the fermentation process the product is usually more nutritious, more digestible, has improved flavour and is toxicologically and microbiologically safer.

Table 12.1. Biotechnology at all levels of the food chain.

Food Chain	*Potential biotechnological impact*
Biological living raw materials Food raw materials Food ingredients	*Agronomic*: increased yield, extend geographical and environmental range, all year growing *Non-agronomic*: increased benefit to processor by lowering the costs of manufacturing operations, keep fresh longer, improve texture and taste, phyto-production of flavours, colours and other more natural additives, using tissue culture, single cell protein
↓	Improving processing and reducing product manufacture costs, e.g. starter cultures, enzyme treatments, genetic engineering of micro-organisms, detoxification of food 'toxins', upgrading of waste materials, analytical applications and modification of fatty acids, carbohydrates and proteins
Food products at the factory gate	
↓	Improving distribution and product quality by inhibiting physical, chemical and microbiological deterioration, introducing less harsh processes and new preservation regimes
Food products at the point of consumption	Ensuring products meet the consumer's expectations of texture, flavour, nutrition, preservation, wholesomeness, and being more natural
Products consumed	

Fermented foods and beverages derived from plant and animal materials are an accepted and essential part of the diet in almost all parts of the world, involving a wide diversity of raw materials as substrates, using technology from the most primitive to the most advanced and achieving an astounding range of sensory and textural qualities in the final products. Fermented foods include breads, cheeses, yoghurts, sauerkraut, soya sauce, tempeh, mushrooms, etc., while fermented beverages include alcoholic beers, wines, sake, brandy, whisky, and non-alcoholic tea, coffee and cocoa. (Table 12.2). While most of these fermentations remain at the level of village or household arts, others have achieved massive commercial application and play a significant part in most national economies. All such fermentations have been or remain classified as indigenous, native to a country or culture and most were developed before recorded history. However, the very roots of modern biotechnology are to be found in these traditional fermentations. How these fermentations first came about is a question that cannot be completely answered?

Table 12.2. Some traditional processed beverages, foods and ingredients created by biotechnology methods.

Alcoholic beverages			
Beers, wines and spirits			
Food and ingredients			
Cheese	Sauerkraut	Enzymes	Vitamins
Bread	Soya sauce	Flavours	Biopolymers
Vinegar	Tempeh	Organic acids	Sweeteners
Yoghurt	Pickles	Amino acids	Mushrooms

Climate and available raw materials have influenced the types of food and beverage fermentation that developed in different geographical regions, and such products continue to form an enduring part of the cultural background of civilisations. It must be remembered that many of the world's populations are vegetarian, not necessarily by choice, but rather for mainly economic reasons. While a very important reason for the development of such fermentations was to preserve the basic organic components from spoilage, of equal or greater importance were the resulting changes in organoleptic, physical and nutritional characteristics of the relatively bland starting materials, resulting in products of enhanced flavour, improved vitamin content and, in some vegetable products, a meat-like texture and flavour. The nutritional value of these fermentations, in particular to the populations of the developing world, is inestimable and modern fermentation practices (e.g. brewing, cheese-making) are providing increased control, consistency of production and, above all, ensuring improved product safety.

For most of these fermentations the procedures were developed in ignorance of the role of micro-organisms. The original artisans unwittingly controlled and directed microbial activity by purely empirical methods but most often achieved consistent end-products. The Egyptians, Sumerians and Babylonians produced alcoholic beverages from barley, sour-dough bread from rye occurred in Europe in 800 B.C., and accounts of fermented dairy products are found in early Sanskrit and Christian works. It is only in relatively recent times that the microbial nature of most of these fermentations has been recognised and, while some fermentations have been shown to have a relatively simple microbiology, others possess a complexity of microbial involvement that may never be fully unravelled. Fermented foods can be divided into nine groups, namely beverages, cereal products, dairy products, fish products, fruit and vegetable products, legumes, meat products, starch crop products, and miscellaneous products. The relative importance of these fermentations in geographical areas is shown in Table 12.3. These fermentation products represent the largest financial sector of all biotechnologies and are a major aspect of the gross national product of developed nations. Almost 90 per cent of all revenues from biotechnology come from the food and beverage sectors. Too often we laud modern biotechnological innovations and overlook the continued presence of the long-standing traditional biotechnology-based industries. Some of these fermentations will be examined in more detail below and reference made where relevant to the impact of modern biotechnology techniques on their present and future production.

Table 12.3. Production of classes of fermented foods accordingly to geographical region.

World production rate	*Region*	*Importance*	
		Major	*Minor*
High	Europe	Dairy, beverage, cereals, meat	Legumes, starch crops
	North America	Beverages, dairy, meat	Fish, legumes, starch crops
	Africa south of Sahara	Starch crops, cereals, beverages	Dairy
Medium	South America	Beverages, dairy	Legumes
	Middle East	Dairy	Legumes, meat
	Indian subcontinent	Cereals, legumes	Meat
	East Asia	Fish, legumes	Dairy
	South-East Asia	Fish, legumes	Dairy
Low	Oceania	Dairy	Legumes
	North Africa	Dairy	Legumes

Alcoholic Beverages

Alcoholic beverages occur throughout the world in many different forms and tastes. The types of beverage produced in any particular region or country almost entirely reflects the crops grown. Thus, the cooler regions of Europe, Scandinavia, Poland and Russia will produce and consume beers and lagers from barley whereas the southern warmer climate of Spain, Greece, Italy and France will have much higher production and consumption of wines derived from grapes. Alcoholic beverages and potable spirit industries world wide represent one of the most economically stable sectors in present-day commerce. Demands on economics and need to increase conversion efficiencies or productivity are all driving forces in the search for new and improved technologies. The main objective is to produce a controlled quantity of alcohol in the liquid to be harvested after the fermentation.

Table 12.4. Substrates for selected alcoholic beverages (non-distilled).

Substrates	*Beverage*	*Country*	*Saccharifying agent*
Starch (barley + cereals)	Ale	Belgium; Germany; Canada; Australia	Barley malto t h e r
	Lager	World wide (industrial countries)	Barley malt
Barley, rye, rice, beet	Kvass	Russia	Barley and rye malt
Millet	Bouza	Russia (Crimea)	Millet malt
	Thumba	India	
Rice	Arrack	India, SE Asia	*Mucor sp.*
	Pachwai	India	*Aspergillus oryzae*
	Sake	Japan	
	Sonti	India	
Rice (red)	Ancu	Taiwan	
	Hung-Chu	China	*Rhizopus sp.*
Sorghum	Kaffir beer	Malawi	*Sorghum malt*
			Aspergillus sp.
			Mucor rouxii
	Merissa	Sudan	*Bacillus spp.*
Sweet potato	Awamori	Japan	
Agave spp. (sap)	Pulque	Mexico	
Apple (juice)	Cider	UK, France, N. America	
Grape (juice)	Wine	Temperate: N and S hemispheres	Not required since sugar is present in the substrate
Honey	Mead	UK	
Pear (juice)	Perry	UK, France	
Palmyra (juice)	Toddy	India, SE Asia	
Palm flower-stalk juice	Tuwak	Indonesia	

The starting material normally comprises either sugary materials (fruit juices, plant sap, honey) or starchy materials (grains or roots) which need to be hydrolysed to simple sugars before the fermentation (Table 12.4). When these substrates are incubated with suitable micro-organisms and allowed to ferment, the end-product is a liquid containing anything from a few per cent up to 16 per cent or more of alcohol, with an acid pH and depleted in nutrients for most contaminating micro-organisms; these factors combine to give the product a certain degree of biological stability and safety. The alcoholic beverages can be drunk fresh but normal practice for many requires a period of storage or ageing, leading in many cases to improved organoleptic properties. Further distillation will increase the alcohol strength and produce spirits of many types, e.g. whisky, brandy, vodka, gin, rum, which can contain between 40 per cent and 50 per cent ethanol (Table 12.5). Cordials and liqueurs are sweetened alcohol distillates derived from fruits, flowers, leaves, etc.

The most regularly used fermenting organism is the yeast. *Saccharomyces cerevisiae* or one of its closely related forms. This organism can assimilate and utilise simple sugars such as glucose and fructose and metabolise them to ethanol, for which it has a high tolerance. The process details of wine and beer production are briefly examined here since their productions represent major world-wide biotechnological industries.

Table 12.5. Potable alcohol production from sugar or starch-containing raw materials.

Sugar	*Product*	*Starch*	*Product*
Molasses	Rum Cognac	Barley	Whisky
Agave	Tequila	Maize and rye	Bourbon whiskey
Pear	Pear brandy	Potatoes and barley	Aquavit
Cherry	Kirsch	Potatoes, rye and wheat	Vodka
Plums	Slivovitz	Rice	Chinese brandies

Wines

Historically, wine is a Middle Eastern and European drink, and although other parts of the world such as the USA and Australia are now large producers, France, Italy and Germany still produce over half the total world output of approximately 10^{10} litres annually.

Most commercial wines use the wine grape *Vitis vinifera*, and cultivars of this species have been transported throughout the world to establish new wine-producing areas. Soil quality can have an important and subtle effect on the eventual quality of the wine.

Red wine is formed when black grapes are crushed and fermented whole. In contrast, if the skins are removed from black grapes or when white grapes are used, white wine is the final product. Hundreds of different wines are recognised in the many producing areas of the world. Rose wine results from some limited contact with the skins of black grapes, dry wine is the end-product of complete sugar utilisation whereas sweet wine will still retain some residual sugars.

Harvesting time of the grapes is judged largely by artisan skills, and the grapes, containing 15–25 per cent sugar, are then crushed mechanically or by treading of feet. The juice (now termed *must*) is the substrate for the truly biotechnological stage of the production. Since the *must* will contain many contaminating yeasts and bacteria it is usual practice to add sulphur dioxide to control or abolish this natural fermentation capacity. In large-scale wine production the must is partially or completely sterilised, inoculated with the desired strain of yeast, *Saccharomyces cerevisiae* var. *ellipsoideus*, and subjected to controlled fermentation in suitable tanks or bioreactors. The dryness or sweetness of the wine will depend on the degree of sugar conversion, glycerol levels, level of bacterial contamination, etc.

Fermentation conditions such as time and temperature will depend on the type of wine desired. After fermentation, the wines are run into storage vats or tankers where the temperature drops quickly, precipitates form and subtle chemical changes take place. Many wines undergo a spontaneous secondary bacterial (*Leuconostoc* spp.) or malolactic fermentation, converting residual malic acid to lactic acid. The final alcoholic content of wines ranges between 10 per cent and 16 per cent.

Modern scientific research now supports the view that moderate wine consumption is associated with lower mortality from coronary heart disease. As Louis *Pasteur* repeatedly stated, 'Wine is the most healthful and most hygienic of beverages'.

Fortified wines, such as sherry, port and vermouth, are wines to which additional alcohol is added after fermentation, raising the alcohol level to about 20 per cent.

Beers

Beers, ales and lagers are produced mostly from starchy cereals such as barley. Additional carbohydrate sources, known as adjuncts, are normally added in varying proportions. In practice, there are five major steps in the manufacture of beers from grains: malting, mashing, fermentation, maturation and finishing.

Malting

Dried barley is soaked or steeped in water and then spread out on the malthouse floor or in revolving drums, where the seeds germinate, with the formation of starch-degrading (amylase) and protein-degrading (protease) enzymes. The germinated seeds are then killed by kilning (slow heating to 80°C) while still retaining most of the enzyme activity (*malt*).

Mashing

In this stage the malt is mixed with hot water (55–65°C), and the starches and proteins break down to produce dextrins, maltose and other sugars, protein breakdown products, minerals and other growth factors (the *wort*). This is the medium for the beer fermentation. Hops may be added prior to the fermentation to give characteristic flavour and some antiseptic properties.

Fermentation

The wort is transferred to open bioreactor systems and inoculated with pure strains of yeast. In Britain a top-fermenting *Saccharomyces cerevisiae* is used at 20–28°C to produce beers, ales or stouts. In continental Europe a bottom-fermenting yeast. *Saccharomyces uvarum* ferments the wort at a lower temperature (10–15°C) to produce lager.

Maturation and finishing

Beer is usually matured in casks at 0.°C for several weeks to improve flavour, settle out the yeasts and remove haze. Bottled or canned beers are usually pasteurised at 60–61°C for 20 minutes. The alcohol content of beer is usually between 4 per cent and 9 per cent; with ales it is somewhat higher.

While European-type lagers and beers are now produced world wide the traditional beer in India and Asia is rice beer and in Africa sorghum beer. Sorghum beer is a very crude material rich in solids and vitamins and is a valuable source of nutrition to those who drink it.

Traditional applied genetics, together with protoplast fusion and recombinant DNA technology, are constantly improving the yeast strains used in these fermentations. In particular, there has been a vast upsurge in genetic engineering knowledge of *Saccharomyces cerevisae*. A new commercial brewing yeast has been developed and approved using recombinant DNA techniques.

Coffee, Tea and Cocoa

In Asia, India, Africa and South America, non-alcoholic fermented beverages are derived from coffee, tea and cocoa plants. These beverages have gained world-wide approval and high commercial value. Tea is derived from the enzymic activity released after the crushing of the leaves, while for coffee and cocoa the pulp surrounding the beans is removed in part by a natural fermentation with bacteria, yeast and fungi, which is critically important for full flavour and aroma development. The dried products, namely tea leaves and coffee and cocoa beans, can then be shipped throughout the world and the final beverage formed by the addition of water. Little is known about the exact microbial contribution to these fermentation processes. The processes are still empirical, with little exact science. Huge quantities of these products are consumed world wide and form the economic basis of several multinational companies.

Dairy Products

The origin of the development of dairy products such as fermented milk, butter and cheeses, is lost in antiquity. Such fermentations are related to areas with high numbers of lactating animals, cows, goats and sheep and Europe is the major world area of production (Table 12.3). World wide, fermented dairy products account for about 10 per cent of all fermented food production. It is now known that these fermentations result largely from the activity of a group of organisms called lactic acid bacteria. Fermentation by lactic acid bacteria results in preservation and transformation of milk and has been used unknowingly for thousands of years. In the past, these fermentations arose directly from the natural occurrence of lactic acid bacteria, but gradually it was recognised that a portion of a previously successful 'ferment', when added to milk, gave better results. Nowadays, an inoculum (a pure starter culture) of selected bacteria is generally added to the milk to be fermented. The modern world-wide dairy industry owes much to the development of pure starter cultures, good fermentation practices and strict adherence to hygienic protocol.

The lactic acid bacteria can have many beneficial effects in the foods in which they grow:

1. They have an inhibitory effect (bacteriocins) on many undesirable bacteria while they themselves are generally harmless; in this way they preserve the milk.

2. They produce highly acceptable texture and flavour modifications in the milk.
3. Reputedly, they have beneficial health effects on intestinal microflora (probiotics).

When growing in milk, these beneficial bacteria break down lactose to lactic acid; however, many other reactions can occur, depending on the composition of the substrate, types of additive and mode of fermentation. These can result in many other metabolites being formed, giving distinctive flavour and appearance to the milk products, e.g. buttermilk, sour cream, yoghurt and the vast range of cheeses.

One of the largest activities of the dairy industry is cheese production. Cheese is made by separating the casein of milk from the liquid or *whey*. Over 900 individual types of cheese are recognised; yet they could all be prepared from any given batch of milk by proper control of the fermentation and by correct selection of the promoting micro-organisms.

The discovery of the role of *rennet* is believed to have arisen from the use of animal stomachs by nomadic sheep herders for carrying liquids. When milk was transported in this way, it would become heated from the sun, soured by naturally occurring bacteria and contaminated with enzymes (*rennet*) from the stomach lining. The consequence of this interaction would be the transformation of the milk into solid curds and liquid whey. The curds were then eventually drained, salted and could be used later – an early example of basic food preservation.

In Europe the Romans were the first to really document this process, which used sheep's or goat's milk. Cow's milk was a very much later innovation. The first industrial production of calf rennet essence or enzymes was in Denmark in 1874.

Cheese production from milk is essentially a dehydration process in which the milk protein (casein) and fat are concentrated 6 to 12 fold. The common, basic steps in most cheese productions are:

1. Acidification of the milk by the conversion of sugar lactose into lactic acid and by the lactic acid bacteria.
2. Coagulation of the casein by a combination of proteolysis and acidification.

Proteolysis is started by the rennet (chymosin enzyme) (animal or fungal origin) and the coagulated caseins form a gel which entraps any fat present (Fig. 12.1). The separated curd is cut into blocks, drained and pressed into shapes, matured and made into cheeses. The details of cheese production are very complicated and involve many individual strains of bacteria and in some cases filamentous fungi, (e.g. Camembert, blue-cheese), special milks and related additives and differing process techniques that cannot be covered here.

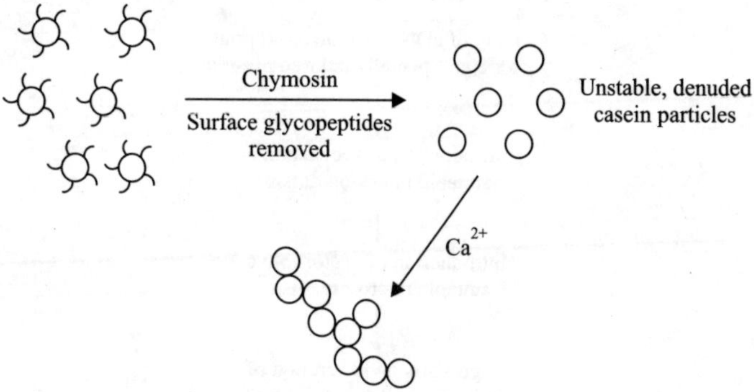

Fig. 12.1. Mode of action of chymosin (rennet).

However, an important recent biotechnological innovation in cheese production has been the use of recombinant DNA techniques for chymosin production and commercial use. At present there are six sources of commercial rennet: three from animals (veal calves, adult cows and pigs) and three fungal sources. The fungal sources are almost identical in function to the animal chymosins and account for approximately one-third of world cheese production, particularly in the USA and France. However, they can on occasion cause yield reductions and poor flavour when compared to animal chymosins.

Within the last decade, generically modified micro-organisms have been produced that can yield chymosin identical to the animal chymosin. Several industrial companies have now produced pure animal-derived chymosin by such methods and these products are now available world wide. The enzyme behaves in exactly the same manner as normal calf chymosin, it has fewer impurities and its activity is more predictable. In the UK it is sold by a major cooperative as generically modified cheese; contrary to some pessimistic forecasts, it has been well received by the public and also by the Vegetarian Society. Expert tasters cannot detect any derence between cheeses made using recombinant chymosin and calf chymosin. Its commercial success is ensured. The production of calf chymosin by genetically modified microorganisms is shown in Fig. 12.2.

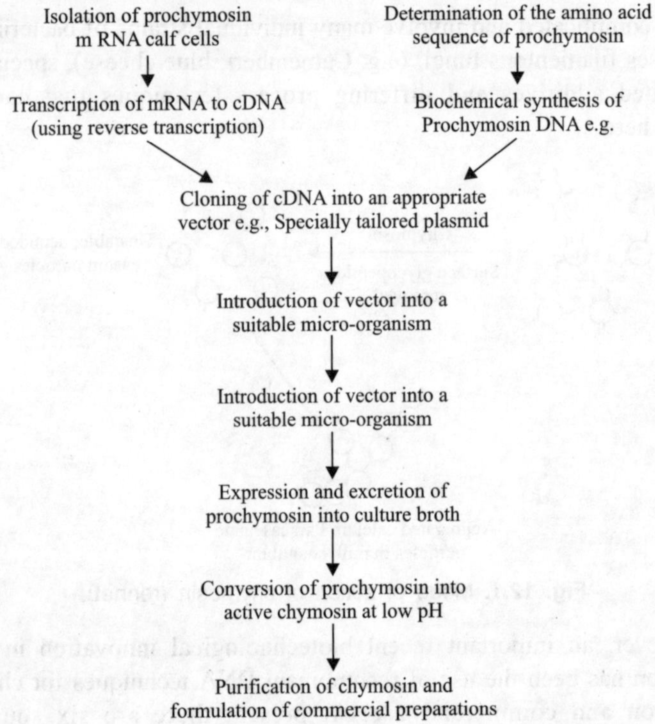

Fig. 12.2. Production of calf chymosin by genetically modified microorganisms.

The flavour of raw cheese, such as Cheddar, is bland and the texture rubbery. It is the period of ripening or maturation, when other micro-organisms such as bacteria and fungi can have pronounced effects, that causes the development of distinctive flavours and aromas as well as major textural changes (Table 12.6).

Table 12.6. Principal types of cheeses.

Unripened cheeses
Low fat (cottage cheese)
High fat (cream cheese)
Ripened cheeses
Hard cheese (internal ripening)
Ripened by bacteria (Cheddar, and Swiss cheese)
Ripened by mould (Roquefort and other blue cheeses)
Soft cheeses (ripening proceeds from outside)
Ripened by bacteria (Limburger)
Ripened by bacteria and moulds (Camembert)

The second major group of dairy products are the yoghurts. They are major foods consumed world wide and represent one of the fastest growing food products in the food industry. Claims are now being made that live yoghurt bacteria can become established in the human gut, with benefits to the digestive and other systems. This growing field of *probiotics* is still to be definitively proved in humans but is increasingly being seen as an important future area of medical advance.

Traditionally, yoghurt is fermented whole milk; the process uses a mixed culture of *Lactobacillus bulgaricus* and *Streptococcus thermophilus*. The characteristic flavour compound, acetaldehyde, is produced by *L. bulgaricus* while *S. thermophilus* generates the fresh acid taste by the conversion of lactose to lactic acid. Both bacteria produce extracellular polymers that give the characteristic viscosity of the product. Incubation is at 30 or 45°C. Set yoghurt is packed into the container after inoculation and allowed to ferment in the container. Frozen yoghurt is gaining increasing popularity as an alternative for ice cream.

Vegetable Fermentations

In various ways throughout the world, fruits and vegetables can be preserved using salt and acid, the acid being derived largely from bacteria in the form of lactic acid. Of principal Western interest are the fermentation preservation of cabbage to give sauerkraut and the pickling of cucumbers and olives.

In sauerkraut production shredded cabbage is packed anaerobically with salt, the salt reducing the water activity and promoting the leakage of sugars from the cabbage leaves (Fig. 12.3). Subsequently, lactic acid bacteria proliferate, releasing lactic acid, lowering the pH and preventing the growth of putrefying bacteria. Accurate control of temperature (7.5°C), salt concentration (2.25 per cent) and the anaerobic state will produce excellent, long-lasting sauerkraut – a nutritious, tasteful food. Large-scale production of sauerkraut can be traced back in Germany to the year A.D. 800.

In cucumber and olive fermentations especially in Greece and Spain, the fermentations are carried out at much higher salt concentrations (5–8 per cent) and the microbial sequences are relatively similar to those of sauerkraut fermentation.

Cereal Products

In almost all parts of the world cereals are produced and are the main class of food consumed by people; a considerable proportion of these cereals will be fermented into solid foods or into alcoholic beverages.

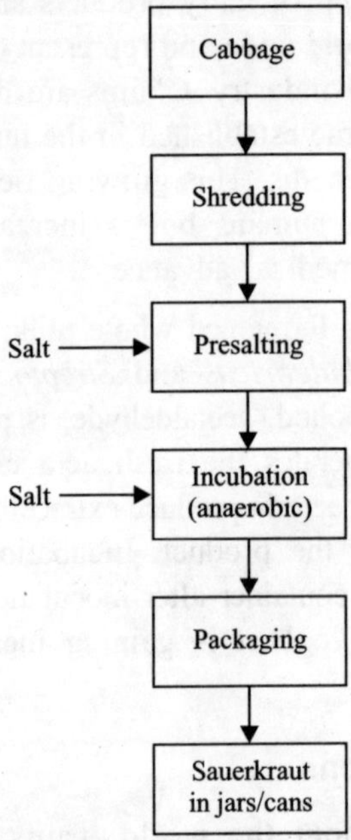

Fig. 12.3. The production of sauerkraut.

Bread in its many local forms is the principal fermented cereal product and has been known since Roman times. In Europe, wheat and rye are two widely used cereal flours and are usually mixed with water or milk, salt, fat, sugar and other varied ingredients, together with the yeast *Saccharomyces cerevisiae*. As the fermentation proceeds the dough rises due to the formation of CO_2. The expansion and stretching of the dough, particularly with wheat, is due to the unique extensible and elastic protein gluten. In this way dough rises and retains its shape on oven baking.

Bread texture is affected by fats, emulsifiers and oxidising agents while the speed of bread-making (of commercial importance) is affected by fats, oxidising and reducing agents and soya flour. While the yeast enzymes have an important role, additional enzymes, e.g. amylases, are added to assist mixing, fermentation, baking and eventual storage characteristics of the bread. Modern biotechnology will increasingly supply improved enzymes to bring even greater control over this complex process. Overall the fermentation achieves three primary objectives: Leavening CO_2 production), flavour development and texture changes in the dough. At the end of the fermentation process the risen dough is baked in an oven, giving a final product free from living micro-organisms and an extended shelf-life.

Modern applied genetics seeks always to improve the quality of the yeast organism, leading to improved activity, better flavour and improved texture of the product. A genetically engineered *Saccharomyces cerevisiae* with improved fermentation properties has been produced and has passed all regulatory requirements for safety. However, as yet, the producer company has not put it into commercial operation!

In other parts of the world sour-dough breads use the yeast *Candida milleri* and the bacterium *Lactobacillus sanfrancisco* for the fermentation stage while *Streptococcus* and *Pediococcus* species are used in the India subcontinent to ferment mixtures of cereal and legume flours to produce idli and popadoms. Rice is widely used in Asia for legume and fish fermentations. In South America corn bread from the cereal maize is a staple food.

Legume Fermentations

The soya bean *Glycine max* is the main legume used for fermentation and products derived are of particular importance in diets in East Asia, South-East Asia and India. Fermentation improves digestibility of the beans by breaking down antinutritional factors and compounds that cause flatulence in the intestine. In Indonesia cooked soya beans are fermented with the fungus *Rhizopus oligosporus* and the product, tempeh, may be fried in oil and eaten as a snack or in soups. Over 250,000 people are involved in the production of tempeh in Indonesia. There is a growing interest in this product in Europe and the USA.

The major fermentation of soya beans is in the production of soya sauce and miso. The production of soya sauce has three phases: the *koji*, the *moromi* and *maturation*. The koji is a solid substrate fermentation in which cooked soya beans and wheat flour are fermented with *Aspergillus oryzae* to breakdown starches, proteins and pectins. The moromi is a liquid/slurry fermentation under anaerobic conditions with *Candida* and *Pediococcus* as the fermentation agents. Maturation gives the final full flavour spectrum.

The Asiatic countries enjoy a rich range of legume and rice fermented products with high annual per capita consumption.

In all of the discussed food and beverage fermentation specific micro-organisms play an indispensable function in achieving the final product. Starter cultures are now used in all these fermentations and bring control and greater uniformity to the end-product(s). In most processes the micro-organisms become part of the food and are consumed intact. In others such as wine, beer, vinegar, and soya sauce, the cells are removed by filtration or centrifugation to eliminate turbidity.

ENZYMES AND FOOD PROCESSING

In recent years, one of the main applications of modern biotechnology to food production has been the use of enzymes. Enzymes are an essential part of most food and beverage fermentations and while most of the enzymes will be derived from the specific micro-organisms, increasingly there is the opportunity to improve the processes by the addition of exogenous enzymes (Table 12.7).

Table 12.7. Use of enzymes in food processing.

Industry	*Enzymes*
Brewing	α-Amylase, β-amylase, protease, papain amyloglucosidase, xylanase
Dairy	Animal/microbial chymosins, lactase, lipase, lysozyme
Baking	α-Amylase, xylanase, protease, phospholipases A and D, lipoxygenase
Fruit and vegetable processing	Pectin esterase, polygalacturanase, pectin lyase, hemicellulases
Starch and sugar	α-Amylase, β-amylase, glucoamylase, xylanase, pullulanase, isomerase, oligoamylases

The role of exogenous enzymes to facilitate or even replace mechanical processes is well demonstrated in fruit and vegetable processing, while in industrial starch transformation chemical processing has yielded almost completely to enzyme processing.

A major development that will revolutionise the use of enzymes in the food industry involves the near elimination of water from enzyme reaction media. Thus, hydrolytic enzymes can be reversed so that with no metabolic energy input the same enzymes that degrade biomolecules can now synthesise them. A wide range of food-related compounds have now been produced by this navel approach and include polyglycerol esters (emulsifiers), chiral flavour esters and oligopeptides, and structural polymers.

The new concept of protein engineering will facilitate the design or alteration of food enzymes at a molecular level, allowing minor modifications or the design of completely novel enzyme catalysts. An important application has involved the enzyme phospholipase A2 currently used as a food emulsifier. There is little doubt that the process of protein engineering coupled to gene cloning technology will be extensively applied to many enzymes used in food processing, allowing greater accuracy and selectivity of action.

Enzymes will also be extensively used in the design of novel and functional foods. Particularly in Japan there has been considerable research into oligosaccharides for insulin-low calorie response, designer fats, and special food fibre ingredients. Much effort is now being directed to the role of nutrition and the ageing process. The affluent, ageing populations of the West will wholeheartedly support such research efforts!

SWEETENERS

In most societies there is a considerable need for sweeteners to accompany food intake. Until the late 1960s, sweeteners were mainly cane and beet sugar, sucrose. In the 1970s, enzyme technology created a new class of sweeteners derived from starch. Saccharin, which is chemically derived, has been widely used as a sweetener for many years but is now being increasingly challenged by new, natural, low calorific sweeteners, while biotechnological methods have been used to develop one of the most important additions to this market, aspartame. Aspartame (trade name Nutra-sweet) is used extensively in many low calorie ('diet') soft drinks. Thaumatin, a protein extracted from berries of the plant *Thaumatococcus danielli*, is the sweetest compound known (Table 12.8). It is marketed extensively in Japan and now Europe and considerable effort is being made to produce the protein in genetically engineered micro-organisms.

Table 12.8. Traditional and alternative sweeteners.

Product	*Relative sweeteners*
Sucrose	1.0
55% HFCS	1.4
Cyclamate	50
Aspartame	150
Saccharin	300
Thaumatin	3000

Sweeteners find extensive markets and applications in soft drinks, confectionery, jams and jellies, ice cream, canning, baking, fermentation, pickles and sauces, and meat products – truly an immense market that will benefit from biotechnological innovations.

The most expensive component in the synthesis of aspartame is the amino acid phenylalanine, which is now produced on a large scale by fermentation methods. Extensive toxicological studies were necessary before aspartame was permitted to be marketed. All new biotechnology-derived food products most undergo a programme of regulatory approval similar to that demanded for new pharmaceuticals. Approval for aspartame took 10 years.

FOOD WASTES

Wastes from food and drinks industries are becoming an increasing problem, particularly to large production centres, because of increasingly strict legislation on the dumping of high-biological oxygen demand wastes. Much effort is now made to use such organic wastes to generate valuable by-products, while also achieving active waste removal. There is a large market for biological waste treatment systems.

MISCELLANEOUS MICROBIALLY DERIVED FOOD PRODUCTS

These products are derived from microbial fermentations and are used as ingredients in food production.

Vinegar

Vinegar is an aqueous solution containing at least 4 per cent by volume acetic acid and small amounts of esters, sugars, alcohol and salts. It is usually derived from wine, malt or apple cider. The fermenting bacteria are normally species of *Acetobacter*. It is widely used as an acidulant and flavouring compound in processed liquid foods such as sauces and ketchups.

Organic Acids

Citric acid is widely used in the food-related industries in fruit drinks, confectionery, jams, preserved fruits, etc. Over 100,000 tonnes of citric acid are manufactured annually by fermentation processes involving the fungus *Aspergillus niger* and molasses as substrate. The fermentation can be as static liquid surface cultures in trays, or in deep-tank, large-scale bioreactors. Citric acid is used in foods to enhance the flavour, to prevent oxidation and browning and as a preservative. Lactic acid can be produced by fermentation (40 per cent) or by chemical synthesis (60 per cent) and is used largely as an acidulant.

Amino Acids and Vitamins

Amino acids are widely used in the food and beverage industries as flavour enhancers, as seasonings, or nutritional additives. Glutamic acid and lysine are two amino acids produced by fermentation processes involving the bacteria *Corynebacterium glutamicum* and *Brevibacterium flavum*, respectively. Extensive mutant selection has produced micro-organisms that overproduce these primary metabolites. DNA technology will further improve the production capabilities. Vitamins are usually used as dietary supplements. However, vitamin C (ascorbic acid) is used as a food ingredient. Several micro-organisms can be used in the production processes.

Polysaccharides

Extracellular microbial polysaccharides are produced copiously by many micro-organisms and have been used in foods to enhance thickening and to form gels. They can stabilise food structure and improve appearance and palatability. The bacteria species used mainly are *Pseudomonas* spp. (xanthan gums) and *Leuconostoc mesenteroides* (dextrans). Species of *Acetobacter* can produce cellulose which forms the basis of certain oriental foods.

Flavours/Flavour Enhancers

The best-known flavour or sensory enhancer is monosodium glutamate, now largely made by fermentation using natural or engineered micro-organisms. Enzymic degradation of yeast RNA can produce nucleotide derivatives that are powerful flavour enhancers for meat. Biotechnology will have a profound influence on this market using gene manipulation and improved enzyme procedures.

RAPID DIAGNOSTICS

It is a major requirement of all food processes to supply a product free from dangerous microbial contamination of toxins. A wide array of traditional, time-consuming procedures have routinely been applied to identify, control, reduce or remove such contaminants. Major food pathogens include *Listeria*, *Salmonella* and *Campylobacter*, while the bacterial endotoxins and the fungal mycotoxins are the principal toxins considered.

Developments in biotechnology now permit a complete reshaping of many of these testing protocols by the application of antibody detection systems and DNA and RNA probe technology. These procedures are much shorter in time, generally cheaper, and often do not need highly skilled operators. By means of these new technologies there will be major improvements in safety standards in the food supply.

The use of immunoassays in the food industry is a recent development but they are rapidly becoming an accepted analytical tool. The enzyme-linked immunosorbent assay (ELISA) based on a 96-well microtitration plate is the most widely used format by the food scientist. It is extremely versatile in that a wide range of handling methods can be used, from manual operation to completely automated assays. Time and labour savings offered by ELISAs over conventional assays can be considerable. The detection of *Salmonella* bacteria in foods is a very complex procedure and can take up to five days using traditional enrichment procedures. By means of immunoassay procedures this can be as little as one day. Similarly, the mycotoxin aflatoxin requires many hours of extraction, purification and concentration for accurate traditional chemical identification. However, using affinity columns containing the specific antibodies to the toxin, the whole process of extraction and quantification can be carried out within 30 minutes.

Food immunoassay procedures are now available for many potential food analytes, e.g. trace residues, mycotoxins, antibiotics, hormones, bacterial toxins, etc. With the increased development and availability of such diagnostic procedures in user-friendly kit form new options will be available for food safety control that will achieve obvious benefits to manufacturer, consumer and regulatory authorities.

In food biotechnology, fermentation is the main means of producing a wide range of products. Improved bioprocess technology in food production will increase productivity, lower costs, improve nutrition and reduce environmental damage presently occurring in many food processes.

13

Biotransformations

INTRODUCTION

Biotransformation deals with the use of biological catalyst to converts a substrate into a product in a limited number of enzymatic steps. The establishment of an efficient biotransformation process requires the extensive examination of factors affecting the development of optimal biocatalysts, reaction media and bioreactors (Fig. 13.1).

There are many opportunities for industrial use of biological catalysts for biotransformations. These include not only the traditional hydrolytic (e.g., starch and protein hydrolysis) and isomerisation (e.g., glucose conversion to fructose) reactions but, more recently, synthesis of chiral compounds, reversal of hydrolytic reactions, complex synthetic reactions such as aromatic hydroxylations and enzymatic group protection chemistry and degradation of toxic and environmentally harmful compounds.

Biological catalysts when compared with chemical catalysts have the advantages of their regioselectivity and stereospecificity which lead to single enantiomeric products with regulatory requisites for pharmaceutical, food and agricultural use. They are also energy effective catalysts working at moderate temperatures, pressures and pH values.

Biotransformations have been performed by a variety of biological catalysts, such as isolated enzymes, cells, immobilised enzymes and cells. The developments of recombinant DNA technology have led to improvements in the enzyme production in different host organisms giving the bioprocess engineer a greater choice of biocatalyst option.

The optimal biocatalyst must be selective, active and stable under operational conditions in the bioreactor, which may not be necessarily conventional in terms of composition, concentration, pressure and temperature. In particular it is necessary to evaluate the biocatalyst performance in non-conventional media (e.g., organic solvents and supercritical fluids).

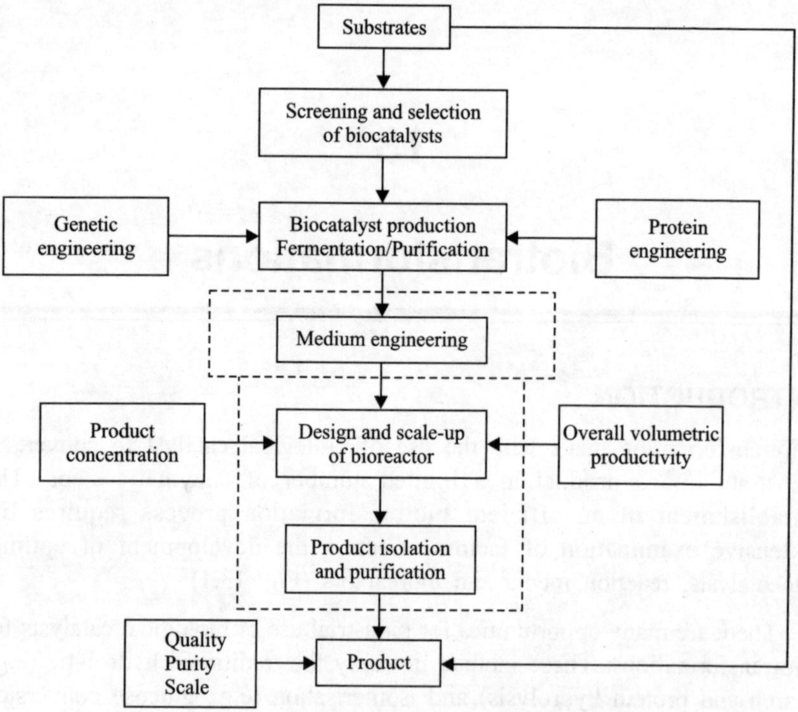

Fig. 13.1. Biotransformation process development.

A key issue is the availability of suitable biocatalysts. More rational screening and selection techniques are required to: (i) isolate biocatalysts, e.g., enzymes and cells, able to catalyse novel reactions of industrial interest; and (ii) select and design catalysts suitable for industrial use with improved operational stabilities and kinetic properties. This requires a much greater understanding of the mechanisms of protein denaturation and decay of catalystic activities under process conditions and an evaluation of methods to maintain and improve biocatalyst stability, e.g., chemical modification, immobilisation and protein engineering.

In the optimisation of the overall process it is also important to enhance the predictability and performance of the biocatalyst in the reaction media in particular in multiphasic media, for example involving a solid phase, e.g., immobilised biocatalyst and one (aqueous) or two (aqueous and organic) liquid phases. It is very important to obtain reliable data and models on physical/chemical transport and interfacial phenomena. Medium engineering plays an important role in the definition of the optimal biocatalyst operation and to evaluate the effect of medium composition on the biocatalysts.

The optimal bioreactor should be simple, safe, well controlled, easy to design and flexible. The design of bioreactors requires knowledge or reaction kinetics as well as fluid dynamics, substrate dispersion and mass transfer. In addition for multiphase bioreactions, interfacial phenomena, substrate and product partitioning and separation of two liquid phases should also be taken into account.

BIOCATALYST SELECTION

After selecting an appropriate starting material to be converted into the product, it is necessary to select the appropriate biocatalyst with suitable activity, selectivity and stability to work under the required operational conditions (temperature, salt concentration, pH, organic solvents, substrate and product concentrations). Several strategies can be followed to obtain the biocatalyst for the pertinent biotransformation: (i) screening for novel biocatalysts; (ii) use of existing biocatalysts; and (iii) genetic modification of existing biocatalysts.

Screening for Novel Biocatalysts

Selection of new micro-organisms with novel activities is still worthwhile taking into account the overwhelming biochemical diversity present in nature. To screen large numbers of organisms requires that cheap simple, rapid and selective detection methods, preferably capable of some automation, should be available to facilitate this usually tedious process.

Selective selection methods for colonies on plates can be very useful, as shown for the isolation of micro-organisms able to hydroxylate L-tyrosine to L-DOPA, a drug used in the treatment of Parkinson's disease. The colonies which produce L-DOPA turn violet-blacks as a result of the reaction of L-DOPA with ferrous ions added to the agar plates.

Microbial selection has also been performed in the presence of high concentrations of the target compound. This approach was used to isolate benzoic acid-assimilating strains for the production of *cis*, *cis*-muconic acid from benzoic acid. Similar approaches have been followed to isolate nitrile-hydrolysing enzymes, such as nitrile hydratase, nitrilase and amidase, which have great potential as catalysts for producing high value amides and acids from the corresponding nitriles.

The resistance to organic solvents is often an important criterion in the selection of a suitable biocatalyst. *Pseudomonas* strains have been isolated with the ability to grow in the presence of toluene and aromatic and aliphatic hydrocarbons and long chain alcohols. These strains and their enzymatic activities are therefore important biocatalytic sources for the degradation of harmful compounds as well as for the synthesis of important chiral compounds.

Use of Existing Biocatalysts

A well-known way to accomplish a desired biotransformation is the use of existing biocatalysts (e.g., commercial enzymes) on natural and unnatural substrates. The substrate specificities of lipase and proteases are currently under intense investigations. The hydrolytic capacity of lipases is not restricted only to triacylglycerols. This type of enzyme is also able to hydrolyse mono-, di- and triacyl esters with different chain lengths of the various acyl groups.

The exploitation of existing enzymes under different reaction conditions could lead to the finding of a biocatalyst for the desired biotransformation. For example, lipases have been used to perform synthetic reactions in media under controlled water activity, e.g., esterification, inter-esterification and trans-esterification reactions. Methods to optimise the enantioselectivity of lipases have been reported, namely the non-covalent modification of lipase and the control of the surface tension of an emulsion.

Genetic Modification of Existing Biocatalysts

A distinct way to obtain a biocatalyst is by *in vivo* (metabolic pathway engineering) and *in vitro* (protein engineering) construction of a novel biocatalyst. *In vivo* genetic engineering has been applied in large scale to obtain a recombinant organism with the desired enzymatic activity. Mutational events leading to the novel enzyme activities include transfer of genes, gene duplication, gene fusion, recombination between genes, deletion or insertion of gene segments and one or more single site mutations, or combination of these activities. An example of this metabolic pathway engineering is the production of dyes, such as the indigo biosynthesis in *E. coli* (Fig. 13.2). By assembling, on a single operon, genes encoding for tryptophan formation, the gene specifying tryptophanase and a fragment of the NAH plasmid of a *Pseudomonas* encoding the napththalene dioxygenase, a recombinant *E. coli* was obtained which was able to synthesise indigo from simple starting compounds.

Another approach is the use of protein engineering to modify an existing protein/enzyme or create *de nova* a protein of pre-specified properties. The protein engineering process can be viewed as an interactive cycle of several interconnected steps (protein engineering cycle). The aim of protein engineering has been to elucidate the structure–function relationship of proteins and to use this information to develop novel/modified proteins (enzymes) with improved characteristics for process applications. An elucidative example is the design of subtilisin mutants with altered properties (substrate specificity and pH activity profile) and improved thermal and oxidative stabilities. For example in subtilisin BPN′, two methionines, Met^{124} and Met^{222}, are especially susceptible to oxidation. To prevent the negative influence caused by the formation of methionine sulphoxide, Met can be replaced, using site-directed mutagenesis, by a non-oxidative amino acid, such as Ala, Ser or Leu, without losing more then 12–53 per cent of the initial activity. The mutant Met^{222}–Ala^{222} is currently in use as a commercial detergent enzyme: 'Durazyme'.

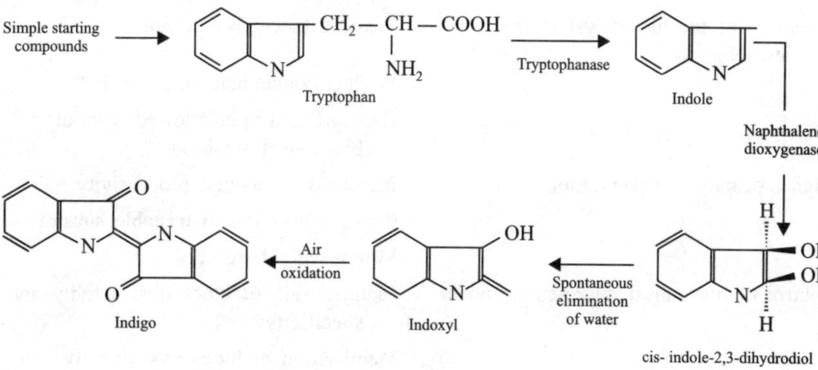

Fig. 13.2. Metabolic pathway engineering for indigo biosynthesis.

BIOCATALYST IMMOBILISATION AND PERFORMANCE

Biocatalyst Immobilisation

The immobilisation of biocatalysts for laboratory studies, analytical and medical applications and large-scale industrial processes is presently a widespread technique. Immobilisation can be defined as the confinement of a biocatalyst inside a bioreaction system, with retention of its catalytic activity and stability and which can be used repeatedly and continuously. Table 13.1 lists some advantages and limitations which can arise from the use of immobilised biocatalysts.

The biocatalysts which can be immobilised range from purified enzymes to viable microbial cells, animal and plant tissues. Isolated enzymes can give high activities per unit mass or mole, high specificity and minimum side reactions. They are, however, often difficult and costly to prepare. In addition, they are frequently unstable and, in many cases, require parallel co-factor regenerating systems. Due to their relatively simple chemical nature, as compared to organelles or whole cells, isolated or partially purified enzymes are the biocatalysts most extensively studied in relation to immobilisation. Immobilised, purified enzymes find suitable applications in developing biosensors and preparing high added-value substances, such as chiral compounds. In more crude forms, immobilised enzymes are also used in large-scale applications in the carbohydrate, food and pharmaceutical industries.

Table 13.1. Advantages and limitations on the use of immobilised biocatalysts.

General aspects		*Specific aspects*
Advantages		
Retention of the biocatalyst in the bioreactor		Possible biocatalyst re-use
		Product contamination avoided
		High dilution rates allowed without biocatalyst wash-out
High biocatalyst concentration		Increased volumetric productivity
		Rapid conversion of unstable substrates
		Minimised side-reactions
Control of biocatalyst micro-environment		Manipulation of biocatalyst activity and specificity
		Stabilisation of biocatalyst activity
		Protection of shear-sensitive biocatalysts
Facilitated separation of the biocatalyst from the product		Precise control of bioreaction time
		Minimisation of further product transformation
Limitations		
Increased costs of biocatalyst production		Increased requirements of materials and equipment
		Need for specific reactor configurations.
Loss of biocatalyst activity during immobilisation	Biocatalyst-related	Exposure to pH and temperature extremes
		Exposure to toxic reactants
		Exposure to high shear or mechanical strain

(Contd...)

	Micro-environment related	Exclusion of macromolecular substrates
		Blocking of the enzymatic active site
		Local pH shifts
		Mass transfer limitations
Loss of biocatalysts activity during bioreactor operation	Leakage of bio-catalyst	Matrix erosion or solubilisation
		Small support particles carried in the outflow
		Cell growth inside the matrix
		Broad pore-size range
	Matrix poisoning or fouling	Build-up of inhibitors in the micro-environment
		Retention of suspended solids
		Growth of contaminating species (biofilms)
		Need for a stricter control of feed composition
Empiricism		Need for case specific, multi-parameter optimisation
		Difficult process modelling and control

Multi-enzyme systems, such as organelles, whole cells or cell tissues, have some clear advantages for immobilisation over isolated enzymes. They can be efficiently retained by mild, physical means, preserving, in adequate conditions, the enzyme-synthesising and co-factor regenerating capabilities and producing a suitable micro-environment for single and multiple enzymatic activities. However, the efficient use of immobilised cells relies on the control of metabolic and physiological alterations throughout the retention produce and the subsequent catalytic process. The major large-scale utilisations of immobilised cell systems take advantage of the natural tendency of many microbial species to flocculate or to adhere to solid surfaces. Other applications are restricted to single-enzyme transformations with non-growing cells in the manufacture of pharmaceuticals and amino acids.

Methods for Biocatalyst Immobilisation

A wide range of basic immobilisation procedures with their specific variations has been described in a large number of reviews. Several classification schemes have also been proposed, one of which is given in Fig. 13.3.

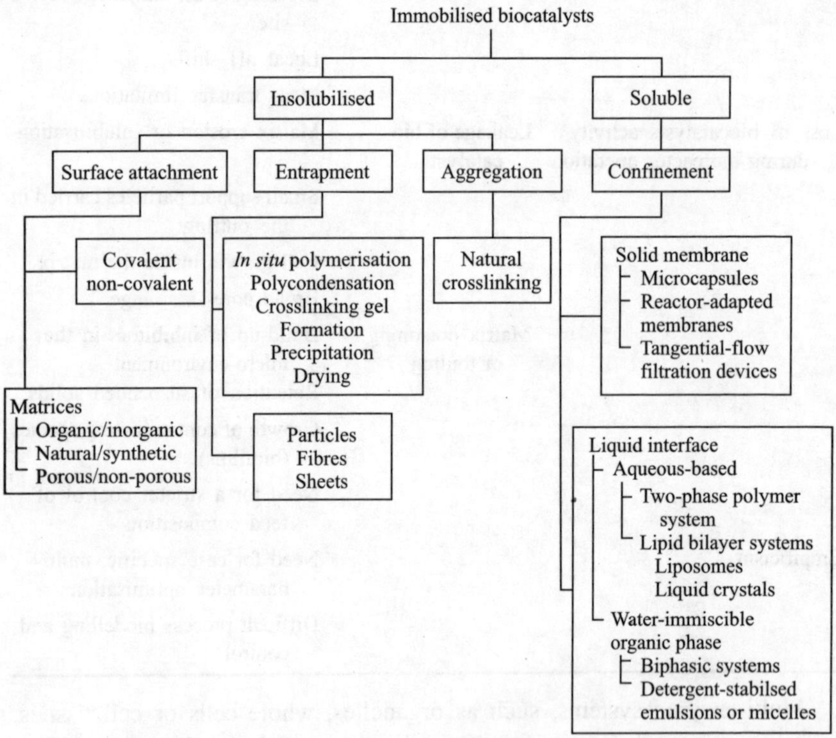

Fig. 13.3. General methods for biocatalyst immobilisation.

Cross-linking with bi-functional reagents

Both cells and enzymes can be covalently cross-linked with bi- or multi-functional reagents, such as aldehydes or amines. However the toxicity of these reagents limits their applicability to the immobilisation of non-viable cells and enzymes. This method produce three-dimensional, crosslinked enzymes aggregates, which are then insoluble in water. Glutaraldehyde has been the most extensively used crosslinking reagent which reacts with the lysyl residues of the enzyme forming a Schiff's base:

$$\text{Enzyme-NH}_2 + \text{OHC(CH}_2)_3\text{CHO} + \text{H}_2\text{N-Enzyme} \rightarrow$$

$$\text{Enzyme-N} = \text{CH(CH}_2)_3\text{CH} = \text{N-Enzyme}$$

The linkages formed between enzyme and glutaraldehyde are irreversible and survive extreme values of pH and temperature, which suggests that the aldimine bond is stabilised.

Biocatalyst crosslinking with glutaraldehyde is critically dependent on a delicate balance of factors such as the concentration of the biocatalyst and crosslinking reagent, pH and ionic strength of the aqueous solution, temperature and time of reaction. The most important advantage of this method is that only a single reagent is required and the reaction is easy to carry out. This method has been used successfully to immobilise industrial biocatalysts such as glucose isomerase and penicillin amidase.

Supported immobilisation methods

The available methods for biocatalyst immobilisation, involving solid supports, fall into two general categories: surface attachment and lattice entrapment. By surface attachment, the enzyme, organelle or cell is bound to a solid interface through interactions which range from weak van der waals forces to essentially irreversible, covalent bonding. The milder interactions can result from direct contact, in suitable conditions, between the biocatalyst and a natural, unmodified surface. However, the versatility and effectiveness of surface immobilisation have been greatly increased by introducing synthetic carriers and chemical modifications to natural and fabricated matrices. In lattice entrapment, a chemical or physical solidification process is induced in a solution containing the biocatalyst, ideally resulting in a water-insoluble lattice retaining the biocatalyst in its active or viable form. Mechanisms like polymerisation, thermal gelation or precipitation can be employed in this type of procedure. Figs. 13.4 and 13.5 give a general overview of biocatalysts, supports and retention methods used in surface attachment and entrapment.

Supports for biocatalyst immobilisation

The development of a useful, support-immobilised biocatalyst necessarily involves a choice of a solid support. Ideally, this selection step should be based on established structural and activity data for the biocatalyst and the general immobilisation method to be used and process conditions. The important factors for examining a broad range of possible supports are summarised in Tables 13.2 and 13.3. Because several factors have usually to be considered when integrating biocatalyst immobilisation in a process, the optimal solution is frequently a compromise. In view of this, a support with a more flexible character is most often used. For example, a porous support can immobilise large biocatalyst loads, however, to avoid diffusional limitations, this same support should be used as very small particles. In other cases, a single or few factors determine the choice of the support. Such is the case of systems where the aim is to preserve biocatalytic activity in the presence of aggressive components in the reaction medium, such as toxic species, organic solvents or strong inhibitors. With these systems, a porous matrix entrapping the biocatalyst and excluding the inhibitor is often the only efficient choice, regardless of the diffusional hindrances slowing down the reaction. Here, the possibility of changing the shape, porosity or hydrophobicity of the support can be advantageous in fine-tuning substrate or non-substrate size exclusion and external and internal mass transfer rates.

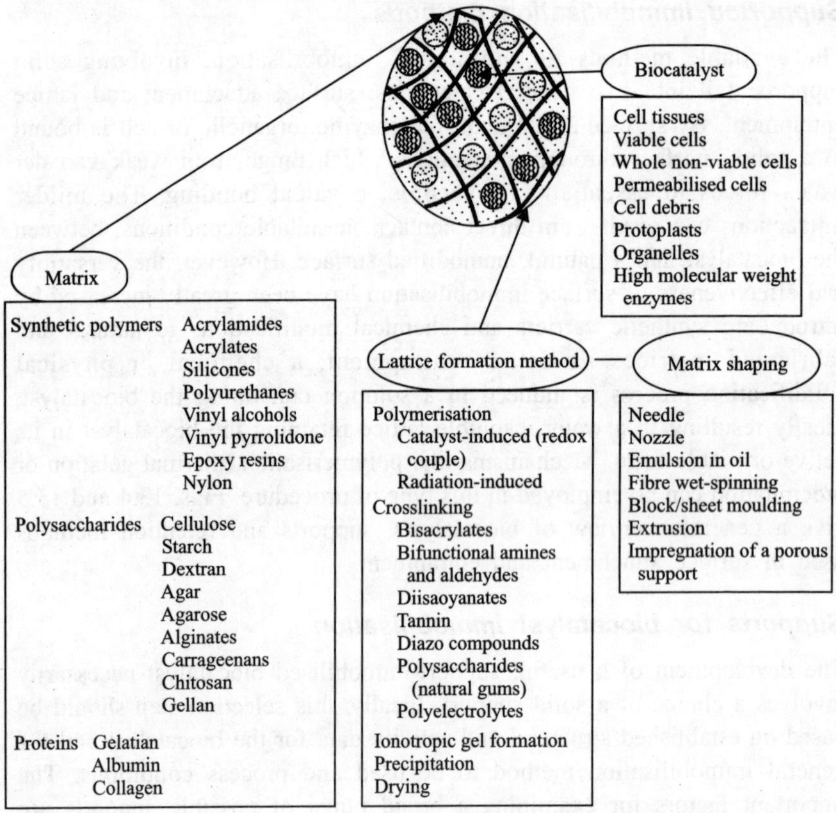

Fig. 13.4. Overview of surface attachment of biocatalysts.

Organic polymers are the most widely employed supports for biocatalyst immobilisation (for examples, see Figs. 13.4 and 13.5). This preference derives from the adaptability of these supports to nearly all kinds of surface-binding or entrapment techniques and to the broad variety of their chemical and physical characteristics. Their major drawbacks come from insufficient mechanical and chemical resistances, which limit both their use under harsher conditions and their regenerability.

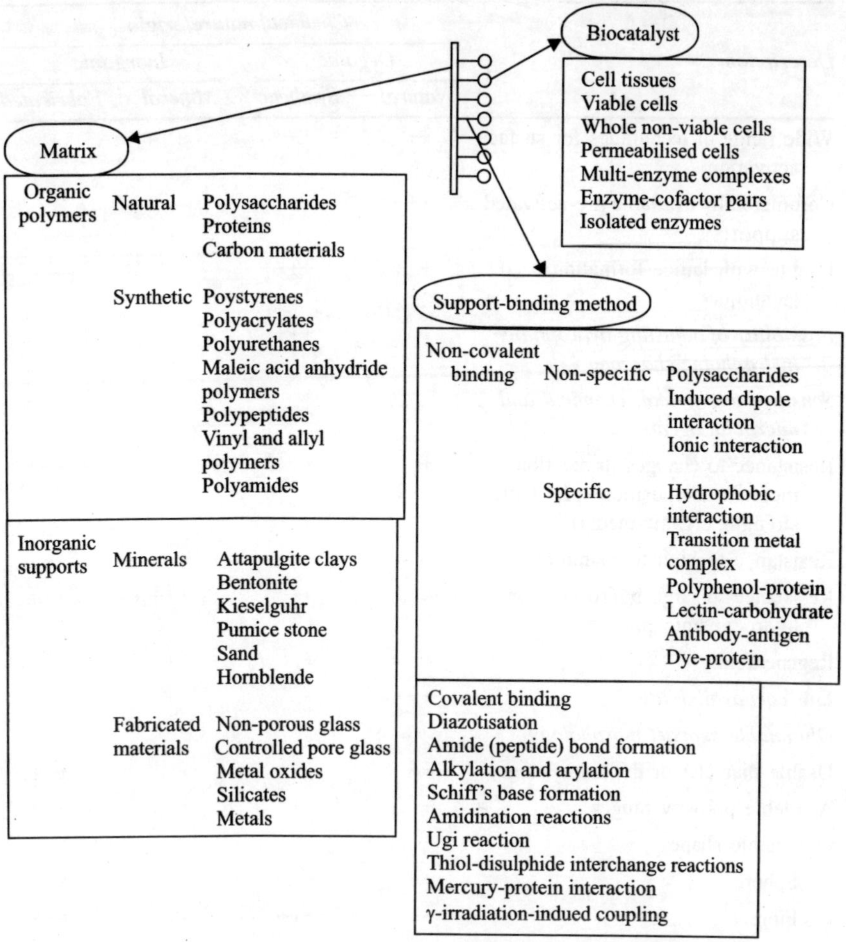

Fig. 13.5. Overview of lattice entrapment of biocatalysts.

Table 13.2. Important aspects in the chemical nature of potential supports for biocatalyst immobilisation.

Description	*Chemical nature/origin*			
	Organic		*Inorganic*	
	Natural	*Synthetic*	*Mineral*	*Fabricated*
Availability of reactive functional groups				
Usable with large variety of biocatalysts	++	+++	+	+

(Contd...)

Description	Chemical nature/origin			
	Organic		Inorganic	
	Natural	Synthetic	Mineral	Fabricated
Wide range of techniques for surface activation	++	+++	+	+
Commercially available pre-activated supports	+++	++	–	+
Usable with lattice formation techniques	+++	+++	–	–
Possibility of adjusting hydrophilic/ hydrophobic character	+	+++	+	+
Sensitive to physical, chemical and microbial agents				
Resistance to changes in reaction medium composition (pH, ionic strength, organic media)	+	++	+++	+++
Resistance to high temperatures	–	+	+++	+++
Resistance to large hydrostatic or hydrodynamic pressures	–	++	+++	+++
Regenerability	–	+	+++	++
Low cost/availability	+++	+	+++	+
Obtainable support morphology				
Usable diameter or thickness ranges	++	+++	+	+++
Available porosity ranges	+	++	++	++
Obtainable shapes				
Sphere	++	++	–	+++
Fibre	++	+++	–	+++
Sheet/membrane	++	+++	–	+++

Notes: –, inadequate; +, poor; ++, fair, +++, good.

Table 13.3. Important aspects of the morphology of potential supports for biocatalyst immobilisation.

Characteristics	Morphology	
	Porous	Non-porous
Total surface area available per unit weight	+++	+
Low incidence of diffusional limitations	+	+++
Attainable biocatalyst load	+++	+
Biocatalyst protection from external aggressions	++	+
Usable with macromolecular substrates	+	+++

(Contd...)

	Fabricated			
	Mineral	*Inorganic*	*Gel*	
Pore size uniformity	+	+++	++	Not applicable
Pore size stability	++	+++	+	Not applicable
Low cost	+++	–	++	+++

Notes: –, inadequate; +, poor; ++, fair, +++, good.

Adsorption and Ionic Binding

Adsorption on a support is the oldest and simplest retention method for biocatalysts, involving no previous modification of the solid surface and relaying on weak interactions of van der Walls, electrostatic, hydrophobic or hydrogen-bond types. This is low-cost procedure largely retaining the native conformation of an immobilised enzyme and its intrinsic catalytic activity. The general application of physical adsorption methods is, however, severely limited by the reversible nature of the biocatalyst-support bond, which is critically dependent on process conditions, such as temperature, pH, ionic strength and dielectric constant.

This factor makes it difficult to operate large scale bioreactors without significant biocatalyst leakage which then result in loss of productivity and product contamination. On the other hand, reversible adsorption methods sallow straightforward regeneration of the supports. After adsorption, the enzyme may be cross-linked, however, this limits the possibility of reusing the support.

A growing field for applying physically adsorbed enzymes is non-aqueous biocatalysis. By using an organic solvent in which the protein is insoluble, an enzyme bound to a carrier by simple adsorption undergoes virtually no desorption in prolonged processes. A lipase from *Rhizomucor miehei* was adsorbed on a polyacrylate support with a very high activity retention (90 per cent) and used in hydrolytic and synthetic reactions. An industrial example is the use of a *Rhizopus* lipase adsorbed onto Celite for the continuous production of cocoa butter-like fats in organic media.

A slightly stronger biocatalyst-support linkage can be achieved by the use of ionic supports. Their advantages and limitations are mostly the same as those of physical adsorption. The stability of the established ionic bonds is particularly sensitive to pH and ionic strength. Glucose isomerase from *Streptomyces rubiginous* was adsorbed onto an anion-exchange resin consisting of DEAE-cellulose agglomerated with polystyrene and TiO_2. This process was industrially implemented for the isomerisation of glucose into fructose.

Covalent Coupling of Enzymes

Probably the most thoroughly investigated approach to enzyme immobilisation involves covalent binding of amino acid residues in the protein to reactive groups in the support. In principle, the wide variety of surface activation and coupling reactions available makes it a generally applicable method. However, the high cost of materials, the often complicated procedures and the almost unavoidable loss of part of the catalytic activity, restrict practical applications of covalent surface immobilisation to specific cases with outstanding advantages.

Covalent coupling of enzymes to supports produces highly stable conjugates with no protein leakage over a wide range of operational conditions. Trypsin, penicillin acylase and lipase immobilised by multipoint covalent attachment onto CNBr-activated agarose yielded derivatives much more stable (300- to 50000-fold) than their free counterparts. Ideally, the binding reaction should not interfere with the amino acids residues at the active site of the enzyme and should not significantly distort the native protein conformation nor alter its flexibility, either as a result of single or multipoint linkages. To meet these requirements, methods include multi-stage activation steps. Although ten different amino acids residues from enzymes can, in principle, be used for covalent coupling, most procedures are targeted at amino, thiol, phenolic and hydroxyl groups. Some of the basic coupling reactions are presented in Fig. 13.6.

Lattice Entrapment

The immobilisation of biocatalysts within the lattices of solid matrices aims to take the advantage of the size difference between the substrates, or products and the biocatalyst so as to achieve total retention of the latter which the former move freely between the bulk medium and the catalytic site. In lattice entrapment, the biocatalyst is not generally subject to strong binding forces; no structure distortion or active site blocking take place. Some deactivation can, nevertheless, occur during the immobilisation process due to pH and temperature changes and contact with aggressive monomers or solvents. Matrices are formed in the presence of the biocatalyst by *in situ* polymerisation, starting from the appropriate monomers, or by solidifying polymer solutions through ionic (alginates) or covalent (polyurethanes) crosslinking, cooling (agarose, gelatin), drying or induced (carrageenan) precipitation (Fig. 13.7). Critical parameters in such processes are always related to optimising pore size and its influence on support rigidity and substrate diffusion within the lattice.

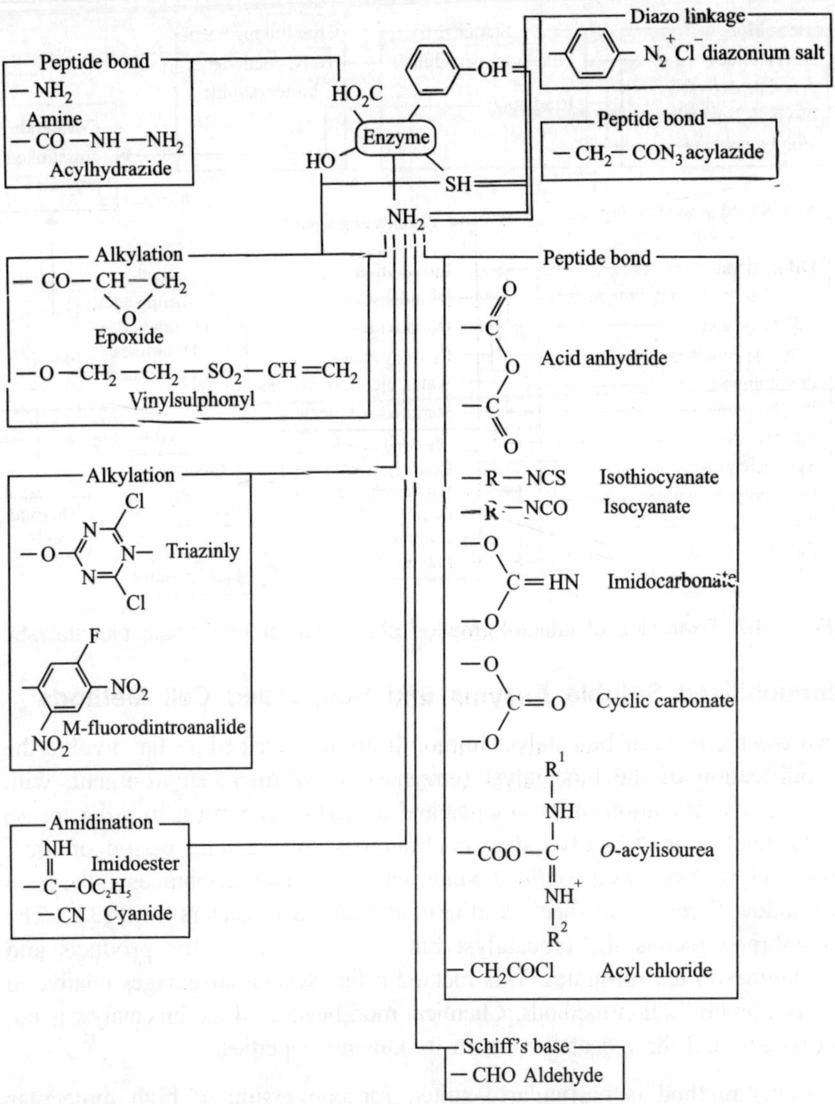

Fig. 13.6. Examples of covalent coupling of enzymes to active groups on supports.

Several attempts have been made to entrap isolated enzymes, however, due to enzyme leakage, this immobilisation method is mainly used for whole cell immobilisation. Several industrial examples include the use of entrapped cells for the production of amino acids (L-aspartic acid, L-isoleucine), L-malic acid, hydroquinone and acrylamide.

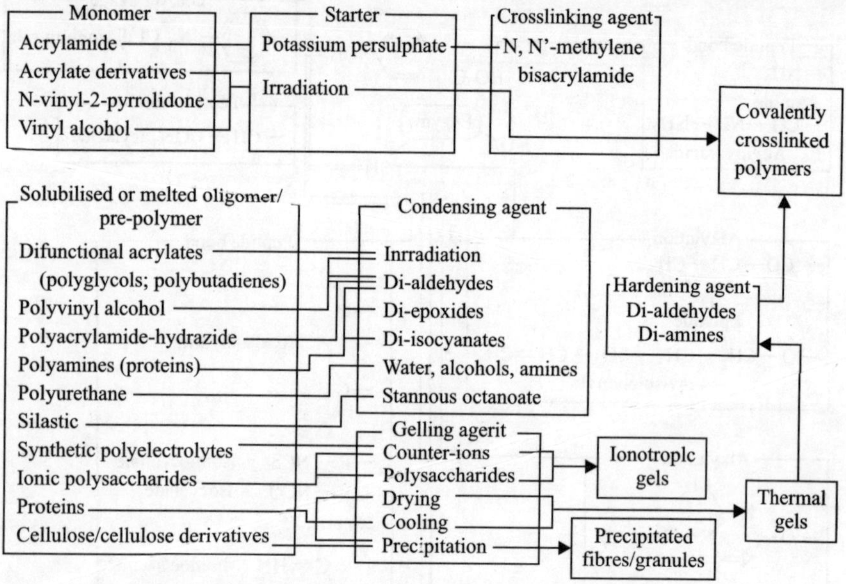

Fig. 13.7. Examples of lattice-formation strategies for entrapping biocatalysts.

Immobilised Soluble Enzyme and Suspended Cell Methods

All the methods of biocatalyst immobilisation described so far involve the modification of the biocatalyst (enzyme) or its micro-environment, with subsequent alteration of its kinetics and catalytic properties. In order to use a biocatalyst in its native state continuously over a long period of time, biocatalysts have been confined within semipermeable membranes in the form of hollow fibres or flat sheet ultrafiltration membrane reactors (Fig. 13.8). The membrane retains the biocatalyst but is permeable to the products and sometimes to the substrates. This method offers several advantages relative to other immobilisation methods. Chemical modification of the biocatalyst is not necessary and the biocatalyst retain its kinetic properties.

This method is particularly suited for conversion of high molecular weight or insoluble substrates, such as starch, cellulose and proteins, as it allows the intimate contact of the biocatalyst with the substrate achieving an efficient conversion of the substrates. However, some disadvantages are inherent in the method: the possible decrease in the reaction rate as a result of the permeability resistance of the membrane; and the adsorption of the biocatalyst and/or substrates and products on the membrane surface. This type of immobilisation has found applications on the modification of fats and oils (e.g., olive oil, palm oil) by lipases and dipeptide synthesis (acetyl-phenylalanine-leucinamide) by proteases in organic media.

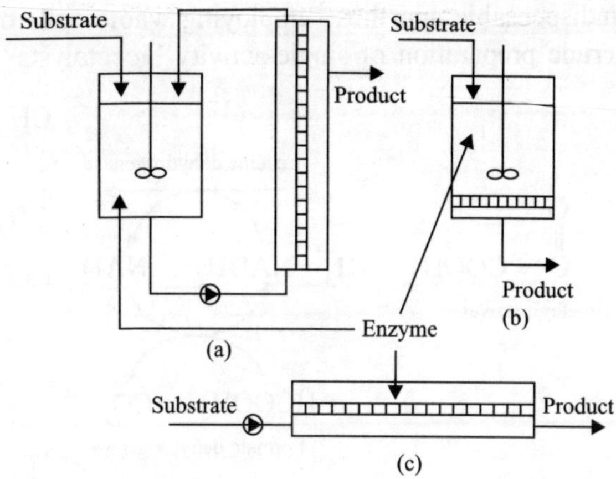

Fig. 13.8. Enzyme membrane reactors: (a) continuous stirred tank reactor with recirculation; (b) dead-end cell; and (c) tubular.

Immobilisation of Multi-Enzyme Systems and Cells

One of the strong disadvantages of single enzyme systems, free or immobilised, is their limitation to single step transformations. This limitation is particularly acute with thermodynamically unfavourable conversions and those requiring the regeneration of enzyme co-factors, such as redox reactions or phosphorylations. Among the possible solutions is coupling the intended enzymatic reaction to a chemical one or to a second enzymatic reaction. The latter alternative has been investigated for several NAD(P)- or ATP-dependent systems, using a pair of enzymes and two substrates with the co-factor shuttling between them. When two or more biotransformations are carried out simultaneously in the same vessel to regenerate co-factors, to shift thermodynamic equilibria, or to favour process economics – the intermediate products should be rapidly converted. In this context, co-immobilisation of the involved enzymes is likely to minimise the diffusion paths of the intermediates between active sites, thus, accelerating the potential rate-limiting steps. However, such systems suffer from severe problems related to the correct relative positioning of the immobilised enzymes and co-factor which are extremely difficult to achieve in practice. An example of this type of immobilisation method is the synthesis of L-tertiary leucine (Fig. 13.9), a chiral intermediate for chemicals.

The immobilisation of the multienzyme systems contained in whole cells or cell particles can lead to marked improvements in a process when compared to using free cells or immobilised single or multiple enzymes. However, a clear distinction should be made between the cases in which cell viability is indispensable and those employing whole cells or cell parts as non-viable, crude preparation of single-activity biocatalysts.

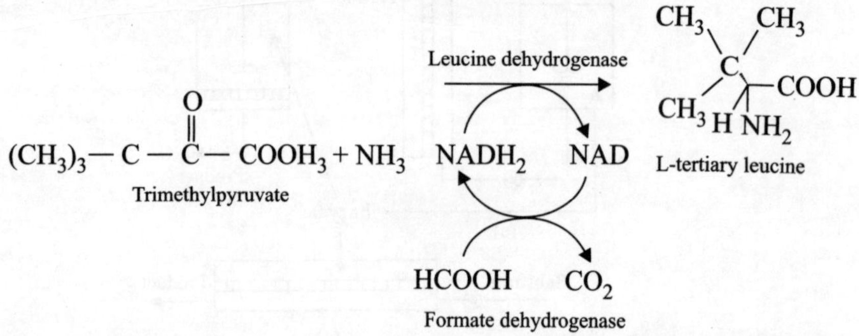

Fig. 13.9. Synthesis of L-tertiary leucine from trimethylpyruvate and ammonia.

In the first situation, one deals with sensitive catalytic forms which require very mild immobilisation and close control of operating conditions in order that cell viability is preserved. This situation corresponds, for example, to the immobilised cell fermentations and the culture of anchorage-dependent mammalian cells.

Immobilised, non-viable cells systems are sometimes preferred to immobilised single enzymes to avoid costly purification processes, or to increase catalytic stability and to retain lattice-entrapped enzymes more efficiently without the need for tight control of matrix porosities. With this type of biocatalyst, entrapment or attachment procedures designed for enzymes can be safely used, through lower productivities per unit weight of biocatalyst are to be expected.

In addition, diffusional resistances are enhanced by the cell membrane and permeabilising treatments with heat, surfactants or solvents are often required. Such treatments can also be required to inactivate contaminating enzyme activities in the cells. On the whole, single-activity, immobilised cell preparations are convenient for industrial applications, where cost reduction is necessary and process control as well as sanitisation procedures are feasible.

One important and industrial example is the immobilisation of glucose isomerase, being an expensive intracellular enzyme, the producing cells were successfully immobilised and used continuously at industrial scale in packed bed reactors. A heat treatment was included which degraded other enzymes and thus avoided substrate (glucose) and product (fructose) degradation. Other examples are listed in Table 13.4.

Table 13.4. Examples of immobilised whole cells for single- or two-enzyme conversions of industrial interest.

Microbial biocatalyst	*Immobilisation method*	*Application*
Escherichia coli	Entrapment	L-Tryptophan production from indole and DL-serine
Bacteria and yeasts	Entrapment	Biosensors
Escherichia coli	Entrapment	L-Aspartic acid production from fumaric acid and ammonia
Arthrobacter simplex	Entrapment	Prednisolone production from hydrocortisone
Rhodococcus rhodocrous	Entrapment	Acrylamide production from acrylonitrile
Pseudomonas chlororaphis		
Saccharomyces cerevisiae	Surface attachment	Sucrose hydrolysis
Humicola sp.	Entrapment	Conversion of Rifamycin B to Rifamycin S
Zymomonas mobilis	Entrapment	Sorbitol and gluconic acid production from glucose and fructose
Pseudomonas AMI	Entrapment	L-Serine production from glycine and methanol

Effect of Immobilisation on the Enzyme Kinetics and Properties

Although enzyme immobilisation can be very useful, immobilisation may also change the kinetics and other properties of the enzyme, usually with a decrease of enzyme-specific activity. This may be ascribed to several factors; (i) conformational and steric effects; (ii) partitioning effects; and (iii) mass transfer or diffusional effects.

Conformational and steric effects

The decrease of specific activity of enzymes, which occurs on their binding either to solid supports or upon inter-molecular cross-linking, is usually attributed to conformational changes in the tertiary structure of the enzymes. For instance, covalent bonds between the enzyme and the matrix can stretch the enzyme molecule and thus the three-dimensional structure at the active site. Denaturation of the enzyme can arise by the action of reagents used in entrapment methods.

The specific activity decrease may also be attributed to steric hindrance resulting in limits on the accessibility of the substrate. In these two cases, the decrease in enzyme activity can be minimised or prevented by choosing suitable conditions for immobilisation. Thus the active centre of the enzyme can be protected with a specific inhibitor, substrate, or product and the shielding effect of the support that causes steric hindrances can be reduced by the introduction of 'spacers' that keep the enzyme at a definite and certain distance from the support. By introducing bi-functional compounds, such as 1,6 diaminohexane, as 'spacers', the specific activity of covalent coupled glucoamylase on porous silica was increased by a factor of 10-fold.

In addition to their influence on the enzyme activity, any physical or chemical matrix-enzyme interactions may additionally modify the selectivity and stability of the bound enzyme from that which it normally possesses in free solution.

Partition effects

In the support binding method, when the support is charged or has a hydrophobic character, the kinetic behaviour of the immobilised enzyme may differ from that of the free enzyme even in the absence of mass transfer effects. This difference is commonly attributed to partition effects that cause different concentrations of charge species, substrates, products, hydrogen ions, hydroxyl ions and so on, in the micro-environment of the immobilised enzyme and in the domain of the bulk solution, owing to electrostatic with fixed charges on the support.

The main consequences of these partition effects is a shift in the optimum pH, with a displacement of the pH-activity profile of the immobilised enzyme towards more alkaline or acidic pH values for negatively or positively charge carriers, respectively. For example, chymotryspin immobilised on a polyanionic support – ethylene/maleic anhydride co-polymer – shifted 1 pH unit to the alkaline side, while immobilised on a polycationic support, polyornithine, a shift of 1.5 pH units to the acid side occurred.

By similar considerations, the partitioning of charged compounds, substrate or product, between a charged enzyme particle and the bulk solution can also be evaluated. For a positively charged substrate, when using a negatively charged immobilised enzyme, a higher concentration of substrate is obtained in the local environment or micro-environment than in the bulk solution and a higher value of relative activity is obtained than with a neutrally charged matrix. However, when effects other than partitioning are present, it is possible to have no shift of the enzyme's pH optimum on charged supports.

Mass transfer effects

When an enzyme is immobilised on or within a solid matrix, mass transfer effects may exist because the substrate must diffuse from the bulk solution to the active site of the immobilised enzyme. If the enzyme is attached to non-porous supports there are only external mass transfer effects on the catalytically active outer surface; in the reaction solution, being surrounded by a stagnant film, substrate and product are transported across the Nernst layer by diffusion. The driving force for this diffusion is the concentration difference between the surface and the bulk concentration of substrate and product. When an enzyme is immobilised within a porous support, in addition to possible external mass transfer effects, there could also be resistance to the internal diffusion of the substrate (as it must diffuse through the pores in order to reach the enzyme) and resistance of product for its diffusion into the bulk solution. Consequently, a substrate concentration gradient is established within the pores, resulting in a concentration decreasing with increased distance (in depth) from the surface of immobilised enzyme preparation. A corresponding product concentration gradient is obtained in the opposite direction.

Unlike external diffusion, internal mass transfer proceeds in parallel with the enzyme reaction and takes into account the depletion of substrate within the pores with increasing distance from the surface of the enzyme support. The rate of reaction will also decrease, for the same reason. The overall reaction is dependent on the substrate concentration and the distance from the outside support surface.

Miscellaneous effects

Other properties of the enzyme can change upon immobilisation. The substrate specificity alters, particularly when using a substrate of high molecular weight by the effect of steric hindrance and diffusional resistances. The kinetics constants K_m and V_m of the immobilised enzyme are different from the free enzyme as a consequence of conformational changes of the immobilised form, which affect the affinity between enzyme and substrate. The increase of activity energy for some immobilised enzymes may be attributed to diffusional resistances, mainly in porous supports.

IMMOBILISED ENZYME REACTORS

Classification of Enzyme Reactors

Among the applications of immobilised enzymes, their utilisation in industry is perhaps the most important and consequently the most frequently discussed. The use of immobilised enzymes in industrial processes is performed in basic chemical reactors. A classification of enzyme reactor based on the mode of operation and the flow characteristics of substrate and product is presented in Table 13.5. The configurations of the different reactor types are shown in Fig. 13.10.

Table 13.5. Classification of enzyme reactors.

Mode of operation	*Flow pattern*	*Type of reactor*
Batch	Well mixed	Batch stirred tank reactor (BSTR)
	Plug flow	Total recycle reactor
Continuous	Well mixed	Continuous stirred tank reactor (CSTR)
		CSTR with ultrafiltration membrane
	Plug flow	Packed bed reactor (PBR)
		Fluidised bed reactor (FBR)
		Tubular reactor (other)
		Hollow fibre reactor

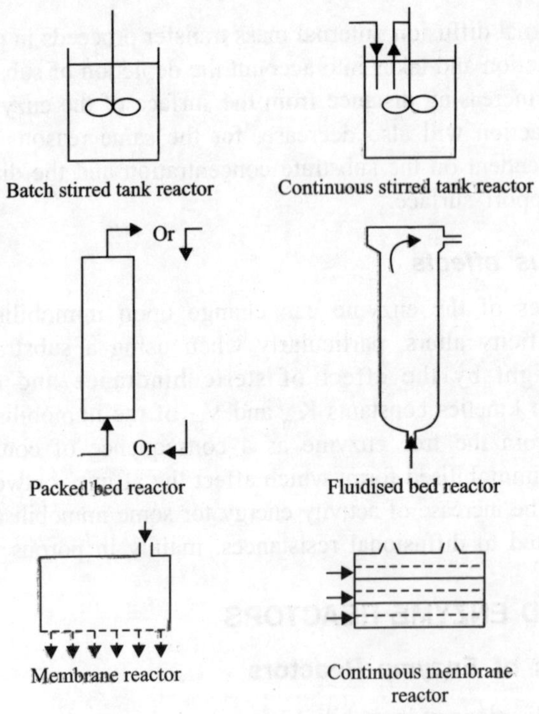

Fig. 13.10. Examples of immobilised biocatalyst reactors.

Batch Reactors

Batch reactors are most commonly used when soluble enzymes are used as catalysts. The soluble enzymes are not generally separated from the products and consequently are not recovered for re-use.

Since one of the main goals of immobilising an enzyme is to permit its re-use, the application of immobilised enzymes in batch reactors requires a separation (or an additional separation) to recover the enzyme preparation. During this recovery process, appreciable loss of immobilised enzyme material may occur as well as loss of enzyme activity. Traditionally, the stirred tank reactor has been used for batchwise work. Composed of a reactor and a stirrer, it is the simplest type of reactor that allows good mixing and relative ease of temperature and pH control. However, some matrices, such as inorganic supports, are broken by shearing in such vessels and alternative designs have therefore been attempted. A possible laboratory alternative is the basket reactor, in which the catalyst is retained within a 'basket' either forming the impeller 'blades' or the baffles of the tank reactor.

Another alternative is to change the flow pattern, using a plug flow type of reactor: the total recycle reactor or batch recirculation reactor, which may be a packed bed or fluidised bed reactor, or even a coated tubular reactor. This type of reactor may be useful where a single pass gives inadequate conversions. However, it has found greatest application in the laboratory for the acquisition of kinetic data, when the recycle rate is adjusted so that the conversion in the reactor is low and it can be considered as a differential reactor. One advantage of this type of reactor is that the external mass transfer effects can be reduced by the operational high fluid velocities.

Continuous Reactors

The continuous operation of immobilised enzymes has some advantages when compared with batch processes, such as ease of automatic control, ease of operation and quality control of products. Continuous reactors can be divided into two basic types: the continuous feed, stirred tank reactor (CSTR) and the plug flow reactor (PFR).

In the ideal CSTR the degree of conversion is independent of the position in the vessel, as a complete mixing is obtained with stirring and the conditions within the CSTR are the same as the outlet stream, that is, low substrate and high product concentrations. With the ideal PFR, the conversion degree is dependent on the length of the reactor as no mixing device at all exists and the conditions within the reactor are never uniform.

While a nearly ideal CSTR is readily obtained (since it is only necessary to have good stirring to obtain complete mixing), an ideal PFR is very difficult to obtain. Several adverse factors to obtaining an ideal PFR often occur, such as temperature and velocity gradients normal to the flow direction and axial dispersion of substrate.

Several considerations influence the type of continuous reactor to be chosen for a particular application. One of the most important criteria is based on kinetic considerations. For Michaelis-Menten kinetics, the PFR is preferable to the CSTR as the CSTR requires more enzyme to obtain the same degree of conversion as a PFR. If product inhibition occurs, this problem is accentuated, as in a CSTR high product concentration is always in direct contact with all of the catalyst. There is only one situation where a CSTR may be kinetically more favourable than a PFR, namely, when substrate inhibition occurs.

The form and characteristics of the immobilised enzyme preparations also influence the choice of reactors type and operational requirements are still another factor to be taken into account. Thus, when pH control is necessary, for instance with penicillin acylase, the CSTR or batch stirred tank reactor is more suitable than PFR reactors. Due to possible disintegration of support through mechanical shearing, only durable preparations of immobilised enzyme should be used in a CSTR. With very small immobilised enzyme particles, problems such as high pressure drop and plugging arise from the utilisation of this catalyst in packed bed reactors (the most used type of PFR). To overcome these problems, a fluidised bed reactor, which provides a degree of mixing intermediate to the CSTR and the ideal PFR, can be used with low pressure drop.

Reactant characteristics can also influence the choice of reactor. Insoluble substrates and products and highly viscous fluids are preferably processed in fluidised bed reactors or CSTR, where no plugging of the reactor is likely to occur, as would be the case in a packed bed reactor.

As can be deduced from this outline, there are no simple rules for choosing reactor type and the different factors mentioned must be analysed individually for a specific case.

BIOCATALYSIS IN NON-CONVENTIONAL MEDIA

Water as an essential reaction medium for biocatalysts has been advocated for many years as one of the major advantages of biotransformations. However, this so-called advantages has proved to be one of the severest limitations for broadening the scope of applications of biocatalysts, especially when the reactants are poorly soluble in water. Non-conventional media that have been used include organic solvents, some gases and supercritical fluids. The scopes and limitations of these different systems are described below.

Biocatalysis in Organic Media

The first examples of biocatalyst/enzyme use in organic solvents for the conversion of hydrophobic compounds were presented over 20 years ago. Several examples show the synthesis of peptides (e.g., AcPheAlaNH_2) from amino acids catalysed by proteases, the production of the sweetener, Aspartame (L-aspartyl-L-phenylalanine methyl ester), catalysed by thermolysin using ethyl acetate as solvent and the use of lipases on esterification, transesterification and inter-esterification reactions and the resolution of enantiomers.

The introduction of an organic solvent in the reaction system has several advantages (Table 13.6). The organic solvent increases the solubility of the poorly water-soluble or insoluble compounds, thereby increasing the volumetric productivity of the reaction system. Another important advantages is that the equilibrium of a hydrolytic reaction can be shifted in favour of the product, this being extracted into the organic phase, therefore biocatalyst and product recovery will be facilitated (extractive bioconversions). High product yields may also be achieved by decreasing possible substrate or product inhibition and prevention of unwanted side-reactions. In spite of these advantages of using organic solvents, limitations also exist. The biocatalyst may be denaturated or inhibited by the solvent and in addition, the introduction of an organic solvent leads to an increased complexity of the reaction process system.

Table 13.6. Characteristics of two-liquid phase bioconversion.

Potential advantages
High substrate and product solubilities
Reduction in substrate and product inhibition
Facilitated recovery of product and biocatalyst
High gas solubility in organic solvents
Shift of reaction equilibrium
Potential disadvantages
Biocatalyst denaturation and/or inhibition by organic solvent
Increasing complexity of the reaction

These reaction systems were first applied to enzymatic conversions before their use was extended to whole-cell systems. Recently, the discovery of bacterial strains which are able to grow in the presence of organic solvents has made this field even more promising, opening further possibilities for the understanding of the mechanisms underlying the tolerance or toxicity responses of micro-organisms to organic solvents.

Selection of solvent

Two of the most important technical criteria for solvent selection are high product recovery and biocompatibility, although other characteristics like chemical and thermal stability, low tendency to form emulsions with water media, non-biodegradability, non-hazardous nature and low market price are desirable.

Whereas the other desirable solvent attributes are relatively mild conditions, the requirement of biocompatibility is a particularly restrictive criterion. Several attempts have been made to associate the toxicity of different solvents to some of their physico-chemical properties. The parameters used to classify solvents in terms of biocompatibility have been related to the polarity of the solvent. *Laane* and coworkers have described a correlation between bioactivity and the logarithm of the partition coefficient of the solvent in the octanol/water two-phase system (log P_{oct}), known as the Hansch parameters. Log P_{oct} denotes hydrophobicity, which is not exactly the same as polarity, but it shows as much better correlation with the biocatalytic rates than other models based on solvent polarity. The Hansch parameter has currently been used in the pharmaceutical and medical fields as a part of drug activity studies and can be determined experimentally or calculated by Rekker's hydrophobic fragmental constant approach. Many attempts have been made to explain the empirical correlation between log P_{oct} and the activity retention of cellular biocatalysts but so far the mechanisms of solvent-caused toxicity are poorly understood.

Laane and coworkers observed that a correlation exists between log P_{oct} and the epoxidising reaction activity of immobilised cells and the gas-producing activity of anaerobic cells in various water-saturated organic solvents. When plotting the cellular activity retention against log P_{oct} sigmoidal curves are obtained (Fig. 13.11). A solvent with a log P_{oct} value than the inflection point is usually toxic and one with a log P_{oct} value higher than the inflection point is biocompatible. The inflection point of these curves depends on the micro-organisms studied. In general, solvents having a log P_{oct} lower than 2 are relatively polar solvents not suitable for biocatalystic systems and biological activities vary in solvents having a log

P_{oct} between 2 and 4, being high in apolar solvents having log P_{oct} values above 4 (Table 13.7).

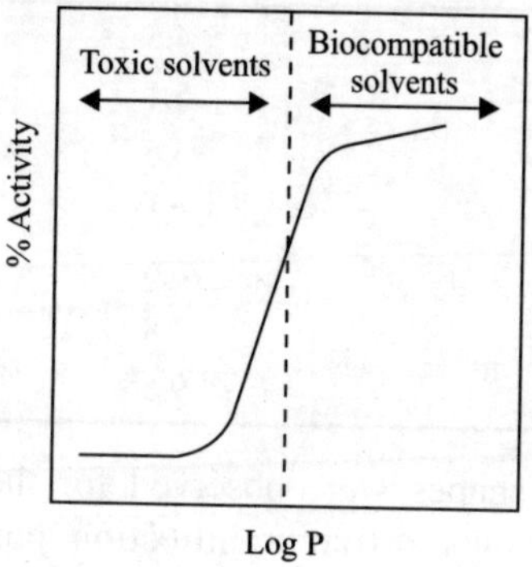

Fig. 13.11. Retention of enzyme activity versus log P.P, partition coefficient.

Table 13.7. Biocompatible organic solvents.

Solvents	*Hansch parameter log P*
Alcohols	
Decanol	4.0
Undecanol	4.5
Dodecanol	5.0
Oleyl alcohol	7.0
Ethers	
Diphenyl ether	4.3
Carboxylic acids	
Oleic acid	7.9
Esters	
Pentyl benzoate	4.2
Ethyl decanoate	4.9
Butyl oleate	9.8
Dibutylphthalate	5.4
Dipentylphthalate	6.5
Dihexylphthalate	7.5
Dioctylphthalate	9.6
Didecylphthalate	11.7
Hydrocarbons	
Heptane	4.0

(Contd...)

Solvents	*Hansch parameter log P*
Octane	4.5
Nonane	5.1
Decane	5.6
Undecane	6.1
Dodecane	6.6
Tetradecane	8.8
Hexadecane	9.6

Similar sigmoidal shapes were observed for the effect of solvents on micro-organisms, however different inflection points were obtained for different micro-organisms which could be due to differences in the characteristics of their cellular membranes. It has also been observed that increasing the agitation rate caused the log P_{oct} curve to shift to the right. A good correlation between the metabolic activity of *Arthrobacter*, *Acinetobacter*, *Nocardia* and *Pseudomonas* and the log P_{oct} of the solvent was found, however, the transition between toxic and non-toxic solvents was observed in the log P_{oct} range from 3 to 5. *Tramper* and coworkers have investigated the relationship between the metabolic activity of cells exposed to organic solvents at 10 per cent (v/v) concentrations and their log P_{oct} values for different homologous series of solvents. They found that the log P_{oct} value, above which all solvents are non-toxic, is different for different homologous series: e.g., *Arthrobacter* and *Nocardia* tolerate alkanols with log P_{oct} above 4 but are only able to tolerate phthalates having a log P_{oct} value higher than 5.

The influence of the characteristics of the cell membrane on the solvent tolerance of the micro-organisms has been evaluated for Gram-negative and Gram-positive bacteria. Gram-negative bacteria and particularly *Pseudomonas*, are in general more tolerant than Gram-positive bacteria. The difference in solvent tolerance is probably due to the presence of the outer membrane of Gram-negative bacteria. This membrane contains major structural proteins, lipoproteins and lipopolysaccharides. A percentage of the total lipoproteins of the outer membrane is bound covalently to the peptidoglycan layer. This linkage might be expected to protect cells from environmental stress. Plant cells appear to be even more sensitive to the presence of organic solvents as shown for cell suspensions of *Morinda citrifolia* which had a biocompatibility limiting log P_{oct} value of 5. Thus the transition range from toxic to non-toxic clearly depends on the type cellular biocatalyst.

The correlation with solvent biocompatibility especially for the effect of solvents on enzyme stability is a 'three-dimensional solubility parameter' or the 'denaturing capacity' used to predict, for example, the concentration of the organic solvent at which half-inactivation is observed.

Classification of organic reaction systems

Biocatalysts can be used in different ways in combination with organic solvents: (i) homogeneous mixture of water and water-miscible solvent; (ii) aqueous/organic two-liquid phase systems; (iii) micro-heterogeneous systems (micro-emulsions and reversed micelles); (iv) enzyme powder and immobilised biocatalyst suspended in solvent without aqueous phase; and (v) covalently modified enzymes dissolved in organic solvent (Fig. 13.12a to g)

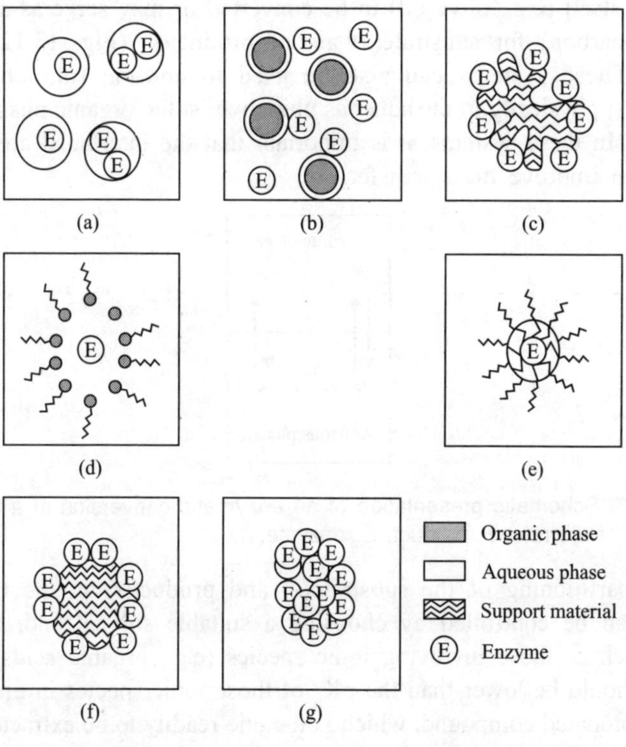

Fig. 13.12. Classification of biocatalysis in organic media systems.

Homogeneous mixture of water and water-miscible solvent

An easy way to increase the solubility of a hydrophobic substrate is to add a water-miscible organic solvent, such as methanol, acetone, ethyl acetate, dimethyl formamide, dimethylsulphoxide etc., to the reaction medium. These systems have the advantages of generally not presenting mass transfer limitations as they are homogeneous systems. However, the biocatalyst in the presence of these systems usually has poor operational stability, particularly if a high concentration of solvent is needed. This results from the fact that water-miscible solvents are polar compounds with log P values lower than 2, being considered as toxic solvents. For example, ribonuclease dissolved in increasing concentrations of 2-chloroethanol undergoes a transition from the native state to an unfolded form.

Aqueous/organic two-phase systems

Two-liquid (aqueous/organic) phase systems are useful when reactants of poor water solubility have to be employed. The organic solvent may be the substrate itself (e.g., olive oil) to be converted or may serve as a reservoir (a hydrocarbon) for substrate(s) and/or product(s) (Fig. 13.12a,b,c and 13.13). These systems can also be used to confine (immobilise) the biocatalyst physically in the aqueous phase whilst the organic phase is being renewed. In these systems, it is important that the interfacial area is large enough to improve mass transfer.

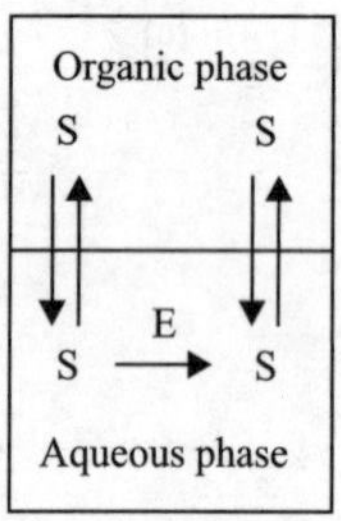

Fig. 13.13. Schematic presentation of an enzymatic conversion in a two-phase system. S, substrate; P, product; E, enzyme.

The partitioning of the substrate(s) and product(s) in the two-liquid phases can be controlled by choosing a suitable solvent and, in certain cases, such as those involving ionic species (e.g., organic acids), the pH (which should be lower than the pK_a of those ionic species in order to get the unprotonated compound, which is the one readily to be extracted) of the water phase. In the latter case, the pH selected should also be compatible with the enzymatic activity. The partitioning of substrate(s) and product(s) is particularly suited when one or both of these compounds is an inhibitor of the enzymatic activity. Its accumulation in the organic phase will alleviate this inhibition. The overall displacement of reaction equilibrium by extraction of the products is also an advantage of this type of systems. The phase ratio can be varied over a wide range leading to an optimisation of reactor capacity. Applications of two-liquid phase systems include: olive oil hydrolysis by lipases, peptide synthesis, extractive ethanol fermentations by coupling an extractant (e.g., dodecanol) with fermentation step, steroid transformations (e.g., prednisolone dehydrogenation by adsorbed and gel-entrapped *Arthrobacter simplex* cells etc.).

Micro-heterogeneous systems

Micro-emulsions and reversed micelles represent a special case of two-liquid phase systems in which the aqueous phase is no longer macroscopically distinguishable from the continuous organic phase. These systems usually contain surfactants to stabilise the distribution of water and its contents in the continuous organic phase.

Reverse micelles are aggregates formed by surfactants in apolar solvents. Surfactants are amphipathic molecules that possess both a hydrophilic and a hydrophobic part (Fig. 13.12d). The hydrophobic tails of the surfactant molecules are in contact with a polar bulk solution; the polar head groups are turned towards the interior of the aggregate forming a polar core. This core can solubilise water (water pool) and host macro-molecules such as proteins. The group of amphipathic molecules used in the formation of reversed micelles in hydrocarbons solvents include both natural membrane lipids and artificial surfactants (Table 13.8).

Table 13.8. Example of surfactant-forming reversed micelles.

Surfactant	*Solvent*
Sodium dioctyl sulphosuccinate (AOT)	*n*-Hydrocarbon (C_6–C_{10})
	Iso-octane
	Cyclohexane
	Carbon tetrachloride
	Benzene
Cetyltrimethylammonium bromide (CTAB)	Hexanol/iso-octane
	Hexanol/octane
	Chloroform/octane
Methyltrioctylammonium chloride (TOMAC)	Cyclohexane
Brij 60	Octane
Triton X	Hexanol/cyclohexane
Phosphatidylcholine	Benzene
	Heptane
Phosphatidylethanolamine	Benzene
	Heptane

The amount of water solubilised in the reversed micellar systems is commonly referred to as w_0, the molar ratio of surfactant to water (w_0 = H_2O/surfactant). This is an extremely important partameter, since it will determine the number of surfactant molecules per micelle, the availability of water molecules for protein hydration and biocatalysis and is the main factor affecting the micelle size.

The formation of reversed micelles depends largely on the energy change due to dipole-dipole interactions between the polar head groups of the surfactant molecules. The solubilisation properties of surfactants are often expressed by a three or four component phase diagram, the reversed micelles being identified by the regions of optical transparency. Most of the work performed with reversed micelles in biological systems used AOT (sodium dioctyl sulphosuccinate), an anionic surfactant. AOT forms stable micellar aggregates in organic solvents, the most used being isooctane. The maximum amount of solubilised water for an AOT/iso-octane/H_2O) system is around w_0 = 60. Above this value the transparent reversed micelle solution becomes a turbid emulsion and phase separation occurs.

Examples of biocatalysis in reversed micelles include the controlled hydrolysis of lipids and vegetable oils, the esterification and trans-esterification reactions catalysed by lipases and peptides synthesis by proteases. The optimum water level can be controlled by the w_0 value, this value being greater then 10 for synthesis reactions. Currently important aspects of research on these systems include the stabilisation of biocatalysts and the development of appropriate reactors to accomplish enzyme retention, product separation and avoid product contamination with the surfactant molecules.

Very low water systems

For many biotransformations it is advantageous to decrease considerably the amount of water in the reaction medium. It is believed that retention of enzyme activity is due to the minimum essential water necessary to maintain the protein structure and the enzymatic function. The selection of the organic solvent is critical as hydrophilic solvents strip the essential hydration shell from the enzyme molecule. The amount of water bound to the enzyme decreases dramatically with the increasing hydrophilicity of the solvent. For example the reactivity of α-chymotrypsin in octane is 10,000-fold higher than that in pyridine.

The amount of water in the reaction medium can be measured in several ways. The most common way is to measure the water concentration (in per cent v/v or mol l^{-1}). This parameter, however, does not describe the reaction conditions for the enzyme. A better way to characterise the degree of hydration of the reaction medium is to use the thermodynamic water activity, a_w, as parameter. This parameter, which is a measure of the amount of water in the system, determines directly the effects of water on the chemical equilibrium. During the reaction the water activity may change, especially if water is formed (esterification reactions) or consumed (hydrolysis). Therefore it is important to keep the water activity constant by controlling its concentration during the reaction, for example by adding salt hydrates directly to the reaction medium which give a constant water activity suitable for the enzymatic conversion. The salt hydrates act as a 'buffer' of the water activity.

Very low water systems involve the use of enzymes as solid suspensions, powders (Fig. 13.12g) or immobilised onto a support, in organic solvents (Fig. 13.12f); and covalently modified enzymes soluble in organic media (Fig. 13.12).

When enzyme powders are used as biocatalysts, problems can occur due to aggregation of the enzyme molecules. A solution to solve these problems is the immobilisation of the enzyme on to an appropriate solid support. The support should be judiciously selected taking into account its surface properties, namely its capacity to attract water (aquaphilicity), as the water in the reaction medium will partition between the enzyme, the support and the reaction medium affecting the enzyme micro-environment.

A very interesting feature of enzyme suspensions in organic media is the phenomenon of 'pH memory'. It has been observed that lyophilised lipases and proteases and several oxido-reductases display pH optima in organic media identical to optima in aqueous solutions. This phenomenon is attributed to the right ionisation state of the enzyme for catalysis. Enzymes suspended in organic solvents have been observed to exhibit altered substrate specificity compared to that in aqueous media. This has been attributed to steric hindrance caused by the lack of conformational mobility of the enzyme (lipases and proteases) in organic solvents and to the substrate's inability to displace water from the hydrophobic binding pockets of the enzyme molecules in organic media.

Since dehydration drastically decreases the conformational mobility of enzymes, it has been observed, as expected, that the thermal denaturation process is slowed down considerably for enzymes suspended in organic solvents.

Suspensions of solid enzymes in organic solvents have been used for a number of biotechnological applications, namely the trans-esterification of fats and oils by lipases, which is an industrial process for upgrading triacylglycerols.

In spite of the advantages of solid enzyme suspensions in organic solvents, these systems are limited by mass transfer. To overcome these diffusional limitations, enzymes have been covalently modified, namely using polyethylene glycol for this purpose, to make them soluble in organic solvents. As the enzyme becomes soluble in the reaction there are no diffusional limitations, however in order to re-use the enzyme, it has to be recovered by precipitation from the reaction mixture with a non-polar solvent such as hexane. Other drawbacks of this type of enzyme are inactivation of the enzyme that may occur during the derivatisation procedure and that the enzyme preparations are soluble only in a limited number of solvents, like aromatic and chlorinated hydrocarbons.

Gas-Phase Reaction Media

It is possible to perform biotransformations with a solid biocatalyst preparation and gaseous substrates. The oxidation of ethanol by alcohol oxidase was a pioneer work. Further applications have involved the use of lipases to catalyse trans-esterifications reactions between short chain alcohols and esters. As the substrates are in the gaseous phase, relatively high temperatures are used, consequently it is advantageous to use thermostable enzymes for these applications. In gas phase reactions, as with using organic media, the water content of the enzyme preparation should be well controlled to achieve the best performance both in terms of activity and stability.

Supercritical Fluids as Bioreaction Media

Enzymatic reactions in supercritical and near critical fluids require pressurised systems. Such systems permit high mass transfer rates and easy separation of reaction products. Due to its non-toxic character and its relatively low critical temperature (31°C) CO_2 has been the most used fluid. In some aspects, supercritical fluids have properties resembling those of non-polar solvents being adequate for biotransformations of hydrophobic compounds. The oxidation of cholesterol by cholesterol oxidase and the stereoselective hydrolysis of racemic glycidyl butyrate by immobilised *Rhizomucor miehei* lipase yielding the homochiral R(-) glycidyl butyrate have been achieved in the presence of supercritical CO_2. The solubility of these compounds may also be improved by the addition of small amounts of co-solvents known as entrainers. For example methanol (3.5 per cent mol) was used as an entrainer to enhance the solubility of cholesterol in supercritical carbon dioxide. The major drawback of this reaction system is the high energy and equipment costs due to the use of high pressures.

To sum up this chapter described the use and performance of biocatalysts in bio-transformations relevant to industrial, analytical and biomedical applications and environment bioremediation.

From a process point of view, there are advantages and limitations for both chemical and biochemical routes. Part of the limitations of the biocatalytic route has been solved through the new developments in the areas of biology, chemistry and process engineering. The recent advances in recombinant DNA technology, metabolic engineering, fermentation and biocatalysis in non-conventional media have also broaden the applications of biocatalysts to synthetic and oxidative/reductive biotransformations. It is also important to emphasise the integration of both processes and disciplines (engineering, biology and chemistry) which is a key feature for the development of competitive biocatalytic routes. New applications of biocatalysts (native or modified) in the fields of chemical synthesis, analysis (biosensors), biomedical and environment are foreseen.

14

Quality Assurance and Quality Control

INTRODUCTION

'Quality is meeting the requirements.' This definition, attributed to Philip B.Crosby, is the essence of quality assurance and quality control. In the pharmaceutical industry, the requirements involve meeting specifications and developing a system to make certain that materials are sampled, tested, and evaluated against those specifications.

Quality control is defined as 'The operational techniques and activities that are used to fulfil requirements for quality.' This is an ISO 9000 definition. (The International Organisation for Standardisation, Geneva, Switzerland, is an organisation that sets international standards. It deals with all fields except electrical and electronics, which is governed by the International Electrotechnical Commission (IEC). The organisation is commonly known as the International Standards Organisation (ISO). Its standards for quality have been issued as policy ISO 9000). These techniques and activities are usually performed by the Quality Control Laboratory.

Quality assurance is defined as 'All those planned and systematic actions necessary to provide adequate confidence that a product or service will satisfy given requirements for quality.' This is an ISO 9000 definition. Quality assurance is performed by the Quality Assurance. Department, which oversees the activities of research and development, manufacturing, and quality control. This function is defined in the current Good Manufacturing Practice Regulations (GMPs) as the Quality Control Unit. The basic principles of quality assurance have as their goal the production of articles that are fit for their intended use. These principles may be stated as follows.

1. Quality, safety, and effectiveness must be designed and built into the product.
2. Quality cannot be inspected or tested into the finished product.
3. Each step of the manufacturing process must be controlled to maximise the probability that the finished product meets all quality and design specifications.

Organic chemicals for commercial use are made by chemical synthesis, recombinant DNA technology, fermentation, enzymatic reactions, recovery from natural materials, or combinations of these processes.

When these chemicals are designated for use as components of drug products, they are typically known as bulk pharmaceutical chemicals (BPCs) or active pharmaceutical ingredients (APIs).

A 'pharmaceutical' is a fancy word for what the Food Drug and Cosmetic (FD&C) Act calls a drug. In section 201(g)(1) of the Act, drugs are defined as:

1. Articles recognised in the official US Pharmacopeia, official Homeopathic Pharmacopeia of the United States, or official National Formulary, or any supplement to any of them.
2. Articles intended for use in the diagnosis, cure, mitigation, treatment, or prevention of disease in man or other animals.
3. Articles (other than food) intended to affect the structure or any function of the body of man or other animals.
4. Articles intended for use as a component of any articles specified in clause (1), (2), or (3).

From this definition, under U.S. law, the component of a drug is also a drug. Thus, any chemical used as an ingredient of a drug product, whether pharmocologically active or not, is also a drug.

CHEMICAL AND PHARMACEUTICAL PRODUCTION

Regardless of whether the product is a chemical or a drug, production typically takes place by using a batch system. In fact, producing a product is very similar to making cookies. To make one batch of cookies, all of the needed raw materials are weighed out individually in the correct proportions. They are then subjected to various operations, such as mixing, homogenising, reacting, granulating, milling, compressing into the correct shape, etc. The specified number of cookies are counted and placed into a container, the cap is screwed on and the label is affixed. The container may then be placed in an outer carton and the flap is closed or sealed.

If a second batch is desired, one would weigh out a second set of raw materials, making more in process items, filling another container, and affixing a new label. The whole process would give us a new batch of finished product. As with cookies, so it is with drug tablets, capsules, vials, etc. A chemical, although produced by a series of chemical or biological transformations, is more often than not produced by a similar batch process. For example, in the preparation of a typical antibiotic by fermentation, an inoculum preparation is made by transferring a carefully preserved bacterium into a vessel containing a mixture of nutrients (chemicals) in solution or suspension (the growth medium) and subjecting it to initial growth. The inoculum preparation is then transferred to fermentation tanks containing greater quantities of the growth medium. The by-product of the bacterial growth is the desired chemical, which is then extracted and purified and the final product is packaged and labeled. Because of the limited size of vessels, the whole process is done as a batch operation.

FIVE VARIABLES

The batch process is built on five types of materials, which can be of five variables:

1. Labels.
2. Packaging materials.
3. Raw materials.
4. In-process items.
5. Finished products.

They are termed variables because minor variations in any one of these areas affects the other areas and in turn can affect the quality control function dramatically.

Quality assurance and quality control are the procedures used to document and control these five variables.

(Labels, packaging materials, and raw materials are usually purchased by a manufacturer and are then used to manufacture and package in-process items and finished products. However, because labels and packaging materials are less likely to be tested for chemical or microbiological properties, the five variables are treated in a different order below from that in which they are listed above.

RAW MATERIALS

Raw materials are the solvents and chemicals, whether active ingredients or inactive excipients, used in the manufacture of the dosage form of a drug product; or the solvents, chemicals, biological materials, and nutrients used to produce chemicals.

Types of Raw Materials

Raw materials may be categorised in several ways: they may be active or inactive, depending upon their physiological action; they may be organic or inorganic, depending upon their chemical composition; they may be solid, liquid, or gas; or they may be classified according to their origin as synthetic, animal, vegetable, or mineral. These classifications are important when assessing microbiological properties.

Method of Manufacture

Raw materials, whether solids, liquids, or gases, all must be purified for use in drug products. Although the bench chemist can ensure purity at each step because of the small scale of such production, the full-scale version requires a great deal of quality control.

For the most part, raw materials are purchased from outside the company. Because of differences in the method of manufacture, handling, and packaging, quality will vary from company to company or even from lot to lot. It is essential to recognise that water used in the process is a raw material and that because it is a hospitable environment for many bacteria, microbiological contamination is a serious concern. For most organic solvents, microbiological contamination is less of a problem. For parenterally used products, liquid purity is crucial; for other products, such a high level of purity might not be necessary. Finally, gases are sometimes used in drug production. Some products are processed under nitrogen to avoid oxidation.

These quality concerns involve testing and inspection of incoming solids, of holding and distribution systems for liquids, and of distillation systems for gases. The goal in each case is to make full-scale batches as similar as possible to the bench chemist's reference standard but at a reasonable cost.

Controls

Raw materials of any form are sampled, tested, and released to the production department by the quality control department. Many suppliers routinely provide raw materials of consistently high quality, in which case the amount of sampling and testing may be reduced or even replaced by a certificate of analysis supplied by the vendor. Just the same, the quality of the raw material is always defined by certain specifications, with tests required to determine its identity, assay, quality, and purity by using physical, chemical, and microbiological procedures.

Many reference compendia supply information on raw materials. The compendia are generally written in 'monograph' format; that is, for each item there is a collection of data under one heading, giving the name of the material, a chemical formula, a description, and specifications and test methods for such parameters as identity, strength, quality, and purity. These items usually include chemical, physical, and microbiological limits for the raw materials. Official compendia include American Chemical Society (ACS) Reagent Chemicals, British Pharmacopeia (BP), British Pharmaceutical Codex (BPC), Deutsches Arzneibuch (DAB), European Pharmacopeia (EP), Food Chemicals Codex (FCC), Japanese Pharmacopeia (JP), Official Methods of Analysis (AOAC), and U.S. Pharmacopeia (USP) including the National Formulary (NF).

IN-PROCESS ITEMS

An in-process item is any item that is produced and held by the manufacturing department for further processing or packaging.

Types of In-Process Items

In-process items include intermediate chemicals, premixes, granulation, bulk tablets or capsules, bulk liquids, ointments, and creams. Also included in this classification are packaged products that require additional processing. For example, sealed vials must undergo terminal sterilisation. Excluding medicinal gases, there are only four types of in-process items: liquids (and solutions), liquid-liquid emulsions, solid-liquid suspensions, and pure solids.

Method of Manufacture

For pharmaceutical liquids, the raw materials are weighed out. The solvent is accurately measured, and the raw materials are added in the order prescribed by the manufacturing instructions. A solution is produced by stirring; the temperature, pH, and other variables are controlled to maintain solubility and to avoid decomposition. Mixing and transfer of liquids are relatively straightforward steps.

Liquid-liquid emulsions, however, require special instructions and care to avoid separation of the phases. The same holds true for solid-liquid suspensions, such as ointments and creams.

Powders and tablets are made by a more complicated procedure than liquids. They are generally mixed, wetted, granulated, milled, dried, and compressed, although advances in the field of direct compression have obviated some of these steps.

Sterile products of all types require special handling techniques; they must be produced in microbiologically clean environments by validated sterilisation procedures under tight microbiological controls.

Before packaging, in-process items are stored in bulk. Solutions and liquids are often held in storage tanks, while bulk powders and tablets are generally held in sealed drums.

Controls

In-process items are subjected to in-process controls and are sampled, tested, and released by the Quality Control Laboratory before being used in the next step. A time limit is set so that in-process items do not remain unused for long periods. In-process items are tested for identity, strength, quality, purity, and other physical properties.

FINISHED PRODUCTS

A finished pharmaceutical product is the article received by the retailer for sale to a consumer. It consists of a dosage form inside a container/closure system. A finished chemical product is usually in the form of a powder or liquid packaged in suitable container.

Types of Finished Products

Finished pharmaceutical products include tablets, capsules, powders, liquids, ointments, creams, and suppositories. These are packed in bottles, jars, vials, ampules, syringes, tubes, canisters, or packets, which are often packed inside a cardboard carton.

A finished chemical product is packaged into plastic or metal drums, large tote bins, or, for bulk shipments, tank cars.

Method of Packaging

If desired, the packaging process for tablets, powders, and liquids can be managed entirely by hand. The procedure, whether done manually or automatically, is as follows.

The material is weighed or counted, containers are filled, closures are applied, labels are affixed, and containers may be placed in cartons. It is usual in modern processing for packaging lines to be partially or fully automated. The only manual operations in automated systems are the filling of hoppers with packaging materials (for example, bottles and caps) and with bulk material (tablets or liquids). High-speed packaging lines are sophisticated and efficient. Indeed, the industry's earliest approaches to robotics came in this area.

Controls

Line clearance checks are performed on the packaging lines before any new product or packaging material is brought into the area. The bill of materials or packaging ticket details the components that are to be assembled. The line is set up, the batch code and expiration date are assigned and checked, and packaging begins.

In-process samples are checked for fill volume or quantity at regular intervals, and samples are taken by the Quality Control Laboratory to check against the specifications for the finished product. The identity, strength, quality, and purity of the finished, packaged product are reconfirmed by a complete records check, and this may be confirmed by additional laboratory testing. The master control documents for the finished product dictate the specifications, methods, and sampling procedures, and the data are recorded in the batch record.

LABELS AND LABELING

In regulatory usage (as defined in the FD&C Act), a label is a display of written, printed, or graphic matter on the immediate container, that is, on the container received by the consumer. The term 'labeling' encompasses all written, printed, and graphic matter accompanying the product. Thus, labeling includes labels, but labels do not include labeling.

The necessity for complete agreement between the label and the product contained within the container is both obvious and essential. The label is therefore closely controlled to prevent a mix-up of products that could harm a patient. While it is just as important to control other kinds of labeling, such as the outside of a shipping case, the potential to harm a patient certainly would not be as great as an error in printing the bottle's label.

Types of Labels and Labeling

The most common type of label is made of paper. This is sometimes preglued like a postage stamp and then attached to the container during the packaging operation. The label, however, may be printed on the bottle, tube, or jar before receipt, in which case the packaging component must be examined for both labeling and packaging defects.

Labeling appears on the following:

Instructions to physicians, patient package inserts, cartons, outer wrapping, cardboard stock for shipping containers, shelf display cartons, and advertising or promotional materials accompanying the product.

Method of Printing

First, the copy (i.e., the text of the labeling) is typed on a sheet of paper. A mechanical art board is then prepared with the required layout. The layout and format are agreed upon by individuals or a group of individuals with knowledge of the regulations, medical aspects of the product, marketing needs, and packaging and printing technology. The printer prepares printing plates and cutting dies to produce labels of the correct size. The printer then prints a few copies, often in black ink. These are called the proofs or stats. After proofreading, the press is set up to produce the correct colour scheme and the material is printed as the final labeling.

Labels are printed in any of three ways. Gang-printing is the term used when many different types of labels are printed on one sheet, which is then cut and sorted into bundles of cut labels. The use of gang-printed labeling for different drug products, or different strengths or net contents of the same drug product, is prohibited unless the labeling from gang-printed sheets is adequately differentiated by size, shape, or colour.

A second, preferable method, is to print only one type of label on each sheet, thereby avoiding errors in sorting. A third printing method, which is the best for drug labels, uses roll labels, where the labels are printed on large roll and not cut into individual labels. Many companies also now use their own computers to generate the labels they need. The cut labels or roll labels are received, sampled, tested, and then released by the Quality Control Laboratory.

Control of Labeling Quality

Sampling a cut label involves fanning the bundle to ensure uniformity in the basic colour and style of the labels. To assist with this, bar codes are usually printed on the label's edge. A few labels are then removed for comparison with the proof and for necessary checks of paper weight, dimensional measurement, and so on.

If acceptable, the labels are signed, dated, and placed in the quality control file for use as references for comparison with future labels, as well as for checking against the final packaged product. The labels are counted by hand, by machine, or by weight and are then released to the packaging department, where they are held under lock and key to prevent mix-ups. Roll labels are similarly evaluated, although they are generally counted electronically. For computer-generated labels, the computer may be programmed to produce the exact number of labels required.

The control of labeling quality varies from extremely detailed examinations of labels, such as would be the case with patient package inserts, to less stringent examinations of packaging materials, such as cardboard for shipping cases, where the printing is limited to such information as name and identification number.

PACKAGING MATERIALS

A packaging material is anything used in packaging a product, including the immediate container and closure; desiccants, cotton, or rayon coil inside the container and cartons, cases, and shipping containers.

Types of Packaging Materials

Packaging materials are generally categorised by their compositions and by the testing methods used to ensure their quality. The following is a list of the basic classifications:

Bottles (glass or plastic), caps (metal or plastic), paper stock, carboard stock, plastic parts, plastic film, metal foil, metal or fiber drums, metal cans, tubes, vials, jars, aerosol cans, and tote bins. (In the cosmetic industry, a greater variety of packaging components is used, including brushes, wipers, plugs, pencils and compacts.)

Methods of Manufacture

Packaging materials are produced in large quantities by highly sophisticated industries. Glass bottles, for instance, are produced in vast quantities by glass manufacturers. The manufacturers begin with sand to produce the glass and then mould and blow a seemingly endless variety of shapes and sizes of bottles to exceedingly precise tolerances.

The manufacturers of plastic containers normally buy the various types of polymers, such as polyethylene, polystyrene, and polypropylene, and mix them with the necessary pigments. They are extruded and pelletised and then moulded into various shapes. Metal cans and drums, by contrast, are produced from sheet steel or aluminium (with or without soldered side seams), and special equipment may be needed for sealing the lid. Paper and cardboard are produced from various grades of pulp.

In essence, each packaging material has its own distinctive source and manufacturing method. The variability possible from lot to lot or manufacturer to manufacturer must therefore be considered in any quality control program.

Controls of Packaging Quality

Packaging materials are produced in large numbers of units, so it is possible to apply known statistical methods in their evaluation. They are accordingly sampled, inspected, and tested for the number of variables by quantitative measurements and for different general defects or attributes. The Quality Control Laboratory must consider the final configuration of the package and assess the quality of the packaging components when they are combined in that configuration.

In other words, a bottle and a cap may each pass individual specifications, but the filled, capped bottle could still leak. The types of defects possible will depend on which category of packaging material they are in and how much information is available from the vendors as an aid to setting the acceptable quality levels (AQLs). All packaging materials are ultimately tested against the master control documents and then released to the 'production department' or rejected as unsuitable. Documentation of the packaging process is discussed below.

DOCUMENTATION

Quality assurance and quality control require the use of many different types of documents, including general procedures, master documents, and batch records, but it still contains basically all the documents needed to manufacture and control a chemical or pharmaceutical product. The documents are listed, in alphabetical order, in Table 14.1.

Table 14.1. Documents needed for manufacture and control of chemical or pharmaceutical product.

No. Document title	*Brief explanation*
1. Alternative Names	List of the various names used with the product (US approved name; generic name; trade name; code number, etc.)
2. Batch Manufacturing Record	Complete record of the manufacturing process
3. Batch Packaging Record	Complete record of the packaging process
4. Bill of Materials	List of all materials used in the finished packaged product
5. Certificate of Analysis	List of the tests done, specifications, and results for the batch
6. Certificate of Disposal	Certification that a rejected product has been properly disposed of
7. Dispostion	Authorised document identifying the batch as accepted, rejected, quarantined, for re-work, etc.
8. Flow Diagram	Schematic showing in simplified form how the product is made
9. General Instructions/Hazards	Special instructions (e.g. 'wear gloves,' flammable material') for the safety of the operator
10. Master Manufacturing Instructions	Preapproved recipe for manufacturing the product
11. Master Packaging Instructions	Preapproved method for packaging the product
12. Quarantine Report	Report of all materials held in quarantine
13. Resample Request	Request to resample from the batch
14. Receiving Report	Record that a material has been received in the company or in the department
15. Sampling Frequency	Preapproved document specifying the frequency of sampling
16. Sample Procedure	Request to resample from the batch
17. Sample Taken	Record that the sample was taken
18. Standard Operating Procedures (SOPs)	Compilation of procedures that cover such things as equipment, facilities, personnel, etc., and the way they are used to manufacture a product according to the approved documentation
19. Specifications: Labels	Preapproved specifications for the labels and labeling
20. Specifications: Packaging	Preapproved specifications for the packaging materials
21. Specifications: Raw Materials	Preapproved specifications for the raw materials

(Contd...)

No. Document title	*Brief explanation*
22. Specifications: In-Process Items	Preapproved specifications for the in-process items
23. Specifications: Finished Products	Preapproved specifications for the finished products
24. Stickers	Stickers physically attached to materials to display their status ('Accepted,' 'Rejected.' 'Quarantine,' etc.)
25. Stability	Documentation confirming the stability of the product
26. Test Frequency	Preapproved document indicating the frequency of testing (e.g., every batch; every fifth batch)
27. Test Methods	Analytical test methods
28. Validation: Cleaning	Validation of the cleaning procedures
29. Validation: Methods	Validation of the analytical methods
30. Validation Process	Validation of the manufacturing process

15

Safety and Ethics in Biotechnology

INTRODUCTION

Society has, for centuries, safely used the products and processes of biotechnology. These processes have employed micro-organisms of known pathogenic potential and, with the exception of vaccine production, all micro-organisms used are non-pathogenic to humans and other animals. Biotechnology has made major advances in public health by the control of communicable diseases with vaccines and the improvement in the quality of the environment by the continued improvements of biological waste treatment processes. However, for all biotechnology processes, safety is of paramount importance. Table 15.1 lists the main areas of consideration, safety aspects specific to biotechnology.

Table 15.1. Safety considerations in biotechnology.

Pathogenicity: potential ability of living organisms and viruses (natural and genetically engineered) to infect humans, animals and plants and to cause disease
Toxicity and allergy associated with microbial production
Other medically relevant effects: increasing the environmental pool of antibiotic-resistant micro-organisms
Problems associated with the disposal of spent microbial biomass and the purification of effluents from biotechnological processes
Safety aspects associated with contamination, infection or mutation of process strains
Safety aspects associated with the industrial use of micro-organisms containing *in vitro* recombinant DNA

Problems of Organism Pathogenicity

Many micro-organisms can infect humans, animals and plants and cause disease. Successful establishment of disease results from interactions between the host and the causal organism. Many factors are involved, only a few of which are well understood.

Most micro-organisms used by industry are harmless and many are indeed used directly for the production of human or animal foods. Many such examples include yeasts, filamentous fungi and many bacteria.

Their safety is well documented from long associations lasting up to hundreds of years. Only a small number of potentially dangerous micro-organisms have been used by industry in the manufacture of vaccines or diagnostic reagents, e.g. *Bordetella pertussis* (whooping cough), *Mycobacterium tuberculosis* (tuberculosis) and the virus that causes foot-and-mouth disease. Stringent containment practices have been the norm.

In recent years there have been many scientific advances permitting alterations to the genetic make-up of micro-organisms. Recombinant DNA techniques have been the most successful but have also been the cause of much concern to the public.

However, this natural anxiety has been ameliorated by several compelling lines of evidence.

1. Risk assessment studies have failed to demonstrate that host cells can acquire novel hazardous properties from DNA donor cells (Table 15.2).
2. More rigorous evaluation of existing information concerning basic immunology, pathogenicity and infectious disease processes has led to relaxation of containment specifications previously set down.
3. Considerable experimentation has shown no observable hazard.

However, care must always be adopted when using recombinant DNA molecules.

Table 15.2. Risk assessment.

Elucidate the capacity of the micro-organism to have an adverse effect on humans or the environment
Establish the probability that micro-organisms might escape, either accidentally or inadvertently, from the production process system
Evaluate the safety of the desired products and the methods for handling by-products

A Classification of the degree of potential hazard of micro-organisms has been drawn up by the European Federation of Biotechnology (Table 15.3). Group E contains those micro-organisms that present risks only to the environment, particularly to animals and plants.

Table 15.3. Classification of micro-organisms according to pathogenicity.

Class 1
Micro-organisms that have never been identified as causative agents of disease in human beings and that offer no threat to the environment
Class 2
Micro-organisms that may cause human disease and might therefore offer a hazard to laboratory workers. They are unlikely to spread in the environment. Prophylactics are available and treatment is effective
Class 3
Micro-organisms that offer a severe threat to the health of laboratory workers but a comparatively small risk to the population at large. Prophylactics are available and treatment is effective.
Class 4
Micro-organisms that cause severe illness in human beings and offer a serious hazard to laboratory workers and to people at large. In general effective prophylactics are not available and no effective treatment is known
Group E
This group contains micro-organisms that offer a more severe threat to the environment than to people. They may be responsible for heavy economic losses. National and international lists and regulations concerning these micro-organisms are already in existence in contexts other than biotechnology (e.g., for phytosanitary purposes)

PROBLEMS OF BIOLOGICALLY ACTIVE BIOTECHNOLOGY PRODUCTS

Vaccines and antibiotics are obvious examples of biologically active products, and care must be taken to prevent their indiscriminate dispersal. Contaminants in otherwise safe processes may produce toxic molecules that could become incorporated into final products, leading to food poisoning. Allergenic reactions to produce formulations must also be guarded against. Overuse of antibiotics in agriculture could lead to carry-over into human foods, resulting in possible development of antibiotic resistance in human disease organisms. Many countries now restrict the use of antibiotics in agriculture.

When properly practised, biotechnology is safe and the benefits deriving from biotechnological innovations will surely lead to major improvements in the health and well-being of the world's population. However, biotechnology must always be subjected to sound regulations for its successful application. The potential risks of biotechnology are manageable, and regulations have been constructed for that management.

HAZARDS

Types of Hazardous Emission

Spillage/leakage

There is a potential for release of liquids at many stages of production. Spillages and leaks can cause gross local contamination of equipment surfaces and of buildings. Personnel exposed to such emissions are generally at risk of being contaminated by ingestion or by skin contact. These can be easily avoided by the provision of protective clothing and by observing appropriate instructions for work.

Effluent discharge

Liquid effluents should be treated before discharge. Effluent treatment usually means the use of heat or chemical treatment. The responsibility for preventing untreated discharge lies with the process operator.

Aerosols

Inhalation of aerosols is a common route whereby process organisms, extrinsic antigens or other products gain access to the body. Aerosols are not easily confined and can also cause environmental contamination or illness in surrounding populations. Airborne transmission can be difficult to detect or prevent since the presence of micro-organisms or products in the atmosphere will not be obvious in most instances. For these reasons, detection and assessment of releases into the air must be considered as an important aspect of biosafety monitoring

Practical Applications

Processes such as fermenter sampling, centrifugation, homogenisation, and cell disruption can generate significant aerosols of process organism or product. Whilst detection of release is important, it is also necessary to have a means of quantifying the significance of that release, particularly for purposes of risk assessment.

Biosafety is not simply an issue for the regulatory agencies of a nation; it also must be taken into account by the management of any transgenic agricultural project. One most important focus of attention is the technology transfer agreement under which a developed and a developing nation cooperate in a research programme, typically through an agreement for insertion of a specified developed-world gene into a specified developing-world variety.

These agreements usually contain some patent provisions and arrangements to protect the commercial interests of both parties. They can also contain trade secrecy provisions.

A crucial aspect in biosafety is to allocate the costs and responsibilities of preparing for biosafety review – particularly when review in the developing nation requires conduct of specific experiments (e.g., demonstration of safety with respect to a local group of beneficial insects) that are not required in the case of the developed world. Similarly, the agreement should ensure that information needed for regulatory purposes is available to the developing nation party – including all information submitted to regulatory authorities in the developed world. Liability questions should also be considered carefully.

The impact of biotechnology in agriculture is just beginning to be felt. Techniques that allow greater precision in developing improved crop varieties are being applied to a number of commercially important crop plants. It may well be that the public's primary interaction with products of biotechnology will not be in the medical field, but with food. It is of the utmost importance that a methodology be developed to assure that foods produced using biotechnology have the same high standards of safety and nutrition we expect from our present-day food supply.

New methods of biotechnology, while enhancing specificity of previous breeding techniques, could also encourage the development of food products which were not possible by classical techniques. Any process for evaluation of safety must be able to effectively evaluate both extremes of the technology. The existing food regulatory policies and practices have been effective in assuring and maintaining a safe food supply. The more closely related a 'novel' food is to a traditional counterpart, the lower the level of safety concern. Also regardless of the process used to develop a product, in the final analysis, the product must be safe for consumption. There are three primary sources of information which can help us in evaluating the safety of a product:

1. Knowledge of the genetic background of the host and donor organisms and the technical procedures used for developing the product.
2. Knowledge of the composition of the food, including any potential toxicants and significant nutrients.
3. Relevant toxicological data on the host organism.

BIOETHICS

Considerable political concerns have been generated over issues such as sustainability, conservation, pollution, and biodiversity; these concerns are increasingly encompassing impacts of modern biotechnology. However very little real progress on legislative and practical fronts has been made so far.

Certain moral and ethical implications of biotechnology and genetic manipulation have generated apprehensions, mostly relating to matters of safety, risk, uncertainty, and the commercial options.

The notion of morality – that certain things are right or wrong – should not be confused with the narrower concept of ethics – the standards by which community decides to regulate its behaviour, so as to distinguish what is legitimate in the pursuit of particular aims from that which is not legitimate. More technically, ethics can also refer to moral philosophy.

Statements that biotechnology is intrinsically wrong need to be disentangled from claims that it is extrinsically wrong because of its consequences. Some workers have considered two intrinsic concerns – that biotechnology is unnatural and that it is disrespectful to living things. Allegations about the unnaturalness of biotechnology have little ethical content, because they rest on unclear language and unsound reasoning. A real biotechnologist shows no less respect for the biosphere than does the person who thwarts the unrestricted growth of grass by mowing the grass of a lawn. It is sometimes stated that riskiness *per se* is not a moral or ethical matter. Some activities are inevitably more risky than others, though none can be totally risk-free, and it does not follow that low-risk activities are morally superior to high risk ones.

The intellectual basis of contemporary concern for the environment has been the ideas of environmental philosophers, four of which have a bearing on biotechnology: (i) the notion of intrinsic value; (ii) mankind's position the biosphere; (iii) the idea of sustainability; and (iv) the practical way forward for bioethics on a global scale.

INTRINSIC VALUE IN NATURE

Humans are by no means the only species capable of valuing. Animals, plants, insects, whole species, and ecosystems are all valuable. The environment has a worth that is totally independent of mankind, and deserves to be treated accordingly. The concept of intrinsic value is illustrated by the fact that nature exists for itself regardless of mankind. In a biotechnological context, particularly with regard to genetic engineering, such ideas seem to conflict with any effort to alter life from what we generally assume to be its 'natural' form. Such activity would, after all, violate the objective characteristic of intrinsic value.

HUMANITY'S POSITION IN THE NATURAL WORLD

The oft-used term *anthropocentrism* has been misused to mean 'human chauvinism' towards nature. In fact, it is rational, reasonable, and logical that humanity should consider humankind as the centre or starting point for its own view of the world. We are at present in a state of confusion caught between our traditional well-entrenched view of humans as 'titular lords' of the Universe and a more modern, scientifically-based vision of the relative insignificance of humankind.

It is here that biotechnology represents a ball in both courts. On the one hand it illustrates our potential power over life, while on the other it highlights our position as just a small part of the natural world, a collection of DNA which in purely biological terms has no particularly privileged place in the biosphere.

SUSTAINABILITY AND ENVIRONMENTALISM

The social scientific definition of sustainability is firstly, that overall capital should not decline, or secondly, that natural capital in particular should not decline. This concept, although allowing some use of non-renewable assets, does not necessarily provide for the demands of environmentalist to protect the natural environment. For instance, a scenario can be envisaged in which a dwindling stock of natural assets is increasingly exploited by technology. This could provide constant overall economic value and human welfare, therefore fitting the criteria for sustainability but not for environmentalism. Such problems in reconciling the goal of sustainability with the aims of environmentalists perhaps stem from the former's emphasis on 'justice for future generations'. This concept fails to capture the meaning of environmentalism, which often has more to do with the notion of intrinsic value. Thus, it is necessary to incorporate into the notion of sustainability the idea that the historical phenomenon of nature is 'good'. What we hand down must retain something of its original form, allowing the preservation of what is good in nature, but not ruling out beneficial human modifications. Genetic engineering may provide a means of exploiting natural capital (e.g., DNA) in a sustainable way.

Need for a New Bioethic

Some feel that there has been the failure of bioethics in its present form to deal with the major environmental crisis of overpopulation. Bioethics, over the years, has centred around health care and the rights of the individual, but individualism and autonomy clash with the cooperation that is required globally if we are to avoid further serious environmental crises. This is particularly seen with respect to reproduction and reproductive technologies. For example, several governments recognise population as a serious problem yet maintain a stance on the autonomous right of individuals to decide the number of children. What we perhaps require is a global bioethic which distinguishes between survival and quality of survival. There should be an emphasis on human mental and physical health, to supplement large-scale environmental protection and controlled reproduction.

Genetic engineering has immense potential to dynamically affect the four chief institutional characteristics of capitalism, surveillance, military power, and industrialism, due to its potential to alter the time/space relationship of biology and life, and the nature of social relations.

Animal productivity vs animal welfare

There are some prospects for improving the productivity of farm animals through genetic engineering by using, e.g., the insertion into sheep of a gene to increase the rate of synthesis of cysteine from serine, thereby removing an important constraint on wool production. This way the synthesis of a specific product can be improved significantly, by modification of a single enzyme to become a commercially interesting prospect.

There may be some potential of genetic engineering in reducing disease in animals, either by the development of improved vaccines or diagnostic aids or by engineering genes for resistance to specific diseases into the animals themselves. The impact of genetic engineering on animal welfare should be considered primarily in terms of the consequence of the process for the animal, rather than the procedure itself. By these criteria, genetic engineering for improved health becomes desirable. Genetic engineering to improve productivity should be judged according to criteria defined by the UK Farm Animal Welfare Council which states: 'Welfare problems may arise through the manipulation of body size, shape or reproductive capacity by breeding, nutrition, hormone therapy or gene insertion in such a way as to increase the risk of injury, metabolic disease, skeletal or obstetric problems, perinatal mortality or psychological distress'.

Any abnormalities in animals forced to rapid growth by natural breeding in the broiler fowl, or by insertion of a gene to synthesise exotic growth hormone in pigs, should be carefully analysed to weigh any suffering to the animal, however slight, against the real benefit to society.

HEALTH BENEFITS FOR HUMANS

The genetic engineering of animals is also being done to achieve improved human health. Genetic engineering of farm animals may be done to secrete pharmaceuticals in their milk. Humans lacking the enzyme α- antitrypsin (AAT) suffer progressive lung degeneration (emphysema) unless treated with 200 gram per year with AAT derived from human blood. Tracey, genetically engineered sheep, has secreted 1.5 kg AAT per lactation in the glycosylated form essential for biological activity in humans. Where complex (e.g., glycosylated) proteins are required in large quantities for the treatment of genetic deficiencies, or the production of recombinant antibodies for diagnostic or therapeutic purpose, the large, lactating, genetically engineered sheep or cow is a better option than fermentation biotechnology.

RISK REGULATION

Several new commercial products being currently developed for use in agriculture, forestry, and as potential pollution control agents are breaking new ground in the regulation of environmental risk. These are the micro-organisms, plants, and animals that express characteristics derived from the genome of other, unrelated species and that well function by being released in the living state, into the environment (genetically manipulated organisms or GMOs).

Apprehension about the possible environmental impacts of some of these products has been expressed by ecologists and environmentalists. Key areas of concern include the fitness of an organism to survive and multiply relative to its unmodified counterpart, its ability to transfer its novel genetic material to other organisms, its physiological tolerance and competitive ability, its potential to behave unpredictably in the environment and, if applicable, its pathogenicity, virulence and host range. In contrast many scientists and industrialists claim that genetic engineering is more precise than classical breeding and that its products are more predictable in their behaviour and hence safer.

In the existing regulatory systems that have been developed to control the risks of potentially hazardous products such as pesticides, the requirement for field trials and subsequent monitoring provides a sufficient safeguard for the environment. However, with GMOs even a trial release might potentially lead to uncontrollable multiplication, destruction of wildlife habitats and the widespread dissemination of alien genetic material among natural populations. Because of this potential, biotechnology has become one of the test cases for the development of precautionary approaches to risk regulation.

The results and applications of biotechnology must be suitably regulated. In most developed countries, such regulation is an important constraint on the activities not only of biotechnology enterprises but also of biotech research laboratories. The issue of regulation is probably more important in the area of biotechnology than in most other industrial fields because biological organisms, which constitute the basic elements of biotechnology, interact in complex ways with the ecosystems of which they form a part. Accordingly, adequate care must be taken to ensure that biotech-related activities do not produce negative effects on the environment.

16

Patenting in Biotechnology

INTRODUCTION

Before discussing 'patenting in biotechnology' it will be appropriate to have a brief look at patenting in general. While industries naturally look to legal protection for their research efforts, society has a stake in encouraging industrial innovation. A strategy that meets both these objectives is the grant of exclusive rights by the government to the inventor for the novel products or processes that they develop. Strategy that decides how the objectives of the user and of the inventor will be best met by sanctioned privileges is 'Intellectual Property Rights'. These sanctioned privileges include trade secrets, copyrights, trademarks, registered designs, plant patents, petty patents and patents.

FUNDAMENTAL REQUIREMENTS OF PATENT

The fundamental requirements of any patent are:

1. Novel (for any country).
2. Not obvious (not merely unknown).
3. Useful.
4. Enabling (implementable).

The discussion on granting of patent in biotechnological areas is not a simple check on the satisfaction of fundamental requirements mentioned above, but is a much more complex issue. Patent laws vary from country to country, although there are ongoing attempts to develop international standards. In the US, for example, the duration of the exclusive rights of a patent is 17 years after it is officially awarded; in Canada it is 20 years; and in many European countries it is 20 years from the date of filing of the patent application (it usually takes 2–5 years following the filling of initial patent application for a patent to be granted).

The holding of a patent can be of considerable economic value, and it is not a trivial matter to receive one. In India, the Indian Patents Act of 1970 allows process patents, but no product patents for foods, chemicals, drugs and pharmaceuticals. Consequently, the same product may be manufactured by an alternative process (which may be cost effective) without any infringement of the patent granted. The duration of a patent in India is five years from the date of grant of patent, or seven years from the date of filling of an application whichever is less. In case of dispute, the burden of proving an infringement lies with the one who has lodged the complaint.

ISSUE OF PATENTING IN BIOTECHNOLOGY

The principal objective of biotechnology is to produce commercial products for economic gain. Bio-technology involves uses of, or organic changes in, animals, plants, micro-organisms and any biological material that can be assimilated by living matter. It involves any technique that uses living organisms (or their parts) to make or modify products, to improve plants and animals or to develop micro-organisms for specific uses. The unforeseen potential uses of biotechnology also contribute to difficulties in resolving issues surrounding patenting of higher life forms.

Patenting in biotechnology has emerged as an issue only in the last two decades. Debates over whether or not a patent should be issued for a higher life form has more facets than only the otherwise appropriateness. These debates focus instead on ethical and economic considerations. They include issues such as biodiversity, human and animal health considerations, environmental safety, animal welfare concerns and economic benefits and risks associated with patented animals.

Important Issues in Patenting

Some of the important issues are:

1. Should genes, DNA and proteins, which are part of living things occurring in nature, be patented?
2. Should DNA sequences (function, position and usefulness not known) discovered as a part of human genome project, be patented?
3. Should higher life forms be patented?
4. Should donor of tissue be rewarded if patent is obtained on such material because of its commercial potential?

5. Should country of origin have right on share in financial benefits if genetic materials from that country is patented and used for commercial gains?
6. Should patents based on products of prior knowledge of other countries be permitted?
7. Should patents be allowed on agricultural sources like seeds, which can cause exploitation of poor farmer and allows monopoly practices?

The challenge remains as to how to reward innovation without compromising fundamental ethical, social and scientific beliefs deeply embedded in our society.

Status Today

Signatories to the GATT and Trade Related Aspects of Intellectual Property Rights (TRIPS) agreement are obliged under the agreement to adopt a patent system for micro-organisms and to establish either patents or some *sui generis* form of IP protection for plants. However, the patenting of higher animal life forms was left unresolved, with signatories having the option to use or not use patents to protect such intellectual property rights. The US, EU, Japan, Australia and a few other countries have legalised the patenting of animals even though none of them has as yet patented a farm animal. Canada is among the countries that do not allow patenting of animals, including farm animals, under its current intellectual property protection laws. There is currently an interest in developing policies that encourage the patenting of animals. The issue, however, is how strengthening IP protection for farm animals will affect the agri-food sector. The economic implications of using patents to protect innovations in animals on various stakeholders in the agri-food sector are being explored.

Patenting of living things

The idea of patenting living forms is not new. In 1877, the German Federation considered plant patents within industrial property legislation and even in 1830s Pope Gregor IV attempted to secure property rights over all plants on all Vatican property in the various Italian states. In spite of the resistance to such efforts of patenting of life forms, the decisions are affected by political and economic factors. A subject matter of an invention shall not be considered unpatentable for the reason only that it is composed of living matter. Shall all life be reduced to the status of a commodity to be sold for profit? Laws of patenting, intellectual property and copyright are being changed to accommodate the ownership of genes, cells, plants, animals, human body cells and human genes.

Changes in patent system in the last decade to accommodate 'biotechnological inventions' have been strongly contested nationally and internationally for development, environmental, spiritual, animal welfare, religious, humanitarian and feminist reasons. Patent legislation was not designed for living organisms. Courts using laws written before the invention of genetic engineering techniques are setting the limits. The results are inconsistent and the implications of patents on living things are not known. The patent profession in Canada and the US is playing a role in experimenting with what is patentable, lobbying the government and bringing test cases to court. However, patenting is still the best incentive or encouraging the pursuit of expensive and risky research, which yields the products society needs.

Whether the patient shares in the rewards for human tissues and cells taken from him is another new issue. In this regard, in a case in the US, rights of a cancer (leukaemia) patient were ruled out, because there were no property rights in discarded human body parts. The commercial potential may be worth billions, but the use of body materials is regarded by some as a new form of slavery and a violation of human dignity. Perhaps the real issue here is about informed consent – whether a patient should be informed about potential commercial interest in his tissues. Many might agree willingly to their body parts being used for research, but not for commercial purposes. Some groups have called for a suspension on animal patenting to provide time for thoughtful reflection and judgement on these issues. In Canada, organisations such as the Canadian Institute for Environmental Law and Policy begun pressing for a top-level government review.

Patenting plants

New plant varieties may be obtained from a number of sources. They may be:

1. Discovered from the field.
2. Hybrid plants through selective breeding.
3. Obtained through cell culture methods like somaclonal variation, gametoclonal variation, and somatogenic embryogenesis, micropropagation.
4. Obtained using recombinant DNA methods to create transgenics with new useful products or characteristics.

Plant varieties obtained by any of the above methods have agricultural and horticultural importance. Transgenic plants can be used to make products of commercial or medicinal value, to get higher yields, to have increased resistance to pesticides etc.

Sexually reproducing plants obtained by human intervention would be considered patentable subject matter in US. The plants themselves, seeds and plant parts, products obtained from new varieties, formulations developed of such products and the processes used to obtain such commercially important plants, are all subjects of patentable matter. IP protection that can be obtained for novel varieties of plants are: plant patents, PVPA certificates and utility patents. The greatest scope of protection is with utility patents, which may be obtained for plants derived by cell culture, plants generated by selective breeding and transgenic plants, but not for plants discovered in the field. The scope of patent protection of novel plant varieties in Europe is different than that in US, and claims to transgenic plants are not patentable but plant cells are. Also, 'essentially biological' (which means natural phenomena such as crossing or selection) processes for obtaining novel plant varieties are to considered patentable. A process for producing a transgenic plant and a process for developing hybrids was also not considered to be 'essentially biological' by the European. Patent Office.

In agriculture patenting genetic innovations gives agribusiness conglomerates more control over independent farmers who buy the products they sell. For instance, Patents may prevent farmers from sowing next year's crops with the seeds collected from this year's harvest. Plant utility patents mean farmers would need to buy new seeds every year. This has become a controversial issue at the WTO negotiations and especially at the NAFTA meetings, as this may hit family-owned farms in poorer countries like Mexico especially hard.

Common Categories of Patents

Product patents

Product patents are the most effective form of protection because proof, that a competitor is infringing the claim is usually apparent from inspection of an analysis of the competitor's product and there is no need to inquire about method of manufacture. This is the only category of patent that has recognition throughout the US, Japan and Western Europe. A new antibiotic produced by any method or new organism degrading certain recalcitrant chemical can be easily patented.

Many of the products produced by recombinant DNA technology are not new—rather they are natural, but now produced in quantity by inexpensive procedure. Still they can be patented (e.g., pure form of vitamin B12 has been allowed for patenting). In recent years, most developed countries have adopted the product claim as an appropriate form of protection for new substances. Before this, in most countries of the world it was necessary, even in case of a new product, to claim the invention in process terms. The unsatisfactory situation, which still exists in Latin America and Eastern Europe, requires alternative methods of preparation to be devised and the use of multiple process claims to cover practically every possible pathway to the product.

Substance: Cloned genes, recombinant proteins, monoclonal antibodies, plasmids, promoters, vectors, cDNA sequences, and monovalent vaccines.

Composition of matter: Multivalent vaccines, biofertilisers, bioinsecticides, pharmaceutical mixtures, micro-organisms and transgenic animals.

Devices: Pulsed field gel electrophoresis apparatus, DNA sequencing apparatus, microprojectile gene transfer machine.

Process patents

The genuine process invention is one where a truly novel route or method has been devised or where new conditions have been found for operating a process, of generally known type, in order to achieve an improved result. The production of indigo by *E.coli* strain carrying a cloned naphthalene dioxygenase gene is one example of a discovery being non-obvious. Another example is of improved substrate utilisation by strains of *Methylophilus methylotropus* carrying a cloned *gdh* gene and which are grown for SCP production. Process patents are weak patents, in that it is difficult to know when a competitor has infringed in your patent. For this reason many companies choose not to patent new processes, but to retain them as trade secrets. The danger here is that a competitor may discover the same process and patent it and then sue you for infringement.

Processes of preparation: Examples are DNA isolation, synthesising double-stranded DNA, vector-insert construction, PCR applications, and purification of recombinant protein.

Method of working: Nucleic acid hybridisation assays, diagnostic procedures, detection systems using PCR, and mutant assays.

Use: Applying biofertiliser or bioinsecticide, fermentation of genetically modified organisms, non-therapeutic animal treatment systems.

The patenting of biotechnological innovations is based on historical experience of patenting inventions by the agricultural, fermentation, pharmaceutical, and medical industries. Today most biotechnology patent applications are straightforward but initially scientists faced many problems to get patent for genetically modified micro-organisms. Dr. Anand Chakraborty could not get a patent from the US Patent Office for a genetically engineered bacterium, which could be used for cleaning of oil-spills on the ground that micro-organisms are products of nature. In 1980, the US Supreme Court in its landmark decision decided that this organism was patentable according to the US patent statute, arguing that 'a live, human-made micro-organism is patentable subject matter–as a manufacture or composition of matter.' In past induced mutations and subsequent selection for novel properties was an acceptable way to create a patentable living organism. Now regardless of method of development, organisms are judged by the standard criteria of novelty, non-obviousness, and utility to determine if they are patentable.

Patenting in different countries can be a different experience, as is borne out from the following example:

1. Human tissue plasminogen activator (t-PA) produced in *E.coli* by Genentech could not get a patent in the UK because it was considered to be novel but the submission was considered obvious.
2. But in US the same product of the same company could get a patent. Genentech even won two lawsuits against two biotechnology companies who were found to be infringing on its tPA patent, although they were selling nonidentical forms of tPA. The US patent not only protects the form of human tPA that was marketed by Genentech, but also gives them exclusive rights to all similar, but not identical, active forms of human tPA.
3. The Japanese version of Genentech's tPA patent is limited to amino acid sequence of the human tPA that was cloned and patented by Genentech, and other companies can sell variant forms of human tPA in Japan.

Thus basically the same patent application was rejected; approved and given broad interpretation; and approved and given a narrow interpretation by three patent offices.

PATENTING OF GENES AND DNA SEQUENCES

Genes, DNA and proteins are part of living things occurring in Nature. However, genes and DNA sequences by themselves are not alive, they can be considered simply carriers of information, in chemical form. As such they fall into the category of composition of matter, which is patentable under the law. The genes that are synthesised artificially normally fulfil the requirements of patents and therefore have no difficulty in getting patent protection. If the protein, that the artificial gene makes and the organism into which the gene is inserted are also novel having desirable properties, such as patent may extend to protein and the organism also. Patenting of naturally occurring useful genes, is debatable and political issue, because they do not fill the requirement of novelty (they are only discovered and not invented). The developing countries are arguing that patenting of naturally occurring genes will allow the developed countries to make use of weed from their land, isolate a desirable gene from this weed, transfer it to commercial crop and then sell it back to the same country (which was the source of the gene). There are counter arguments, that companies should be rewarded for their investment and labour involved in making the natural genes useful, which may otherwise be never exploited for human welfare. The low on this issue is confusing and changing with judgements in the court. However in the US a patent is given to purified form of the chemical although unpurified form of chemical may occur in nature. In UK naturally occurring gene sequences cannot be patented. Patent is however granted to gene sequences, which were obtained as mutation of natural one. Whether a patent claim for a DNA sequence can protect its expression product (proteins) remains to be seen. Scientists involved in the Human Genome Project are bitterly divided over whether the DNA sequences they are discovering should be patented or not. In this regard, complete genes, have been patented in the past and it not controversial because these sequences of DNA have clear uses. But, patenting gene fragments whose function and position in the human genome is not known, is criticised claiming they fail the 'utility' criterion for patentability. These patent applications have not been allowed anywhere in the world.

'Usefulness' is interpreted as practical utility. This may discourage research in this area. However a patent has been awarded to isolated gene, vectors and transformed cells expressing the hormone – 'angiogenesis factor' (AGF) at a time when its use was not known. Many other patents of isolated genes are similar in nature and their validity is questionable. If not patented such isolated genes will remain secret while if patented they will be available to other researchers. Future stand is to be observed. The criteria for granting protection (not necessarily a patent) on plants are that the new variety must be distinct, homogenous and stable.

Criteria such as these prove difficult to assess when applied to patenting life forms. Genes, cells, animals and plants are not inventions: they already exist and are only being modified. They cannot be considered to be non-obvious. Also living objects are not stable. It was not until 1987 that the US Patent office announced that life forms other than micro-organisms were patentable. In 1988 it granted the first patent for genetically engineered animal. This was a transgenic mouse, which contains a gene that makes it extremely sensitive to cancer-producing agents – Oncomouse is sold commercially to test substances suspected to be carcinogenic. In contrast with the decision of the US Patent Office, the European Patent Office (EPO) has rejected the application for a patent for the transgenic mouse. Both the US Patent Office and the EPO have announced that the patenting of humans is prohibited.

Uncertainty of Patent Law

Patent legislation is an inexact science and infringement suits in biotechnology will often be reduced to debates between opposing experts. The situations become further confusing due to variations in patent laws of various countries. A better example of uncertainty concerns the doctrine of equivalence. Thus an analogous process in *Pseudomonas* would not literally infringe a claim of indigo production in *E.coli.* However an appeal against infringement may be upheld under the doctrine of equivalence. The courts generally allow a broader scope and wider range of equivalence to patents that represent a new field or a distinct step in the progress of an art. Another example of uncertainty of patent law is the narrow interpretation of the prohibition on patenting of plant varieties taken by EPO. First, the prohibition does not apply to plant modified other than genetically. Second, inventions relating to plants of one or more species are potentially patentable. Finally, the prohibition does not apply to plants produced by microbiological process. The EPO's interpretation of this term includes virtually all modern genetic engineering methods for creating plants. One patent granted recently concerns a technique to introduce a gene that produces phaseolin, a storage protein, into alfalfa, thereby increasing the crop's protein content and nutrient value. The patent covers the technique and any plants produced by the technique. Since it is not limited to plants of a particular variety, the EPO took the view that prohibition under article 53(b) did not apply.

Advantages of Patenting

Advantages of patenting are same for biotechnology as they are for any other products and processes.

1. Patents are granted in exchange for the disclosure of an invention.

2. Society benefits from the use of inventions in terms of economic growth.
3. Patents reward the inventor and thereby stimulate research and foster technical and scientific progress (research can be and is stimulated in other ways as well).
4. Patent law takes care of disagreement regarding ownership and hence right to control the item, even to destroy it if inventor feels so.
5. Patenting is a blessing in one sense because it promises monopoly to get returns on investments done on research and secures financial interests.
6. One major advantage of patenting is biotechnology companies consider it as an incentive due to proprietary rights possible.

Arguments Against Patenting in Biotechnological Inventions

Arguments against patenting of life forms and genes are many.

1. Patenting is morally repugnant. It reflects exploitative relationship with other living things.
2. Economically it hurts the most vulnerable, e.g., agriculture farmers.
3. Patenting encourages creating of more genetically engineered life forms and their sale as profitable products.
4. Patenting of life forms degrades our status to that equivalent to some commodity.
5. Patents in biotechnology, does not even succeed in increasing creativity and innovations, but on the contrary it limits attention to one kind of the product.

Biotechnological Patents in the US

The US, Patent low was amended on November 1, 1995, by adding new sub-section 35 USC 103(b) to provide that a biological process using or resulting in a composition of matter that is novel and non-obvious will itself be considered non-obvious composition or in a separate application having the same effective filing date as the claims to the new and non-obvious composition made or used in the process. In addition it is required that the composition and the process at the time they were invented were owned by the same person or subject to an obligation of assignment to the same entity.

A biotechnological process is defined in the law as:

1. A process of genetically altering or otherwise inducing a single-or multi-celled organism to:

(a) Express an exogenous nucleotide sequence.
(b) Inhibit, eliminate, augment, or alter expression of an endogenous nucleotide sequence.
(c) Express a specific physiological characteristic not naturally associated with said organism.

2. Cell fusion procedures yielding a cell line that expresses a specific protein, such as a monoclonal antibody.
3. A method of using a product produced by a process defined by (i) or (ii) or a combination of (i) and (ii).

This law was enacted in response to the controversial decision of the Court of Appeals for the Federal Circuit which stated that, although an old process may become a new process when a previously unknown starting material is used and the product produced is new, this does not necessarily mean that the whole process has become non-obvious. Although the claim at issue was for a chemical process, the patenting of biotechnological processes was also affected by the decision. For example, although a host cell for producing a genetically engineered form of protein could be patented, the use of the host cell to produce the protein was not patentable unless there was something inventive about the method used. In addition, the import of finished pharmaceutical products, which were produced abroad using a host cell that was patented in the US, was legal because there had been no use of the patented material in the US, and the Process Patent Amendments Act did not apply because one could not secure a US Patent for the process used. The problem was worsened by the fact that in several countries there has been a problem in securing patent protection for the host cells used since these were regarded as being unpatentable 'living matter'. The Patent Office has now issued a Notice of Treatment of Product and Process Claims that directs examiners in examining claims for a process for making or using a novel and non-obvious composition to consider the claim as a whole and view the making or using of a novel and non-obvious composition as a material limitation.

Biotech Patents Spawn Biopiracy of Genetic Resource of Third World

One of the allegations repeatedly made is that the bio-poor western countries are looting genetic resources of third world poor countries and patenting the products or using prior knowledge, which existed as a 'prior art' in the respective countries. The genetic resources have been maintained and taken care of by the farmers and cultivators of third world poor countries for centuries and their use existed as 'prior art'. Gene banks and botanical gardens have become the routes of biopiracy.

The recent explosive growth in biotech patents in drugs, insecticides, and agro-products in advanced countries is based on the genetic resources of third world. Western countries have patented over 100 plants of Indian origin. Few examples of biopiracy, out of several are cited below:

1. Dr. Aklilu Lemma, an Ethiopean scientist had discovered in 1989 'endod', an African soapberry plant that had remarkable capacity to thwart Schistosomiasis. Although the scientist did not get support for research on this project after some time of his visit to USA a patent was filed on use of endod for control of Zebra muscles, which cling to hulls of ships.
2. An American scientist observed coloured cottonseeds obtained from the Andes, South America. These cottonseeds of green and blue hues produced cotton of short staple but the enterprising Dr. Fox succeeded in breeding the coloured cotton with long staple. This coloured cotton proved to be the goldmine for the world's biggest blue jeans manufacturers who will mass produce coloured jeans with high price tag bearing the Andes logo. The patents and profits from coloured jeans will not go to poor masses of the Andes, who were the pioneers in growing coloured cotton.
3. The same biopiracy of Asian bioresource is reflected in Margosan-O, the patented insecticide from Asia's neem tree extract. The Margosan-O patent is the exclusive property of W.R. Grace Chemical Company (US), although farmers of South Asia have known the insecticidal properties of neem extracts, for generations.
4. On the food front, Lucky Biotech of USA has filed a patent on the modified thaumatin, a powerful sweetener obtained originally from the serendipity plant of West Africa. Thaumatin is 4,000 times sweeter than cane sugar with alluring sweetness. If this patent is granted then it will prohibit some uses of above said plant world-wide and even in countries of its origin.

The WTO dispute is an attempt to put pressure on India to adopt US-style patent laws. However, as the turmeric patent case makes it evident, the US patent system has its own weaknesses, which allow biopiracy to be practised as a rule. The withdrawal of the turmeric patent is only a first step in reversing biopiracy.

The US needs to revoke all patents based on other country's indigenous knowledge and 'prior art'. In addition, the US also needs to change its patent laws, which sanction biopiracy by its non-recognition of foreign 'prior art'.

Biotechnology Patents and India

It emerged very clearly from the survey conducted by US drug multinational Pfizer and Institute of Intellectual Property Development (backed by the Federation of Indian Chamber of Commerce and Industry) for opinion on product patents, that we lack patent literacy. There is a need to learn the tricks of the trade. There is also a need for more technical input and greater awareness of the world market to give our patents a competitive edge. This is holding India from gaining huge benefits of exclusive patents in the global market.

The 1991–92 draft for GATT prepared in by its then Director General, Arthur Dunkel, made the following provisions regarding biotechnology: (i) India will have to alter its patent laws by the year 2007, firstly to introduce product patent and for increase in duration of the patent; (ii) open up its markets to foreign patent holders, and (iii) Open the agriculture sector to patented biotechnology.

Implementation of the Convention on Biodiversity (CBD) and its harmonisation with TRIPs is the only long-term solution to the problem of protecting India's biological wealth. India is thinking about TRIPs but has given less importance to CBD. In fact CBD would provide adequate protection to its plant species. The Union Government is expected to shortly enact the Biodiversity Conservation Act with tough regulatory measures against the commercial exploitation of Indian biodiversity. The Cabinet has also given approval to the Plant Varieties Act, which sets down the varieties, which can be sold or used for research purposes. India's stand as for as patents in biotechnology was made clear in Parliament by stating that India is not going to agree for micro-organisms. It is clear that no public interests will be served by patenting naturally occurring genes or any life forms. Also serious ethical, moral and religious questions are related to patenting of micro-organisms, therefore animal rights groups, environmentalists and religious groups like churches have also spoken against patenting of life forms or naturally occurring genes.

At a national seminar on 'Patenting in Biotechnology' organised by the Department of Biotechnology (DBT), in collaboration with The World Intellectual Property Organisation (WIPO) it was opined that the scope of amendment of the Patents Act of 1970 should be extended to cover biodiversity and the question of compensating the country or region supplying the plant material from which a patentable product is derived.

The country is debating on the crucial issues of making distinction between naturally occurring forms and genetically engineered organisms and on the ethico-legal issues. It is paradoxical to refer to biodiversity as a global heritage and to proclaim simultaneously that biotechnology is the domain of privately-held property.

UNIQUE ASPECTS OF BIOTECHNOLOGY PATENT LAW AND PROCEDURES

Fulfilling the Utility Requirement

Indeed, under the new Utility Examination Guidelines, a patent applicant provides a sworn affidavit or declaration by a competent expert stating the disclosed human therapeutic utility of the invention would be considered credible by a person of ordinary skill in the art (e.g., in view of animal data or even *in vitro* data that are reasonably predictive of human efficacy), that ends the inquiry. The invention will be deemed to meet the utility requirement. Despite this progress, issues about utility continue to arise for biotechnology related inventions, e.g., new DNA sequences, resulting from the Human Genome Project, that are not yet known to encode a useful protein or to have a useful control function. (Several expression sequence tags have been patented by commercial ventures.) Utility questions may also arise about compounds that pass early combinatorial screening assays but are otherwise of unknown therapeutic utility. The patent applicant is advised not to delay filing patent applications until all the therapeutic data are in, but to disclose additional known utilities besides potential therapeutic ones (e.g., use as standards, use in tissue typing, use as probes, use as PCR primers, or some other diagnostic use) in the patent application.

Fulfilling the Novelty Requirement

Biotechnology-related inventions are judged by the same standards as other inventions, including the requirement of novelty. Note, however, that US patent law permits the patenting of naturally occurring micro-organisms and other naturally occurring compounds such as proteins, cofactors, nucleic acids, vitamins, and antibiotics. In one sense, these things are not 'new' because they exist in nature. However, their natural surroundings are such that they are unknown and waiting to be discovered through, e.g., screening and isolation or purification techniques. When a natural product is made to exist in a form other than the one in which it exists in nature, it is considered novel for patentability purposes.

Fulfilling the Nonobviousness Requirement

Many worthy biotechnology-related inventions have been barraged with obviousness rejections by patent examiners. This is due in large part to the standardisation of cloning methodologies and the routine nature of screening cDNA or genomic libraries. Also, the ability to determine the protein sequence from the gene sequence makes certain aspects of biotechnology predictable. For example, examiners were rejecting claims to DNA sequences encoding particular proteins for which partial or complete amino acids sequences of the encoded proteins were known. Examiners reasoned that certain techniques for making hybridisation probes and screening cDNA libraries or for making antibodies to proteins and screening, e.g., expression libraries, had become sufficiently routine that the gene encoding the protein became obvious.

Rejections of claims to nucleic acids (e.g., genes) identified by their specific nucleotide sequences based on the obviousness of the methodologies used to clone genes have largely been vanquished by recent decisions by the federal Circuit.

For many years, biotechnology-related inventions were also plagued by what were known as *'Durden-type'* obviousness rejections. Based on case called *In re Durden,* certain processes (e.g., expressing as gene and making the encoded protein) were considered obvious even though the reactants and resulting products were novel and nonobvious. Relief from such rejections has come in the form of a new statute, which provides that, if certain PTO procedural requirements are met, a biotechnological process using or resulting in a composition of matter that is novel and non-obvious shall be considered non-obvious. 'Biotechnological process' is specifically defined as meaning:

1. A process of genetically altering or otherwise inducing a single or multicelled organism to
   - (a) Express an exogenous nucleotide sequence.
   - (b) Inhibit, eliminate, augment, or alter expression of an endogenous nucleotide sequence.
   - (c) Express a specific physiological characteristic not naturally associated with said organism.
2. Cell fusion procedures yielding a cell line that expresses a specific protein, such as a monoclonal antibody.
3. A method of using a product produced by a process defined by subparagraph A or B, or a combination of subparagraphs A and B.

Fulfilling the Sufficiency of Disclosure Requirements

It is difficult to provide specific recommendations as to how to ensure that your biotechnology related invention is adequately described, because descriptions vary on a case by case basis. Even though a written description is called for, sometimes words are inadequate to convey completely the intended scope of the invention. Always consider using figures and photographs. For example, suppose you are claiming a plant that is identifiable by some special physical trait, e.g., flower, leaf or seed shape, or colour. Including a photograph can often easily solve the inherent ambiguities and vagaries of the English language. Other pictorial examples include (i) photomicrographs or electron microscope images revealing particular characteristics of the thing claimed; (ii) photographs of protein or nucleic acid gel analyses (e.g., sodium dodecyl sulphate polyacrylamide gel electrophoresis or Southern blot results) to demonstrate size distributions; or (iii) photographs of treated and untreated conditions to show the effect of a claimed cosmetic or therapeutic compound.

For many biotechnology related inventions, nucleotide sequences and amino acid sequences are very frequently presented in a patent application. The presentation of a complete sequence does not necessarily provide adequate written description for gene or protein fragments that fulfill the same function as the complete gene or protein. Therefore, if you wish to claim less than the entire gene or protein sequence, part of your written description should provide some explicit guidance as to what the specific fragments are. Do not rely on your entire sequence to adequately describe these smaller nucleic acid sequences or polypeptides. In the same fashion, make sure any mutations, deletions, or insertions are adequately described. If you wish to claim a genus of nucleic acids, do not rely solely on functional language (e.g., the type of protein encoded). Be sure to describe the essential chemical structural features, or, better yet, give examples of as many complete sequences within the genus as possible.

A truly unique aspect of certain biotechnology related inventions is the necessity of making a 'micro-organism deposit' to fulfill the enablement requirement. Recognising that words may not be adequate to enable someone else to make and use a new life form.

It is recommended that biological materials be deposited before the patent application is filed, although this is not always required under U.S. law. The patent rules in many foreign countries, however, strictly require that the deposit of biological material be made before filing the application to which the foreign patent application claims priority and that, within a specified time period, the application be updated to contain the accession number. For example, if you wish to file for a patent in the European Patent Office (which examines for many European countries), claiming priority of your US, application, the deposit of biological material under the Budapest

Treaty must be made before filing the U.S case. It is preferable to include the accession number in the specification and claims as of the time of filing, although the accession number can be added to the application for a limited time after filing. Also, it is a good idea to make the deposit sufficiently far ahead of filing the application so that the viability of the culture can be determined by the depository. Certain foreign countries require that the deposit be viable as of the filing date of the application. If the culture is not viable, and the application was already filed, the priority date may be lost. However, if the application has not yet been filed, there will be time to redeposit a viable culture with the depository.

New procedures recently went into effect in the European Patent Office for depositing biological materials in connection with patent filings. A competent European patent practitioner should be consulted when applying for European patent protection on a biotechnology-related invention. European patent applications do not have to be filed in the name of the inventor (as in the United States) and are usually filed in the name of the company that owns the application. To avoid various procedural requirements, it is recommended that the applicant for the European patent and the depositor of the biological material be one and the same. Thus, if the European patent claims priority to a US, application should be the depositor. Also, you should be aware that in the European Patent Office, deposited biological materials become publicly accessible to authorised individuals when the patent application is published. (Most foreign jurisdictions publish patent applications within a certain number of months of the claimed priority date.) There are procedures that the applicant can adopt of limiting the accessibility of the cultures to specified experts. Obviously, the goal is to provide public disclosure of the complete invention, yet prevent misappropriations.

Since deposits of biological material are usually necessary to fulfill the enablement requirement for biotechnology related inventions, the question often arises. Must I deposit the biological material that I consider to be my best mode? The answer is, not necessarily. Enablement and best mode are distinct requirements. However, an inventor must not only disclose the best mode, but must disclose it in a way that skilled artisans can understand it and practice it. That is, the best mode has to be enabled. Thus, it may not be necessary to deposit your best genetic construct if the materials you made it from are publicly available or if you deposit the necessary starting biological materials and you provide enabling guidance or instruction in your specification for constructing what you consider to be the biological material that is or best performs the claimed invention. If you cannot adequately describe how to make the best mode, then you have not enabled it and you should deposit the biological material you consider to be your best mode.

Thus from the discussions of general aspects of US, patent law and unique aspects of patent protection for biotechnology related inventions is by no means exhaustive. Furthermore, because of global trade concerns and other influences (e.g., health care concerns about the patenting of medical procedures), US, patent laws are probably in a greater state of flux than at any other point in history. Consequently, you should work closely with your patent attorney to plan your company's or university's patent strategy with respect to obtaining patent protection on biotechnology-related inventions as well as licensing and litigating your issued patents. Your company's or university's financial well being could depend on it.

17

Environmental Biotechnolgy

INTRODUCTION

'Environmental Biotechnology' involves specific applications of biotechnology to the management and related socio-economic and development issues, keeping in view the concept of sustainable development. 'Environmental biotechnology' encompasses issues like: (i) Environmental monitoring; (ii) restoration of environmental quality; (iii) resource/residue/waste-recovery/utilisation/treatment through application of r-DNA technology; (iv) substitution of non-renewable resources by renewable ones; (v) strain improvement for degradation of highly-toxic pollutants with the production of chemicals; (vi) global changes; (vii) biological diversity; and (viii) risk management.

Industrial pollution management is thus one amongst the many issues that environmental biotechnology addresses to. It is important to note here that biotechnological applications to environment protection relate to all the five options such as – (i) In-process treatment; (ii) end-of-pipe treatment; (iii) remediation of polluted sites; (iv) modification of existing processes; and (v) introduction of new processes and products. A cleaner technology is a means of providing a human benefit which, overall, uses fewer resources and causes less environmental damage than alternative means with which it is economically competitive. Pollution prevention and waste minimisation are important parts of cleaner technology, but more fundamental is the shift from selling products to providing services. Cleaner technology is a way of thinking, rather than a set of 'hard technologies'. One productive approach is through dematerialisation.

Pollution prevention is major economical and environmental issue in the chemical processing industries. Since waste regulations became more stringent, new technologies for treating these streams became necessary. To this end, technical solutions for recovering some valuable raw materials form these streams were developed or applied to terminal streams for the first time. Once these materials were recovered, they could be reused in the process.

These recycle/reuse techniques kept the industries in good standing for many years, but the outcomes they developed were still based on an end-of-pipe approach. In order to eliminate pipe treatment, the push is to look within the core of the plant itself, and attempt to reduce the amount of waste being produced. This may be accomplished through modifications within the process where waste is created. This approach is generally referred to as 'source reduction'.

While source reduction is an extremely important goal, recycle/reuse strategies and destructive end-of-pipe treatments are still useful techniques. In fact, while both source reduction and recycle/reuse techniques may achieve the desired reduction in environmental emissions, these approaches may yield vastly different economical consequences when implemented.

METHODS FOR ACHIEVING CLEANER PRODUCTION

Following are few sources of reduction techniques, not necessarily in the order of their hierarchy:

1. Good housekeeping practices.
2. Equipment modification.
3. Process modification/retrofitting.
4. Raw material substitution.
5. Product modification.
6. Innovative manufacturing technology.

However it is important to note that source reduction is one of the means to achieve a reduction in waste generation from industrial operations and thus cleaner production. There are other ways to reduce waste and their undesirable impacts on the environment.

These include:

1. By-product recovery from waste.
2. Recycling waste (with or without treatment) in the production process.
3. Reuse of waste (with or without treatment) in some other process.

All of these techniques can result in substantial reduction in the quality, toxicity and the pollution strength of the waste. The residual waste could then be treated and disposed off in a more safe and environmental-friendly manner while meeting the required effluent/emission standards.

BIOTECHNOLOGY IN MEETING ENVIRONMENTAL PROBLEMS

Today biotechnological processes are playing a significant role in the protection and rehabilitation of environment and are advantageous alternative or adduct to physical and chemical treatment technologies. Biotechnology in environment invites its specific use in the management of environmental problems including waste treatment and pollution control and their integration with non-biological technologies.

With regard to wastes produced by different industries, the following strategies are adopted.

1. To minimise the amount of waste generated.
2. To reduce the toxicity of wastes that are generated.
3. To find more satisfactory ways of disposing off and degrading wastes which have been generated.
4. To find new alternatives for treatments causing pollution.

Point Source Reduction

Much of the waste minimisation resources have been focused on the source of waste. Several examples of enzyme application in waste minimisation by acting on sources especially in food processing, animal feed, leather treatment and detergent industries have been established. These include:

1. Hemicellulose and cellulose are being tested to hydrolyse hemicellulose in pulp in order to reduce consumption of bleaching chlorine in the pulping process. This reduces the generation of chlorinated organic wastes in bleached pulp product.
2. Invertases are employed in recycles of effluents from food and dairy industries where residential starch content is 5–10 per cent which allows the growth of various harmful microbes.

   Effluents → biofilters → starch → microbial invertase

   → glucose ⟶ animal feed.
3. A major disadvantage of use of coal as a fuel is that the pollutants are released into environment during combustion. SO_2 and nitrogen oxides affect air quality and acid rain control.

By-Product Utilisation

Various by-products obtained during usual processes are released as such which can cause pollution of air, water and soil by acting as potential source for the growth of various harmful micro-organisms. By using various biotechnological methods, these by-products can be used up for various purposes. Most of its applications are in chemical, food and dairy industries.

End of pipe treatment (waste destruction technique)

This treatment involves the removal of waste or pollutants at the stage of release in the environment. This depends on the type of wastes encountered whether gaseous, liquid or solid.

BIOLOGICAL TREATMENT OF WASTE-WATER

The principal objective of waste-water treatment is generally to allow human and industrial effluents to be disposed of without danger to human health or unacceptable damage to the natural environment. Irrigation with waste-water is both disposal and utilisation is indeed an effective form of waste-water disposal (as in slow-rate land treatment). However, some degree of treatment must normally be provided to raw municipal waste-water before it can be used for agricultural or landscape irrigation or for aquaculture. The quality of treated effluent used in agriculture has a great influence on the operation and performance of the waste-water-soil-plant or aquaculture system. In the case of irrigation, the required quality of effluent will depend on the crop or crops to be irrigated, the soil conditions and the system of effluent distribution adopted. Through crop restriction and selection of irrigation systems which minimise health risk, the degree of pre-application waste-water treatment can be reduced. A similar approach is not feasible in aquaculture systems and more reliance will have to be placed on control through waste-water treatment.

The most appropriate waste-water treatment to be applied before effluent use in agriculture is that which will produce an effluent meeting the recommended microbiological and chemical quality guidelines both at low cost and with minimal operational and maintenance requirements. Adopting as low a level of treatment as possible is especially desirable in developing countries, not only from the point of view of cost but also in acknowledgement of the difficulty of operating complex systems reliably. In many locations it will be better to design the reuse system to accept a low-grade of effluent rather than to rely on advanced treatment processes producing a reclaimed effluent which continuously meets a stringent quality standard.

Nevertheless, there are locations where a higher-grade effluent will be necessary and it is essential that information on the performance of a wide range of waste-water treatment technology should be available. The design of waste-water treatment plants is usually based on the need to reduce organic and suspended solids loads to limit pollution of the environment. Pathogen removal has very rarely been considered an objective but, for reuse of effluents in agriculture, this must now be of primary concern and processes should be selected and designed accordingly. Treatment to remove waste-water constituents that may be toxic or harmful to crops, aquatic plants (macrophytes) and fish is technically possible but is not normally economically feasible. Unfortunately, few performance data on waste-water treatment plants in developing countries are available and even then they do not normally include effluent quality parameters of importance in agricultural use.

The short-term variations in waste-water flows observed at municipal waste-water treatment plants follow a diurnal pattern. Flow is typically low during the early morning hours, when water consumption is lowest and when the base flow consists of infiltration-inflow and small quantities of sanitary waste-water. A first peak of flow generally occurs in the late morning, when waste-water from the peak morning water for use reaches the treatment plant and a second peak flow usually occurs in the evening. The relative magnitude of the peaks and the times at which they occur vary from country to country and with the size of the community and the length of the sewers. Small communities with small sewer systems have a much higher ratio of peak flow to average flow than do large communities. Although the magnitude of peaks is attenuated as waste-water passes through a treatment plant, the daily variations in flow from a municipal treatment plant make it impracticable, in most cases, to irrigate with effluent directly from the treatment plant. Some form of flow equalisation or short-term storage of treated effluent is necessary to provide a relatively constant supply of reclaimed water for efficient irrigation, although additional benefits result from storage.

Conventional Waste-Water Treatment Processes

Conventional waste-water treatment consists of a combination of physical, chemical, and biological processes and operations to remove solids, organic matter and, sometimes, nutrients from waste-water. General terms used to describe different degrees of treatment, in order of increasing treatment level, are preliminary, primary, secondary and tertiary and/or advanced waste-water treatment. In some countries, disinfection to remove pathogens sometimes follows the last treatment step.

Preliminary treatment

The objective of preliminary treatment is the removal of coarse solids and other large materials often found in raw waste-water. Removal of these materials is necessary to enhance the operation and maintenance of subsequent treatment units. Preliminary treatment operations typically include coarse screening, grit removal and, in some cases, comminution of large objects. In grit chambers, the velocity of the water through the chamber is maintained sufficiently high, or air is used, so as to prevent the settling of most organic solids. Grit removal is not included as a preliminary treatment step in most small waste-water treatment plants. Comminutors are sometimes adopted to supplement coarse screening and serve to reduce the size of large particles so that they will be removed in the form of a sludge in subsequent treatment processes. Flow measurement devices, often standing-wave flumes, are always included at the preliminary treatment stage.

Primary treatment

The objective of primary treatment is the removal of settleable organic and inorganic solids by sedimentation, and the removal of materials that will float (scum) by skimming. Approximately 25 to 50 per cent of the incoming Biochemical Oxygen Demand (BOD_5), 50 to 70 per cent of the total Suspended Solids (SS), and 65 per cent of the oil and grease are removed during primary treatment. Some organic nitrogen, organic phosphorus, and heavy metals associated with solids are also removed during primary sedimentation but colloidal and dissolved constituents are not affected. The effluent from primary sedimentation units is referred to as primary effluent.

In many industrialised countries, primary treatment is the minimum level of preapplication treatment required for waste-water irrigation. It may be considered sufficient treatment if the waste-water is used to irrigate crops that are not consumed by humans or to irrigate orchards, vineyards, and some processed food crops. However, to prevent potential nuisance conditions in storage or flow-equalising reservoirs, some form of secondary treatment is normally required in these countries, even in the case of non-food crop irrigation. It may be possible to use at least a portion of primary effluent for irrigation if off-line storage is provided.

Primary sedimentation tanks or clarifiers may be round or rectangular basins, typically 3 to 5 m deep, with hydraulic retention time between 2 and 3 hours. Settled solids (primary sludge) are normally removed from the bottom of tanks by sludge rakes that scrape the sludge to a central well from which it is pumped to sludge processing units. Scum is swept across the tank surface by water jets or mechanical means from which it is also pumped to sludge processing units.

In large sewage treatment plants, primary sludge is most commonly processed biologically by anaerobic digestion. In the digestion process, anaerobic and facultative bacteria metabolise the organic material in sludge, thereby reducing the volume requiring ultimate disposal, making the sludge stable (nonputrescible) and improving its dewatering characteristics. Digestion is carried out in covered tanks (anaerobic digesters), typically 7 to 14 m deep. The residence time in a digester may vary from a minimum of about 10 days for high-rate digesters (well-mixed and heated) to 60 days or more in standard-rate digesters. Gas containing about 60 to 65 per cent methane is produced during digestion and can be recovered as an energy source. In small sewage treatment plants, sludge is processed in a variety of ways including: aerobic digestion, storage in sludge lagoons, direct application to sludge drying beds, in-process storage (as in stabilisation ponds), and land application.

Secondary treatment

Biological treatment

The uniqueness of micro-organisms and their often unpredictable nature and biosynthetic capabilities, given a specific set of environmental and cultural conditions, has made them likely candidates for solving particularly difficult problems in the life sciences and other fields as well. The various ways in which micro-organisms have been used over the past 50 years to advance medical technology, human and animal health, food processing, food safety and quality, genetic engineering, environmental protection, agricultural biotechnology, and more effective treatment of agricultural and municipal wastes provide a most impressive record of achievement. Many of these technological advances would not have been possible using straightforward chemical and physical engineering methods, or if they were, they would not have been practically or economically feasible.

Nevertheless, while microbial technologies have been applied to various agricultural and environmental problems with considerable success in recent years, they have not been widely accepted by the scientific community because it is often difficult to consistently reproduce their beneficial effects. Micro-organisms are effective only when they are presented with suitable and optimum conditions for metabolizing their substrates including available water, oxygen (depending on whether the micro-organisms are obligate aerobes or facultative anaerobes), pH and temperature of their environment. Meanwhile, the various types of microbial cultures and inoculants available in the market today have rapidly increased because of these new technologies. Significant achievements are being made in systems where technical guidance is coordinated with the marketing of microbial products. Since micro-organisms are useful in eliminating problems associated with the use of chemical fertilisers and pesticides, they are now widely applied in nature farming and organic agriculture.

Environmental pollution, caused by excessive soil crosion and the associated transport of sediment, chemical fertilisers and pesticides to surface and groundwater, and improper treatment of human and animal wastes has caused serious environmental and social problems throughout the world. Often engineers have attempted to solve these problems using established chemical and physical methods. However, they have usually found that such problems cannot be solved without using microbial methods and technologies in coordination with agricultural production.

Concept of useful and effective micro-organisms

The concept of effective micro-organisms (EM) was developed by Professor *Teruo Higa*, University of the Ryukyus, Okinawa, Japan. EM consists of mixed cultures of beneficial an naturally-occurring micro-organisms that can be applied as inoculants to increase the microbial diversity of soils and plant. Research has shown that the inoculation of EM cultures to the soil/plant ecosystem can improve soil quality, soil health and the growth, yield, and quality of crops. EM contains selected species of micro-organisms including predominant populations of lactic acid bacteria and yeasts and smaller numbers of photosynthetic bacteria, actinomycetes and other types of organisms. All of these are mutually compatible with one another and can coexist in liquid culture.

Conceptual design is important in developing new technologies for utilising beneficial and effective micro-organisms for a more sustainable treatment of industrial effluents containing high COD and BOD values. The basis of a conceptual design is imply to first conceive an ideal or model and then to devise a strategy and method for achieving the reality. However it is necessary to carefully coordinate the materials, the environment, and the technologies constituting the method. Moreover one should adopt a philosophical attitude in applying microbial technologies for degradation of organics in effluents.

Careful selection of bacterial flora for the Industrial waste-water treatment makes it possible to reuse and recycle technologies ultimately leading to an ZERO discharge situation which proves that the treatment programmes are productive and are not implemented for the statutory requirements of the Pollution Control Boards.

The objective of secondary treatment is the further treatment of the effluent from primary treatment to remove the residual organics and suspended solids. In most cases, secondary treatment follows primary treatment and involves the removal of biodegradable dissolved and colloidal organic matter using aerobic biological treatment processes. Aerobic biological treatment is performed in the presence of oxygen by aerobic micro-organisms (principally bacteria) that metabolise the organic matter in the waste-water, thereby producing more micro-organisms and inorganic end-products (principally CO_2, NH_3, and H_2O). Several aerobic biological processes are used for secondary treatment differing primarily in the manner in which oxygen is supplied to the micro-organisms and in the rate at which organisms metabolise the organic matter.

High-rate biological processes are characterised by relatively small reactor volumes and high concentrations of micro-organisms compared with low rate processes. Consequently, the growth rate of new organisms is much greater in high-rate systems because of the well controlled environment. The micro-organisms must be separated from the treated waste-water by sedimentation to produce clarified secondary effluent. The sedimentation tanks used in secondary treatment, often referred to as secondary clarifiers, operate in the same basic manner as the primary clarifiers described previously. The biological solids removed during secondary sedimentation, called secondary or biological sludge, are normally combined with primary sludge for sludge processing.

Common high-rate processes include the activated sludge processes, trickling filters or biofilters, oxidation ditches and Rotating Biological Contactors (RBC). A combination of two of these processes in series (e.g., biofilter followed by activated sludge) is sometimes used to treat municipal waste-water containing a high concentration of organic material from industrial sources.

Activated sludge

In the activated sludge process, the dispersed-growth reactor is an aeration tank or basin containing a suspension of the waste-water and micro-organisms, the mixed liquor. The contents of the aeration tank are mixed vigorously by aeration devices which also supply oxygen to the biological suspension. Aeration devices commonly used include submerged diffusers that release compressed air and mechanical surface aerators that introduce air by agitating the liquid surface. Hydraulic retention time in the aeration tanks usually ranges from 3 to 8 hours but can be higher with high BOD_5 waste-waters. Following the aeration step, the micro-organisms are separated from the liquid by sedimentation and the clarified liquid is secondary effluent. A portion of the biological sludge is recycled to the aeration basin to maintain a high Mixed-Liquor Suspended Solids (MLSS) level. The remainder is removed from the process and sent to sludge processing to maintain a relatively constant concentration of micro-organisms in the system. Several variations of the basic activated sludge process, such as extended aeration and oxidation ditches, are in common use, but the principles are similar.

Trickling filters

A trickling filter or biofilter consists of a basin or tower filled with support media such as stones, plastic shapes, or wooden slats. Waste-water is applied intermittently, or sometimes continuously, over the media. Micro-organisms become attached to the media and form a biological layer or fixed film. Organic matter in the waste-water diffuses into the film, where it is metabolised. Oxygen is normally supplied to the film by the natural flow of air either up or down through the media, depending on the relative temperatures of the waste-water and ambient air. Forced air can also be supplied by blowers but this is rarely necessary. The thickness of the biofilm increases as new organisms grow. Periodically, portions of the film 'slough' off the media. The sloughed material is separated from the liquid in a secondary clarifier and discharged to sludge processing. Clarified liquid from the secondary clarifier is the secondary effluent and a portion is often recycled to the biofilter to improve hydraulic distribution of the waste-water over the filter.

Rotating biological contactors

Rotating Biological Contactors (RBCs) are fixed-film reactors similar to biofilters in that organisms are attached to support media. In the case of the RBC, the support media are slowly rotating discs that are partially submerged in flowing waste-water in the reactor. Oxygen is supplied to the attached biofilm from the air when the film is out of the water and from the liquid when submerged, since oxygen is transferred to the waste-water by surface turbulence created by the 'discs' rotation. Sloughed pieces of biofilm are removed in the same manner described for biofilters.

High-rate biological treatment processes, in combination with primary sedimentation, typically remove 85 per cent of the BOD_5 and SS originally present in the raw waste-water and some of the heavy metals. Activated sludge generally produces an effluent of slightly higher quality, in terms of these constituents, than biofilters or RBCs. When coupled with a disinfection step, these processes can provide substantial but not complete removal of bacteria and virus. However, they remove very little phosphorus, nitrogen, non-biodegradable organics, or dissolved minerals.

Tertiary and/or advanced treatment

Tertiary and/or advanced waste-water treatment is employed when specific waste-water constituents which cannot be removed by secondary treatment must be removed. Individual treatment processes are necessary to remove nitrogen, phosphorus, additional suspended solids, refractory organics, heavy metals and dissolved solids. Because advanced treatment usually follows high-rate secondary treatment, it is sometimes referred to as tertiary treatment. However, advanced treatment processes are sometimes combined with primary or secondary treatment (e.g., chemical addition to primary clarifiers or aeration basins to remove phosphorus) or used in place of secondary treatment (e.g., overland flow treatment of primary effluent).

An adaptation of the activated sludge process is often used to remove nitrogen and phosphorus. Effluent from primary clarifiers flows to the biological reactor, which is physically divided into five zones by baffles and weirs. In sequence these zones are: (i) anaerobic fermentation zone (characterised by very low dissolved oxygen levels and the absence of nitrates); (ii) anoxic zone (low dissolved oxygen levels but nitrates present); (iii) aerobic zone (aerated); (iv) secondary anoxic zone; and (v) final aeration zone. The function of the first zone is to condition the group of bacteria responsible for phosphorus removal by stressing them under low oxidation-reduction conditions, which results in a release of phosphorus equilibrium in the cells of the bacteria. On subsequent exposure to an adequate supply of oxygen and phosphorus in the aerated zones, these cells rapidly accumulate phosphorus considerably in excess of their normal metabolic requirements. Phosphorus is removed from the system with the waste activated sludge.

Most of the nitrogen in the influent is in the ammonia form, and this passes through the first two zones virtually unaltered. In the third aerobic zone, the sludge age is such that almost complete nitrification takes place, and the ammonia nitrogen is converted to nitrites and then to nitrates. The nitrate-rich mixed liquor is then recycled from the aerobic zone back to the first anoxic zone. Here denitrification occurs, where the recycled nitrates, in the absence of dissolved oxygen, are reduced by facultative bacteria to nitrogen gas, using the influent organic carbon compounds as hydrogen donors. The nitrogen gas merely escapes to atmosphere. In the second anoxic zone, those nitrates which were not recycled are reduced by the endogenous respiration of bacteria. In the final re-aeration zone, dissolved oxygen levels are again raised to prevent further denitrification, which would impair settling in the secondary clarifiers to which the mixed liquor then flows.

An experimentation programme on this plant demonstrated the importance of the addition of volatile fatty acids to the anaerobic fermentation zone to achieve good phosphorus removal. These essential short-chain organics (mainly acetates) are produced by the controlled fermentation of primary sludge in a gravity thickener and are released into the thickener supernatent, which can be fed to the head of the biological reactor. Without this supernatent return flow, overall phosphorus removal quickly dropped to levels found in conventional activated sludge plants. Performance data over three years have proved that, with thickener supernatent recycle, effluent quality median values of 0.5-1.38 mg/l Ortho-P, 1.4-1.6 mg/l total nitrogen and 1.4-2.0 mg/l nitrate-N are achievable. This advanced biological waste-water treatment plant cost only marginally more than a conventional activated sludge plant but nevertheless involved considerable investment. Furthermore, the complexity of the process and the skilled operation required to achieve consistent results make this approach unsuitable for developing countries.

In many situations, where the risk of public exposure to the reclaimed water or residual constituents is high, the intent of the treatment is to minimise the probability of human exposure to enteric viruses and other pathogens. Effective disinfection of viruses is believed to be inhibited by suspended and colloidal solids in the water, therefore these solids must be removed by advanced treatment before the disinfection step. The sequence of treatment often specified in the United States is: secondary treatment followed by chemical coagulation, sedimentation, filtration, and disinfection. This level of treatment is assumed to produce an effluent free from detectable viruses. In Near East countries adopting tertiary treatment, the tendency has been to introduce pre-chlorination before rapid-gravity sand filtration and post-chlorination afterwards. A final ozonation treatment after this sequence has been considered in at least one country.

Disinfection

Disinfection normally involves the injection of a chlorine solution at the head end of a chlorine contact basin. The chlorine dosage depends upon the strength of the waste-water and other factors, but dosages of 5 to 15 mg/l are common. Ozone and ultra violet (uv) irradiation can also be used for disinfection but these methods of disinfection are not in common use. Chlorine contact basins are usually rectangular channels, with baffles to prevent short-circuiting, designed to provide a contact time of about 30 minutes. However, to meet advanced waste-water treatment requirements, a chlorine contact time of as long as 120 minutes is sometimes required for specific irrigation uses of reclaimed waste-water. The bactericidal effects of chlorine and other disinfectants are dependent upon pH, contact time, organic content, and effluent temperature.

Effluent storage

Although not considered a step in the treatment process, a storage facility is, in most cases, a critical link between the waste-water treatment plant and the irrigation system. Storage is needed for the following reasons:

1. To equalise daily variations in flow from the treatment plant and to store excess when average waste-water flow exceeds irrigation demands; includes winter storage.
2. To meet peak irrigation demands in excess of the average waste-water flow.
3. To minimise the effects of disruptions in the operations of the treatment plant and irrigation system. Storage is used to provide insurance against the possibility of unsuitable reclaimed waste-water entering the irrigation system and to provide additional time to resolve temporary water quality problems.

   Advanced waste-water treatment in these plants follows high rate secondary treatment and includes addition of chemical coagulants (alum + polymer) as necessary followed by filtration through sand or activated carbon granular medium filters.
4. To provide additional treatment. Oxygen demand, suspended solids, nitrogen, and micro-organisms are further reduced during storage.

Reliability of conventional and advanced waste-water treatment

Waste-water reclamation and reuse systems should contain both design and operational requirements necessary to ensure reliability of treatment. Reliability features such as alarm systems, standby power supplies, treatment process duplications, emergency storage or disposal of inadequately treated waste-water, monitoring devices, and automatic controllers are important. From a public health standpoint, provisions for adequate and reliable disinfection are the most essential features of the advanced waste-water treatment process. Where disinfection is required, several reliability features must be incorporated into the system to ensure uninterrupted chlorine feed.

Natural Biological Treatment Systems

Natural low-rate biological treatment systems are available for the treatment of organic waste-waters such as municipal sewage and tend to be lower in cost and less sophisticated in operation and maintenance. Although such processes tend to be land intensive by comparison with the conventional high-rate biological processes already described, they are often more effective in removing pathogens and do so reliably and continuously if properly designed and not overloaded. Among the natural biological treatment systems available, stabilisation ponds and land treatment have been used widely around the world and a considerable record of experience and design practice has been documented. The nutrient film technique is a fairly recent development of the hydroponic plant growth system with application in the treatment and use of waste-water.

Waste-water stabilisation ponds

Stabilisation ponds are the preferred waste-water treatment process in developing countries, where land is often available at reasonable opportunity cost and skilled labour is in short supply.

Waste-water stabilisation pond systems are designed to achieve different forms of treatment in up to three stages in series, depending on the organic strength of the input waste and the effluent quality objectives. For ease of maintenance and flexibility of operation, at least two trains of ponds in parallel are incorporated in any design. Strong waste-waters, with BOD_5 concentration in excess of about 300 mg/l, will frequently be introduced into first-stage anaerobic ponds, which achieve a high volumetric rate of removal. Weaker wastes or, where anaerobic ponds are environmentally unacceptable, even stronger wastes (say up to 1000 mg/l BOD_5) may be discharged directly into primary facultative ponds. Effluent from first-stage anaerobic ponds will overflow into secondary facultative ponds which comprise the second-stage of biological treatment. Following primary or secondary facultative ponds, if further pathogen reduction is necessary maturation ponds will be introduced to provide tertiary treatment. Anaerobic ponds normally have a depth between 2m and 5m and function as open septic tanks with gas release to the atmosphere. The biochemical reactions which take place in anaerobic ponds are the same as those occurring in anaerobic digesters, with a first phase of acidogenesis and a second slower-rate of methanogenesis. Ambient temperatures in hot-climate countries are conducive to these anaerobic reactions and expected BOD_5 removals for different retention times in treating sewage have been given. More recently, some scientists have suggested conservative removals of BOD_5 in anaerobic ponds as 40 per cent below 10°C, at a design loading, λ_v , of 100 g/m^{3}d, and 60 per cent above 20°C, at a design loading of 300 g/m^{3}d, with linear interpolation for operating temperature between 10 and 20°C. Higher removal rates are possible with industrial wastes, particularly those containing significant quantities of organic settleable solids. Of course, other environmental conditions in the ponds, particularly pH, must be suitable for the anaerobic micro-organisms bringing about the breakdown of BOD.

In certain instances, anaerobic ponds become covered with a thick scum layer, which is thought to be beneficial but not essential, and may give rise to increased fly breeding. Solids in the raw waste-water, as well as biomass produced, will settle out in first-stage anaerobic ponds and it is common to remove sludge when it has reached half depth in the pond. This usually occurs after two years of operation at design flow in the case of municipal sewage treatment.

Facultative ponds

The effluent from anaerobic ponds will require some form of aerobic treatment before discharge or use and facultative ponds will often be more appropriate than conventional forms of secondary biological treatment for application in developing countries. Primary facultative ponds will be designed for the treatment of weaker wastes and in sensitive locations where anaerobic pond odours would be unacceptable. Solids in the influent to a facultative pond and excess biomass produced in the pond will settle out forming a sludge layer at the bottom. The benthic layer will be anaerobic and, as a result of anaerobic breakdown of organics, will release soluble organic products to the water column above.

Organic matter dissolved or suspended in the water column will be metabolised by heterotrophic bacteria, with the uptake of oxygen, as in convential aerobic biological waste-water treatment processes. However, unlike in convential processes, the dissolved oxygen utilised by the bacteria in facultative ponds is replaced through photosynthetic oxygen production by microalgae, rather than by aeration equipment. Especially intreating municipal sewage in hot climates, the environment in facultative ponds is ideal for the proliferation of microalgae. High temperature and ample sunlight create conditions which encourage algae to utilise the Carbon Dioxide (CO_2) released by bacteria in breaking down the organic components of the waste-water and take up nutrients (mainly nitrogen and phosphorus) contained in the waste-water. This symbiotic relationship contributes to the overall removal of BOD in facultative ponds, as shown in Fig. 17.1.

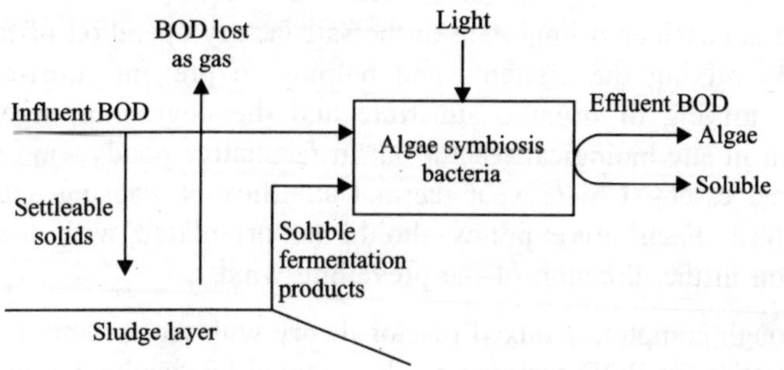

Fig. 17.1. Removal of BOD in facultative ponds.

To maintain the balance necessary to allow this symbiosis to persist, the organic loading on a facultative pond must be strictly limited. Even under satisfactory operating conditions, the dissolved oxygen concentration (DO) in a facultative pond will vary diurnally as well as over the depth. Maximum DO will occur at the surface of the pond and will usually reach supersaturation in tropical regions at the time of maximum radiation intensity, as shown in Fig. 17.2.

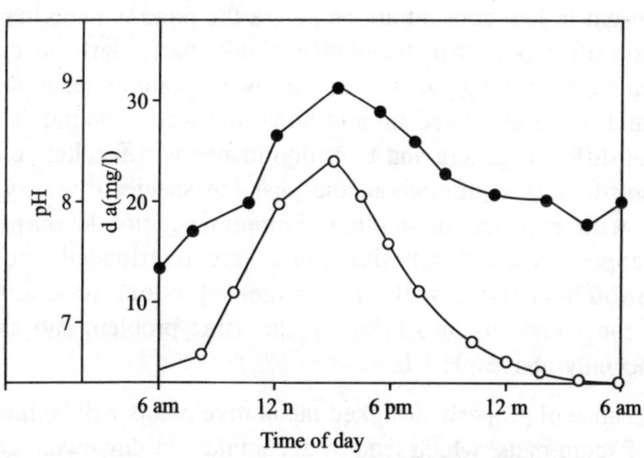

Fig. 17.2. Diurnal variation of dissolved oxygen and pH in facultative pond, pH: dissolved oxygen.

From that time until sunrise, DO will decline and may well disappear completely for a short period. For a typical facultative pond depth, D_f, of 1.5m the water column will be predominantly aerobic at the time of peak radiation and predominantly anaerobic at sunrise. As illustrated in Fig. 17.2, the pH of the pond contents will also vary diurnally as algae utilise CO_2 throughout daylight hours and respire, along with bacteria and other organisms, releasing CO_2 during the night.

Wind is considered important to the satisfactory operation of facultative ponds by mixing the contents and helping to prevent short-circuiting. Intimate mixing of organic substrate and the degrading organisms is important in any biological reactor but in facultative ponds wind mixing is considered essential to prevent thermal stratification causing anaerobiosis and failure. Facultative ponds should be orientated with the longest dimension in the direction of the prevailing wind.

Although completely-mixed reactor theory with the assumption of first-order kinetics for BOD removal can be adopted for facultative pond design such a fundamental approach is rarely adopted in practice. Instead, an empirical procedure based on operational experience is more common. The most widely adopted design method currently being applied wherever local experience is limited is that introduced by *McGarry* and *Pescod.*

Efficiently operating facultative ponds treating waste-water will contain a mixed population of flora but flagellate algal genera such as *Chlamydomonas*, *Euglena*, *Phacus* and *Pyrobotrys* will predominate. Non-motile forms such as *Chlorella, Scenedesmus* and various *diatom* species will be present in low concentrations unless the pond is underloaded. Algal stratification often occurs in facultative ponds, particularly in the absence of wind-induced mixing, as motile forms respond to changes in light intensity and move in a band up and down the water column. The relative numbers of different genera and their dominance in a facultative pond vary from season to season throughout the year but species diversity generally decreases with increase in loading. Sometimes, mobile purple sulphur bacteria appear when facultative ponds are overloaded and sulphide concentration increases, with the danger of odour production. High ammonia concentrations also bring on the same problem and are toxic to algae, especially above pH 8.0.

Maintenance of properly designed facultative ponds will be limited to the removal of scum mats, which tend to accumulate in downwind corners and the cutting of grass on embankments. To ensure efficient operation, facultative ponds should be regularly monitored but, even where this is not possible, they have the reputation of being relatively trouble-free.

Maturation ponds

The effluent from facultative ponds treating municipal sewage or equivalent input waste-water will normally contain at least 50 mg/l BOD_5 and if an effluent with lower BOD_5 concentration is required it will be necessary to use maturation ponds. For sewage treatment, two maturation ponds in series, each with a retention time of 7 days, have been found necessary to produce a final effluent with $BOD_5 < 25$ mg/l when the facultative pond effluent had a $BOD_5 < 75$ mg/l.

A more important function of maturation ponds, however, is the removal of excreted pathogens to achieve an effluent quality which is suitable for its downstream reuse. Although the longer retention in anaerobic and facultative pond systems will make them more efficient than conventional waste-water treatment processes in removing pathogens, the effluent from a facultative pond treating municipal sewage will generally require further treatment in maturation ponds to reach effluent standards imposed for reuse in unrestricted irrigation. Faecal coliform bacteria are commonly used as indicators of excreted pathogens and maturation ponds can be designed to achieve a given reduction of Faecal Coliforms (FC). *Protozoan cysts* and *helminth ova* are removed by sedimentation in stabilisation ponds and a series of ponds with overall retention of 20 days or more will produce an effluent totally free of cysts and ova.

Reduction of faecal coliform bacteria in any stabilisation pond (anaerobic, facultative and maturation) is generally taken as following first-order kinetics:

A suitable design value of N_i in the case of municipal sewage treatment is 1×10^8 faecal coliforms/100ml, which is slightly higher than average practical levels.

The value of N_e should be obtained by substituting the appropriate levels of variables assuming a retention time of 7 days in each of two maturation lagoons (for sewage). If the calculated value of N_e does not meet the reuse effluent standard, the number of maturation ponds should be increased, say to three or more each with retention time 5 days, and N_e recalculated. A more systematic approach is now available whereby the optimum design for maturation ponds can be obtained using a simple computer programme.

Polprasert has developed an approach to the assessment of bacterial die-off which attempts to take into account the complex physical characteristics of ponds and biochemical reactions taking place in them. A multiple-regression equation involving parameters such as retention time, organic loading, algal concentration and ultra-violet light exposure has been suggested. The Wehner and Wilhelm non-ideal flow equation, including the pond dispersion number, was adopted to predict bacterial survival, in preference to the first order rate equation.

Maturation ponds will be aerobic throughout the water column during daylight hours and the pH will rise above 9.0. The algal population of many species of non-flagellate unicellular and colonial forms will be distributed over the full depth of a maturation pond. Large numbers of filamentous algae, particularly blue-greens, will emerge under very low BOD loading conditions. Very low concentrations of algae in a maturation pond will indicate excessive algal predation by zooplankton, such as *Daphnia sp.* and this will have a deleterious effect on pathogen die-off, which is linked to algal activity.

Saqqar in his analysis of the performance of the Al Samra stabilisation ponds in Amman, Jordan, has shown that the coliform and faecal coliform die-off coefficients varied with retention time, water temperature, organic loading, total BOD_5 concentration, pH and pond depth. Total coliform die-off was less than the rate of faecal coliform die-off, except during the cold season. For the series of ten ponds, including at least the first five totally anaerobic, the faecal coliform die-off coefficient, k, for the temperature range 12 – 15°C increased through the pond sequence from 0.11 per day in the first anaerobic pond to 0.68 per day in the final two ponds, which operated as facultative ponds.

Apart from the use of effluent for irrigation of crops, termed 'slow rate' land treatment in the US Environmental Protection Agency's Process Design Manual for Land Treatment of Municipal Waste-waters (EPA) and 'rapid infiltration' or 'infiltration percolation' of effluent discussed as soil-aquifer treatment in a later section of this document, the EPA manual deals with 'overland flow' as a waste-water treatment method. In overland flow treatment, effluent is distributed over gently sloping grassland on fairly impermeable soils. Ideally, the waste-water moves evenly down the slope to collecting ditches at the bottom edge of the area and water-tolerant grasses are an essential component of the system.

This form of land treatment requires alternating applications of effluent (usually treated) and resting on the land, to allow soil reaction and grass cutting. The total area utilised is normally broken up into small plots to allow this form of intermittent operation and yet achieve continuous treatment of the flow of waste-water. Although this type of land treatment has been widely adopted in Australia, New Zealand and the UK for tertiary upgrading of secondary effluents, it has been used for the treatment of primary effluent in, Australia and is being considered for the treatment of raw sewage in Karachi, Pakistan.

Site Characteristics and Design Features for Overland Flow Treatment of Waste-water

Basic site characteristics and design features for overland flow treatment have been suggested by EPA as shown in Table 17.1. It was pointed out that steeper land slopes might be feasible at reduced hydraulic loadings. The ranges given for field area required and application rates cover the waste-water quality from raw sewage to secondary effluent, with higher application rates and lower land area requirements being associated with higher levels of pre-application treatment. Although soil permeability is not critical with this form of land treatment, the impact on groundwater should not be overlooked in the case of highly permeable soils.

Table 17.1. Site characteristics and design features for flow treatment.

Grade	Finished slope 2–8 per cent
Field area required (ha)	6.55–44
Soil permeability	Slow (clays, silts and soils with impermeable barriers)
Annual application rate (m)	3–20
Typical weekly application rate (cm)	6–40

The application rate for waste-waters will depend principally on the type of soil, the quality of waste-water effluent and the physical and biochemical activity in the near-surface environment. Rational design procedures, based on the kinetics of BOD removal, have been developed for overland flow systems. Slope lengths from 30–60 m are common in the US for overland flow systems.

The cover crop is an important component of the overland flow system since it prevents soil erosion, provides nutrient uptake and serves as a fixed-film medium for biological treatment. Crops best suited to overland flow treatment are grasses with a long growing season, high moisture tolerance and extensive root formation. Reed canary grass has a very high nutrient uptake capacity and yields a good quality hay; other suitable grasses include rye grass and tall fescue.

Suspended and colloidal organic materials in the waste-water are removed by sedimentation and filtration through surface grass and organic layers. Removal of total nitrogen and ammonia is inversely related to application rate, slope length and soil temperature. Phosphorus and trace elements removal is by sorption on soil clay colloids and precipitation as insoluble complexes of calcium, iron and aluminium. Overland flow systems also remove pathogens from sewage effluent at levels comparable with conventional secondary treatment systems, without chlorination. A monitoring programme should always be incorporated into the design of overland flow projects both for waste-water and effluent quality and for application rates.

Macrophyte treatment

Maturation ponds which incorporate floating, submerged or emergent aquatic plant species are termed *macrophyte ponds* and these have been used in recent years for upgrading effluents from stabilisation ponds. Macrophytes take up large amounts of inorganic nutrients (especially N and P) and heavy metals (such as Cd, Cu, Hg and Zn) as a consequnce of the growth requirements and decrease the concentration of algal cells through light shading by the leaf canopy and, possibly, adherence to gelatinous biomass which grows on the roots.

Floating macrophyte systems utilising water hyacinth and receiving primary sewage effluent in Florida have achieved secondary treatment effluent quality with a six day hydraulic retention time, water depth of 60 cm and hydraulic loading 1860 m^3/ha d. Similar results had also been observed for artificial wetlands using emergent macrophytes. In Europe, the land area considered to be necessary for treatment of preliminary-treated sewage is estimated at 2–5 m^2 per population equivalent to achieve a secondary effluent quality.

Floating aquatic macrophyte systems

Floating macrophyte species, with their large root systems, are very efficient at nutrient stripping. Although several genera have been used in pilot schemes, including *Salvinia, Spirodella, Lemna* and *Eichornia*, *Eichornia crassipes* (water hyacinth) has been studied in much greater detail. In tropical regions, water hyacinth doubles in mass about every 6 hrs. and a macrophyte pond can produce more than 250 kg/ha d (dry weight). Nitrogen and phosphorus reductions up to 80 per cent and 50 per cent have been achieved. In Tamil Nadu, India, studies have indicated that the coontail, *Ceratophyllum demersum,* a submerged macrophyte, is very efficient at removing ammonia (97 per cent) and phosphorus (96 per cent) from raw sewage and also removes 95 per cent of the BOD_5. It has a lower growth rate than *Eichornia crassipes,* which allows less frequent harvesting.

In such macrophyte pond systems, apart from any physical removal processes which might occur (especially sedimentation) the aquatic vascular plants serve as living substrates for microbial activity, which removes BOD and nitrogen achieves reductions in phosphorus, heavy metals and some organics through plant uptake. The basic function of the macrophytes in the latter mechanism is to assimilate, concentrate and store contaminants on a short-term basis. Subsequent harvest of the plant biomass results in permanent removal of stored contaminents from the pond treatment system.

The nutrient assimilation capacity of aquatic macrophytes is directly related to growth rate, standing crop and tissue composition. The potential rate of pollutant storage by an aquatic plant is limited by the growth rate and standing crop of biomass per unit area. Water hyacinth, for example, was found to reach a standing crop level of 30 tonnes (dry weight)/ha in Florida, resulting in a maximum storage of 900 kg N/ha and 180 kg P/ha.

Fly and mosquito breeding is a problem in floating macrophyte ponds but this can be partially alleviated by introducing larvae-eating fish species such as *Gambusia* and *Peocelia* into the ponds. It should be recognised that pathogen die-off is poor in macrophyte ponds as a result of light shading and the lower dissolved oxygen and pH compared with algal maturation ponds. In their favour, macrophyte ponds can serve a useful purpose in stripping pond effluents of nutrients and algae and at the same time produce a harvestable biomass. Floating macrophytes are fairly easily collected by floating harvesters. The harvested plants might be fed to cattle, used as a green manure in agriculture, composted aerobically to produce a fertiliser and soil conditioner, or can be converted into biogas in an anaerobic digester, in which case the residual sludge can then be applied as a fertiliser and soil conditioner. Maximum removal by water hyacinth was 5850 kg N/ha year, compared with 1200 kg N/ha year by duckweed.

Emergent macrophyte treatment systems

In recent years, natural and artificial wetlands and marshes have been used to treat raw sewage and partially-treated effluents. Natural wetlands are usually unmanaged, whereas artificial systems are specially designed to maximise performance by providing the optimum conditions for emergent macrophyte growth. The key features of such reed bed treatment systems are:

1. Rhizomes of the reeds grow vertically and horizontally in the soil or gravel bed, opening up 'hydraulic pathways'.
2. Waste-water BOD and nitrogen are removed by bacterial activity; aerobic treatment takes place in the rhizosphere, with anoxic and anaerobic treatment taking place in the surrounding soil.

3. Oxygen passes from the atmosphere to the rhizosphere via the leaves and stems of the reeds through the hollow rhizomes and out through the roots.
4. Suspended solids in the sewage are aerobically composted in the above-ground layer of vegetation formed from dead leaves and stems.
5. Nutrients and heavy metals are removed by plant uptake.

The growth rate and pollutant assimilative capacity of emergent macrophytes such as *Phragmites communis and Scirpus lacstris* are limited by the culture system, waste-water loading rate, plant density, climate and management factors.

Growth rates for emergent macrophytes are also provided in Table 17.2 as well as nutrient contents. High tissue N concentrations have been found in plants cultured in nutrient enriched (waste-water) systems and in plants analysed in the early stages of growth. Maximum storage of nutrients by emergent macrophytes was found to be in the range 200–1560 kg N/ha and 40–375 kg P/ha in Florida. More than 50 percent of the nutrients were stored in below-ground portions of the plants, tissues difficult to harvest to achieve effective nutrient removal. However, because emergent macrophytes have more supportive tissue than floating macrophytes, they might have greater potential for storing the nutrients over a longer period. Consequently, frequent harvesting might not be so necessary to achieve maximum nutrient removal although harvesting above-ground biomass once a year should improve overall nutrient removal efficiency.

Table 17.2. Growth and nutrient (N & P) contents of selected macrophytes.

Description	*Biomass*		*Tissue composition*	
	Standing crop	*Growth rates*	*N*	*P*
	t (dw) ha^{-1}	*t ha^{-1} yr^{-1}*	*g kg^{-1}*	
Floating macrophytes				
Eichhornia crasspipes (water hyacinth)	20.0–24.0	60–110	10–40	1.4–12.0
Pistia stratiotes (water lettuce)	6.0–10.5	50–80	12–40	1.5–11.5
Hydrocotyle spp. (pennywort)	7.0–11.0	30–60	15–45	2.0–12.5
Alternanthera spp. (alligator weed)	18.0	78	15–35	2.0–9.0
Lemna spp. (duckweed)	1.3	6–26	25	4.0–15.0
Salvinia spp.	2.4–3.2	9–45		1.8–9.0

(Contd...)

Description	Biomass		Tissue composition	
	Standing crop	Growth rates	N	P
	t (dw) ha^{-1}	t ha^{-1} yr^{-1}	g kg^{-1}	
Emergent macrophytes				
Typha (cattail)	4.3–22.5	8–61	5-24	0.5–4.0
Juncus (rush)	22.0	53	15	2.0
Scirpus (bulrush)	–	–	8-27	1.0–3.0
Phragmites (reed)	6.0–35.0	10–60	18–21	2.0–3.0
Eleocharis (spike rush)	8.8	26	9–18	1.0–3.0
Saururus cernuus (lizardis tail)	4.5–22.5	–	15–25	1.0–5.0

Nutrient film technique

The nutrient film technique (NFT) is a modification of the hydroponic plant growth system in which plants are grown directly on an impermeable surface to which a thin film of waste-water is continuously applied. Root production on the impermeable surface is high and the large surface area traps and accumulates matter. Plant top-growth provides nutrient uptake, shade for protection against algal growth and water removal in the form of transpiration, while the large mass of self-generating root systems and accumulated material serve as living filters. *Jewell* and *Smith* have hypothesised the following mechanisms, taking place in three plant sections:

1. Roughing or preliminary treatment by plant species with large root systems capable of surviving and growing in a grossly polluted condition. Large sludge accumulations, anaerobic conditions and trace metal precipitation and entrapment characterise this mechanism and a large portion of waste-water BOD and suspended solids would thereby be removed.
2. Nutrient conversion and recovery due to high biomass production.
3. Waste-water polishing during nutrient-limited plant production, depending on the required effluent quality.

A three year pilot-scale study by *Jewell.* proved this to be a viable alternative for sewage treatment. Reed canary grass was used as the main test species and resulted in the production of better than secondary effluent quality at an application rate of 10 cm/d of settled domestic sewage and synthetic waste-water. The highest loading rates achieved were equivalent to treating the sewage generated by a population of 10,000 on an area of 2 ha. Plants other than reed canary grass were also tested and those that flourished best in the NFT system were: cattails, bulrush, strawflowers, Japanese millet, roses, Napier grass, marigolds, wheat and phragmites.

Sewage is the waste-water released by residences, businesses and industries in a community. It is 99.94 per cent water, with only 0.06 per cent of the waste-water dissolved and suspended solid material. The cloudiness of sewage is caused by suspended particles which in untreated sewage ranges from 100 to 350 mg/l. A measure of the strength of the waste-water is biochemical oxygen demand, or BOD_5. The BOD_5 measures the amount of oxygen micro-organisms require in five days to break down sewage. Untreated sewage has a BOD_5 ranging from 100 mg/l to 300 mg/l. Pathogens or disease-causing organisms are present in sewage. Coliform bacteria are used as an indicator of disease-causing organisms. Sewage also contains nutrients (such as ammonia and phosphorus), minerals, and metals. Ammonia can range from 12 to 50 mg/l and phosphorus can range from 6 to 20 mg/l in untreated sewage.

Sewage treatment is a multi-stage process to renovate waste-water before it reenters a body of water, is applied to the land or is reused. The goal is to reduce or remove organic matter, solids, nutrients, disease-causing organisms and other pollutants from waste-water. Each receiving body of water has limits to the amount of pollutants it can receive without degradation. Therefore, each sewage treatment plant must hold a permit listing the allowable levels of BOD_5, suspended solids, coliform bacteria and other pollutants. The discharge permits are called NPDES permits which stands for the National Pollutant Discharge Elimination System.

Preliminary treatment

Preliminary treatment to screen out, grind up, or separate debris is the first step in waste-water treatment. Sticks, rags, large food particles, sand, gravel, toys, etc., are removed at this stage to protect the pumping and other equipment in the treatment plant. Treatment equipment such as bar screens, comminutors (a large version of a garbage disposal) and grit chambers are used as the waste-water first enters a treatment plant. The collected debris is usually disposed off in a landfill.

Primary treatment

Primary treatment is the second step in treatment and separates suspended solids and greases from waste-water. Waste-water is held in a quiet tank for several hours allowing the particles to settle to the bottom and the greases to float to the top. The solids drawn off the bottom and skimmed off the top receive further treatment as sludge. The clarified waste-water flows on to the next stage of waste-water treatment. Clarifiers and septic tanks are usually used to provide primary treatment.

Secondary treatment

Secondary treatment is a biological treatment process to remove dissolved organic matter from waste-water. Sewage micro-organisms are cultivated and added to the waste-water. The micro-organisms absorb organic matter from sewage as their food supply. Three approaches are used to accomplish secondary treatment; fixed film, suspended film and lagoon systems.

Fixed film systems

Fixed film systems grow micro-organisms on substrates such as rocks, sand or plastic. The waste-water is spread over the substrate, allowing the waste-water to flow past the film of micro-organisms fixed to the substrate. As organic matter and nutrients are absorbed from the waste-water, the film of micro-organisms grows and thickens. Trickling filters, rotating biological contactors, and sand filters are examples of fixed film systems.

Suspended film systems

Suspended film systems stir and suspend micro-organisms in waste-water. As the micro-organisms absorb organic matter and nutrients from the waste-water they grow in size and number. After the micro-organisms have been suspended in the waste-water for several hours, they are settled out as a sludge.

Some of the sludge is pumped back into the incoming waste-water to provide 'seed' micro-organisms. The remainder is wasted and sent on to a sludge treatment process. Activated sludge, extended aeration, oxidation ditch, and sequential batch reactor systems are all examples of suspended film systems.

Lagoon systems

Lagoon systems are shallow basins which hold the waste-water for several months to allow for the natural degradation of sewage. These systems take advantage of natural aeration and micro-organisms in the waste-water to renovate sewage.

Final treatment

Final treatment focuses on removal of disease-causing organisms from waste-water. Treated waste-water can be disinfected by adding chlorine or by using ultraviolet light. High levels of chlorine may be harmful to aquatic life in receiving streams. Treatment systems often add a chlorine-neutralising chemical to the treated waste-water before stream discharge.

Advanced treatment

Advanced treatment is necessary in some treatment systems to remove nutrients from waste-water. Chemicals are sometimes added during the treatment process to help settle out or strip out phosphorus or nitrogen. Some examples of nutrient removal systems include coagulant addition for phosphorus removal and air stripping for ammonia removal (Fig. 17.3).

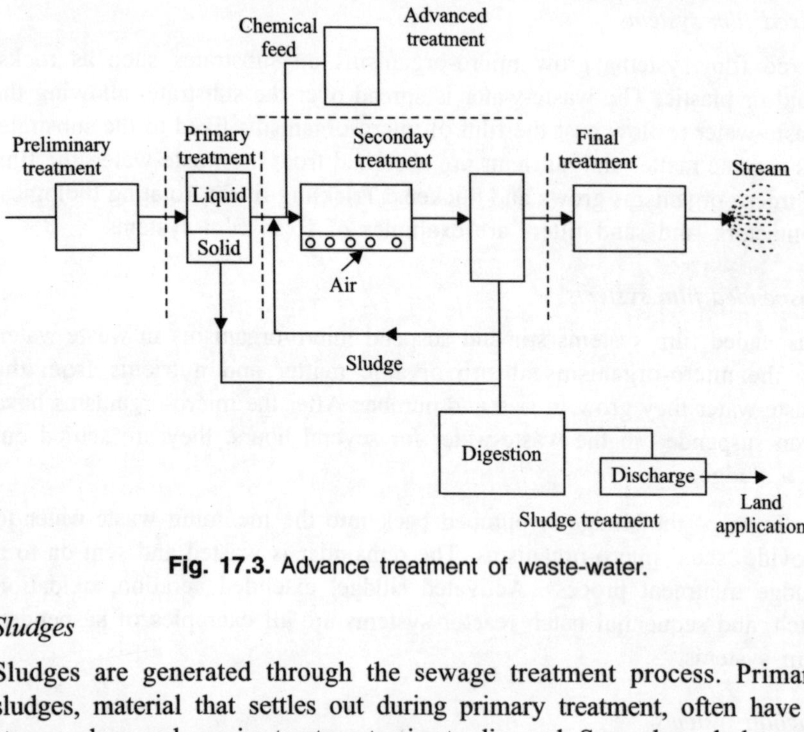

Fig. 17.3. Advance treatment of waste-water.

Sludges

Sludges are generated through the sewage treatment process. Primary sludges, material that settles out during primary treatment, often have a strong odour and require treatment prior to disposal. Secondary sludges are the extra micro-organisms from the biological treatment processes. The goals of sludge treatment are to stabilise the sludge and reduce odours, remove some of the water and reduce volume, decompose some of the organic matter and reduce volume, kill disease causing organisms and disinfect the sludge. Untreated sludges are about 97 percent water. Settling the sludge and decanting off the separated liquid removes some of the water and reduces the sludge volume. Settling can result in a sludge with about 96 to 92 percent water. More water can be removed from sludge by using sand drying beds, vacuum filters, filter presses, and centrifuges resulting in sludges with between 80 to 50 percent water. This dried sludge is called a sludge cake. Aerobic and anaerobic digestion are used to decompose organic matter to reduce volume. Digestion also stabilises the sludge to reduce odours. Caustic chemicals can be added to sludge or it may be heat treated to kill disease-causing organisms. Following treatment, liquid and cake sludges are usually spread on fields, returning organic matter and nutrients to the soil. Waste-water treatment processes require careful management to ensure the protection of the water body that receives the discharge. Trained and certified treatment plant operators measure and monitor the incoming sewage, the treatment process and the final effluent.

UPFLOW ANAEROBIC SLUDGE BLANKET PROCESS (UASB)

The objective of any biological treatment is to coagulate and remove the non settleable colloids and to stabilise organic matter. In case of domestic waste-water, the main purpose is to reduce the organic content and nutrients such as nitrogen and phosphorous, while for industrial waste-water, is to remove or reduce the organic content.

The major groups of bio-conversion of waste-water can be classified as under:

1. Aerobic processes (in presence of Oxygen).
2. Anaerobic processes (in absence of Oxygen).
3. Anoxic processes (obligate microbes in absence or presence of trace amount of oxygen).
4. Combination of aerobic/anoxic or anaerobic processes.

These processes are further divided based on growth system, like suspended growth, attached growth or combination. Various type of anaerobic reactors have been developed in order to obtain the higher organic loading rate as well as efficiency of the system. Anaerobic reactors are combined and equipped with various units like sludge water separator and gas separator. The reactor types are completely mixed, fixed bed, moving bed, fluidised reactors, expanded bed, recycled bed, recycled flocs and sludge blanket reactor (upflow). The upflow anaerobic sludge blanket process (UASB) is the purely anaerobic suspended growth system where the bacterial mass remains suspended in the middle of the reactor and the waste-water with organic content passes through the granulated media of microbes developed in the system.

Depending upon the type of waste-water and the amount of organic waste in terms of COD and BOD levels, aerobic and anaerobic processes are selected. It becomes very expensive to regulate the aerobic system due to oxygen transfer limitations where COD is more than 1500 mg/l. The anaerobic process naturally becomes the choice when the COD is more than 1500 mg/l. Nowadays, anaerobic system have been found to be economical even for lower BOD values. If properly designed and operated anaerobic process specially the upflow anaerobic sludge blanket can yield major benefits in comparison to any other system in addition to production of valuable by-products like methane. Anaerobic conversions are attractive for several reasons listed as below.

1. No oxygen requirement.
2. Low capital cost.
3. Low nutrient requirement.
4. Low production of biological sludge.
5. High treatment efficiency.
6. Low operating cost.

In the UASB system the flocs are kept in suspension by the effect of gas bubbles. In order to avoid mechanical mixing, the waste-water must be evenly distributed over the bottom of the reactor. An often used technique is to distribute the incoming water through bottom inlet pipes. These are by no means mandatory for sludge blanket generation. The flocs have a biofilm structure and are formed in granules of 1–5mm diameter. In the lower part of the reactor this dense layer of sludge granules is present. This results in sludge concentration profiles with considerable concentration variation within the reactor. Some sort of the reactor separation, either internal (in the reactor) or external is needed.

The above reactors are being used in the biological waste treatment, however the upflow anaerobic sludge blanket reactor, is the most recent advancement in the development of bio-reactor which requires lesser power and give better efficiency of removal. Hydroair has improvised the system and developed hydro methane reactor which is described here.

Special Feature of UASB

Considering the various advantages a bio-reactor having features of UASB is designed by Hydroair and is named as 'Hydro Methane Reactor'. In this unit the anaerobic digestion proceeds in absence of oxygen where anaerobic bacteria degrades the organic matter and converts the same into the carbon dioxide and methane.

This process is successfully applied wherever the organic matter is present, due to many advantages which are as follows.

1. No air is required as in case of extended aeration process.
2. No power requirement for mixing and aeration.

3. The loading rate is 5 times hence the area required for treatment is 1/5th.
4. The system can be operated intermittently or continuously without effecting the efficiency.
5. It can also work if there is variation in the effluent characters.
6. More than 90 per cent removal of BOD and COD is expected in very short duration.
7. The solid produced is 1/5th of the conventional process.
8. The solid is easily dewatered when subjected to filteration process.
9. By-product gas is produced which can be used as alternative fuel.

Due to the above advantages this process is suitable for India as the temperature is favourable for the degradation. The 'Hydro Methane Reactor' proposed consists of a diffusing system which diffuses the organic matter in the biomass for active degradation. The mixing takes place with the help of gas produced. The reactor consists of solid liquid separation and gas liquid separation system for maintaining the biomass concentration in the system.

Hydroair has devised new system for the treatment of high and low organic suspended solid/dissolved solid waste with 'Hydro Methane Reactor' based on 'UASB biomethanation process'. The newly developed reactor is the fruit of research and development of few years. The new unit does not require any mechanical mixing and use waste-water with high solids/low solids concentrations (as high as 20 per cent solids as low as few ppm).

The 'Hydro Methane Reactor' is highly efficient and can achieve very high rate of waste stabilisation with practically no power consumption. The reactor works on anaerobic process of organic waste stabilisation.

The unique system 'Hydro Methane Reactor' is compartment-alised into various parts having different functions to work under anaerobic conditions. The anaerobic process is multi-stage ie. acid formation stage where facultative bacteria (acid former) decomposes the organic matter to volatile acids and the second stage is the methane formation stage where methanogens produce the methane gas, CO_2 and water along with stabilised solids. The entire process of digestion can be expressed as shown in Fig. 17.4.

Compartments in the 'Hydro Methane Reactor'

The various compartments in the 'Hydro Methane Reactor' are as follows:

1. The lower part is the main chamber where the fermentation proceeds.

2. The upper chamber is solid liquid separator.
3. The gas dome holds the gas from the upper chamber.

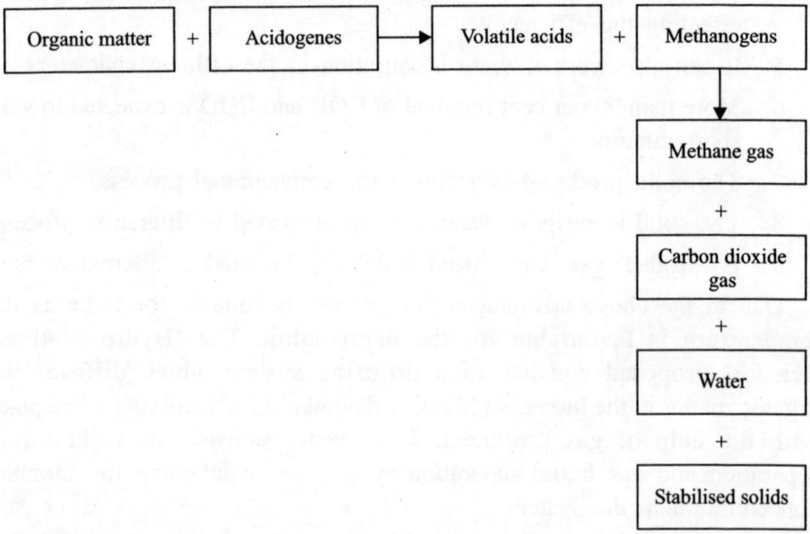

Fig. 17.4. Process of digestion.

The gas is collected in the main chamber by the operation/automatic valves provided in the gas pipe between the chambers.

The gas formed displaces the digester substrate into the upper chamber which results in increase gas pressure in the main chamber due to the level difference.

As soon as the level difference develops the gas pressure is released by opening of the valve in the gas connecting pipe. The substrate starts displacing into the upper chamber and flows back into the main chamber with high velocity. Due to these reflex actions and turbulence created the scum and sediments are perfectly remixed with the existing substrate in the main chamber which gives better stabilisation. The digester is completely automatic and can be operated by two methods.

1. Pressure control method
2. Time control method

The pressure control method is used with substrates having low specific gas production rate. The cycles are determined by gas production and pressure rise. While in time control method small gas blower is added for the microbial gas production to ensure certain fix number of mixing cycles per day.

Advantages of 'Hydro Methane Reactor'

'Hydro Methane Reactor' has unique advantages. There are given below:

1. No mechanical parts or moving parts required.
2. No maintenance cost required.
3. Reduction in the installation cost.
4. No servicing and maintenance cost required.
5. Shorter retention tank are possible by maintaining high concentration of bio-mass.
6. The process is highly reliable as it has good control of scum and sediinent mixing.
7. The net result is reduction in capital cost.
8. No short circuit takes place on account of the separate partitions.
9. Bio-solids do not escape.

Types of wastes for which 'Hydro Methane Reactor' has been successfully adopted

'Hydro Methane Reactor' has been successfully adopted for following types of wastes.

1. Sewage waste-water from township.
2. Pharmaceutical waste from production of protinex.
3. Vegetable waste from kitchen or canteen.
4. Edible oil and fats.
5. Aquaculture, prawns and fish processing waste.
6. Slaughter house waste.
7. Vegetable and food processing waste.
8. Poultry, broiler and mid processing waste.
9. Water hycinth waste, grass and straw waste.
10. Waste from hotel and garden.

The above process can be used for generation of electricity with the bio-gas produced. The Hydro Methane digester is very versatile, robust and easy to operate. This process has been utilised for kitchen, pharmaceutical and sugar distillery type of waste successfully in many of our plants.

The gas containing methane 65 per cent and carbon dioxide 35 per cent with some moisture can be utilised in the process for generation of heat or power as the case may be.

Significance of Microbes in UASB System

As said earlier the objective of waste-water treatment is to remove pollutants from the water which are soluble and insoluble organics in the form of nitrogen, phosphorous, carbon and inert insoluble materials. It is well known if favourable conditions are provided, the soluble and insoluble organic materials can be effectively removed through biological action by micro-organisms. For effective design of biological waste-water treatment process, it is necessary to have understanding of the following.

1. Nutritional requirement of microbes.
2. Environmental factor.
3. Metabolism.
4. Relationship between microbial growth and substrate utilisation.

The biological processes are categorised into aerobic and anaerobic process wherein organisms grow in presence of oxygen or absence of oxygen. The microbes which undertake the microbial reaction based on their nature are divided into various categories like.

1. Autotrophic, (Photosynthetic).
2. Aerobic (Cellular respiration).
3. Hetrotrophic (Anaerobic).

On the basis of energy source required micro-organisms are classified into following:

1. Phototrophs—These are organisms that use light as their energy source.
2. Chemotrophs—These are organisms that employ oxidation-reduction reactions to provide energy. Chemotrophs can be further classified on the basis of the type of chemical compounds oxidised (Electron donor).
3. Chemo-organotrophs—These are organisms that use complex organic molecules as their electron donor, where as chemoautotrophs use simple inorganic molecules such as hydrogen sulphide or ammonia.

Activities of microbial cell depend upon food utilisation and chemical reactions which are controlled by enzymes. Enzymes are proteins produced by living cell that act as catalysts to accelerate reactions like hydrolytic, oxidative and synthetic. Hydrolytic enzymes are used to hydrolyse complex insoluble food substances into simple soluble components which can pass through cell membranes by diffusion. They are also called extracellular enzymes. The hydrolysis reaction involves the addition of water to a complex compound and splitting of compound into simpler soluble compound.

Oxidation Reactions

These are signified by addition of oxygen or removal of hydrogen. Most organisms oxidise their food by enzymatic removal of hydrogen from the molecule. The enzymes known as dehydrogenases remove the hydrogen from the compound. The final hydrogen acceptor is determined by the aerobic or anaerobic nature of the surrounding medium. For aerobic reactions oxygen is the final hydrogen acceptor and water is formed. Under anaerobic conditions oxidised compound accepts the hydrogen and a reduced compound is formed.

Energy is liberated during oxidation and energy is consumed during reduction and the net result is more energy is liberated than the used up which is available for use by the cell. The synthesis of cellular material for both the maintenance of cells and new cell is catalysed by intracellular synthetic enzymes.

The activity of enzymes is affected by environmental conditions like temperature, pH, presence of ions like phosphate, magnesium and calcium. Since enzymes are affected by environmental conditions, hence microbial growth is also affected which is the main basis of design of biological waste treatment. The presence or absence of oxygen divides organisms into three classes.

1. Aerobes—Organism which use molecular oxygen as an electron acceptor.
2. Anaerobes—Organism which do not use molecular oxygen.
3. Facultative—Organism which can use either oxygen or some other compound as their electron acceptor.

Obligate aerobes are not able to grow in absence of oxygen. The principal significance of the electron acceptor used by the microbes is completeness of reaction to enable release of energy for growth and life support. Heterotrophs, aerobic and facultative organisms oxidise their food completely whereas anaerobes do not oxidise completely and only ferment. The anaerobic process proceeds without any extra source of power to inject oxygen as in case of aerobic process and the limiting factor in the aerobic process is the oxygen transfer efficiency of the mechanical systems. The anaerobic waste treatment processes naturally become an effective method for the treatment of organic waste. This treatment is mediated by facultative and anaerobic microbes which in the absence of oxygen convert organic materials into gaseous end products such as carbon dioxide and methane.

Major advantages of this process

The major advantages of this process are as follows.

1. Less bio-mass produced per unit of substrate (organic material) utilised, which means a decrease in the requirement for nitrogen and phosphorus.
2. Economical value of methane gas.
3. Higher organic loading potential because the process is not limited by oxygen transfer capability.

The anaerobic process is successfully utilised in municipal waste and industrial waste of medium strength, brewery, distillery, fatty acid, wood fibre and dairy. The anaerobic treatment of complex waste involved three distinct stages. The first stage complex waste components like fats, proteins and polysaccharides are hydrolysed to sub-units by heterogenous group of facultative and anaerobic bacteria. The second stage of hydrolyses is fermentation of above sub-units by β-oxidation and other metabolic process leading to formation of short chain volatile acids and alcohols. In this stage little stabilisation of BOD and COD is realised. The third stage of the process is the formation of gases like methane and carbon dioxide by several different species of strictly anaerobic bacteria. In this process, the true stabilisation of the organic matter occurs and it is referred as methane fermentation.

The formation of methane by methane bacteria is result of metabolic activity of bacterial group which are terminal members of the food chain and they convert fermentation products formed by other anaerobes. The gas has low solubility in water and it escapes. Methanogenic bacteria are associated with rumenic digestion and divided into four genera:

1. Methano sarcina.
2. Methano bacterium
3. Methano spirillum
4. Methano coccus

The substrates for the bacteria are hydrogen, carbon dioxide, formate, acetate and methanol depending on the genera and ammonia is the preferred nitrogen source.

The bacteria responsible for acid fermentation are relatively tolerant to pH changes and temperature and have a much higher rate of growth than the bacteria responsible for methane fermentation. As a result the methane fermentation is assumed to be rate controlling step (Table 17.3).

Table 17.3. Bacteria responsible for acid fermentation.

Acid producing bacteria	Acid producing bacteria	(Butyric and propionic acid)
	Acetogenic bacteria	(Acetic acid and hydrogen)
Methane producing bacteria	Acetoclastic methane bacteria	(Acetophilic)
	Methane bacteria	(Hydrogenophilic)

Keeping in view of the above mechanism, the rate limiting step approach is used for the anaerobic waste treatment in UASB.

Environmental Factors of Importance for UASB System

Tamperature

Temperature plays an important role in the anaerobic degradation. The most common range for process of operation is between 30–40°C. The yield coefficient, decay rate and specific growth rate of methanogens depends on temperature. Thermophillic processes have higher constant for methane production rate.

The substrate removal rate as compared to mesophillic rate is higher. The major problem with thermophillic processes is low yield resulting in very slow start up and slow accommodation to loading variation, substrate changes or toxic substances. At high temperature few bacterial species are unable to grow. The temperature shocks can be used to facilitate the separation of solids and liquids by suspending gas production temporarily. UASB in tropical country like our use mesophillic bacteria as the temperature are favourable 25 to 35°C hence no heating required for optimum operation.

Nutrients

The theoretical minimum ratio of COD/N has been observed between 350/7. A value of 400/7 is considered to be reasonable good for high load. For low loads less than 0.5 kg COD/kg VSS/d the COD/N ratio of 1000/7 or more is found to be adequate. There are many other nutrients required like phosphorus, iron, nickel, magnesium, calcium, barium, cobalt and sulphate which improves the process performance as mentioned in Table 17.4.

Table 17.4. Nutrients which improve the process performance.

Compound	*Beneficial Conc. G/m³*	*Effects*
Fe^{++}	0.2	Precipitation of sulphide
	12-120	Flocculation/Biofilm structure
Ni^{++}	0.01	Build up of F_{430} co-factor in methanogens
	0.006	Increase in activity
Mg^{++}	0.01–0.02	Flocculation
Ca^{++}	0.01–0.04	Flocculation
Ba^{++}	0.01–0.1	Flocculation
Co^{++}	0.01	Vitamin B_{12}
	0.003	Increase in activity
SO_4^{--}	0.02	Increase in activity

Toxic substances

There are certain toxic substances which will inhibit the process. The bacterial mass may adapt to the toxic substances if they are acclamatised and anaerobic cultures seem to have ability to recover after shock dose exposure to toxicants. The anaerobic microbes are able to dechlorinate chlorophenolic compounds and remove mutagenity and detoxify certain wastes.

Sulphide toxicity

The sulphide and sulphate contents in high concentration like molasses, pulp and paper, pectine, wine, distillery etc. may result in inhibiting process because of the low pH and toxicity of these substances may reduce in presence of iron due to precipitation of ferrous sulphate. Sulphate is reduced to sulphide during the process of fermentation.

It has been observed in the upflow anaerobic process where sulphate bacteria predominates at the bottom of the bed, free hydrogen sulphide can inhibit the process at concentration of 0.1 kg S/m^3. In Indian conditions it have been observed that sulphate concentration in the range of 2000–4000 mg/l can be accepted.

Ammonia

Free ammonia is the most toxic compound and inhibit at a concentration of 0.1–0.2 kg N/m^3. It has been observed that total ammonia concentration of 5–8 kg N/m^3 can be tolerated if the pH of the reactor is low.

Volatile acids

The toxic level of free undissociated volatile acids is observed between 1–2 g/m^3. The propionic acid is more toxic than acetic acid. The toxic level of higher fatty acids have similar inhibitory effect as volatile acids.

pH dependency

The methane bacteria have optimum pH between 6–8 whereas the acid producing bacteria have optimum pH between 5–6. Since methane step is rate limiting the pH should be kept above 6.0. It has been observed that in milk waste if the pH varies occasionally from 5.8–10.2 it has no adverse effect.

Economics and Cost Consideration

As described earlier the UASB process is cost effective w.r.t. capital cost, land cost, operation and maintenance cost etc. It has been observed that the UASB process will be more cost effective as compared to any other process and it will give methane gas as an alternative fuel which gas can be efficiently fired in almost any appliance wherever fuel is used. The composition of gas have 65 per cent methane, 35 per cent carbon dioxide and traces of hydrogen sulphide. The biogas has following characteristics.

Calorific value of methane gas	8500 kcal/m^3
Relative fuel value as compared to coal	1.542
Fuel utilisation efficiency of methane is	90 per cent as compared to 66 per cent of coal
Gross relative value w.r.t. coal	2.1

In addition to monetary savings and the advantages of UASB (HMR) are as below:

1. The proposed anaerobic (UASB) process followed by polishing system will be faster to restart and achieve maximum efficiency within a short time as compared to conventional system.
2. The proposed process can also effectively handle shock load without hampering the process efficiency.

The UASB system has unique advantage as compared to any other process in Indian conditions and it is most attractive as a treatment method. It will be essential to aerate the effluent water to remove the septicity of the effluent for proper disposal in surface drain.

It is undoubtedly, ideally suited process for organic wastes with very high load have energy recovery as compared to other processes which are either power intensive or requires large space of land.

The upflow anaerobic system is found to be versatile in case of dairy, distillery, municipal, pharmaceutical and other related industries. The microbes can be acclamatised and the upflow reactor can be matured to give granulated sludge for very high efficiency of conversion of organic matter. The upflow anaerobic reactor is unique where the organic flocs remain in suspension due to upflow velocities of the effluent water and settle in the specially designed separator for maintaining the microbe concentration in the reactor.

The microbes play a greater role in the UASB system by the property of granulation when subjected to favourable conditions. A very high concentration of microbe can be maintained due to property of granulation thereby accepting higher loading of 300/m^3 of the reactor volume.

Reactor bio-mass can remain in dormant stage for months in case of no effluent flow and can be revived in short period of days to pull operating capacity. Even in case of shock loads the reactor do not get disturbed due to inherent property of granulated bio-mass.

ENZYMATIC PRE-TREATMENT TO ENHANCE THE BIODEGRADABILITY OF INDUSTRIAL WASTE-WATER

Enzymes are components of micro-organisms that catalyse a wide variety of chemical reactions. Evaluationary enzymatic chemistry in highly selective, efficient, temperate and environmental friendly. Though much attention has been paid in the utilisation of biocatalysts in several field, their involvement has been felt very recently in solving environmental problems due to obvious reasons. In this section the role of enzymes in enhancing the biodegradability of waste-water contains xenobiotic compounds and application of enzymatic treatment for various types of wastes are discussed.

Recent advances in waste-water treatment stressed the need for more reliable biological processes, which can enhance the process efficiency. The toxic and inhibitory compounds normally encountered in waste-waters adversely affect the ongoing biological process by inhibiting growth of microbial population and the performance. Pre-treatment option for such kind of waste will enhance the biological process efficiency and due to enhanced biodegradability of the waste-waters. Normally, physico-chemical processes such as adsorption, coagulation, chemical oxidation, etc. are used as pre-treatment options to enhance process efficiency. However, overlying cost, hazardous solid waste generation and regeneration problems are the inherent limitations of these processes.

The recent progress in enzyme science and engineering encouraged waste-water scientists to focus their attention to apply reliable and sophisticated activity of enzymes in the treatment process to enhance the effluent biodegradability. Enzymes (protenaceous material) are defined as temperature sensitive organic catalysts produced by living cells and capable of acting outside or inside the cell. The industrial use of enzymes as biocatalysts dates back to the early 1900s with the development of enzyme processes to manufacture cheese products, bread, beer and wine. Enzymes have capability to differentiate between chemicals with closely related structures and can catalyse reactions at relatively low temperature and in the entire pH range (2–14). This results in high quality products, fewer by-products and simpler purification procedures. In addition, enzymes are non-toxic and biodegradable and can be produced in large quantities without the need for special chemical resistant equipment. Over recent years, the use of enzymes as catalysts has been progressively developed in many different fields of industrial interest. Biocatalysis is now a well-established and active research area. The catalytic action of enzymes is extremely efficient and selective compared to chemical catalyst; these enzymes demonstrate higher reaction rates, milder reaction conditions and greater stereospecificity.

Enzymatic Treatment

Enzymes can act on specific recalcitrant pollutants to remove them by precipitation or transformation to other (innocuous) products and also can change the characteristics of a given waste to render it more amenable to treatment. Enzymatic treatment has been proposed by many researchers as a potential alternative to conventional methods, due to its highly selective nature and relatively less likely inhibited by toxic substance. Enzymes operate over a broad aromatic concentration range and require low retention time with respect to other treatment methods. A large number of enzymes from a variety of different plants and micro-organisms have been reported to play an important role in a array of waste treatment applications. Peroxidases are oxidoreductases produced by a number of micro-organisms and plants (lignin peroxidase, horseradish peroxidase, chloroperoxidase) were reported for treating xenobiotic compounds. Polyphenol oxidase (tyrosenase, laccase, etc.) were also studied for its potential for application. Enzymes were studied for its effectiveness in treating waste containing xenobiotics, cyanades bearing waste, pesticides, food processing waste. Other potential applications for enzymatic treatment are soil decontamination and fly ash and soil stabilisation.

Xenobiotic compounds azo-based synthetic dyes and pesticides are not amenable to conventional biological waste-water treatment. Possible reasons for their non-biodegradability is the lack of requisite enzymes in the conventional biological treatment plant. Anaerobic transformation of azo dyes begins with reductive fission of the azo linkage, resulting in the formation and accumulation of colourless aromatic amines, which can be highly toxic and carcinogenic. Under aerobic condition, azo dyes are considered essentially non-degradable by bacteria. However, using an alternative anaerobic-aerobic treatment, total degradation of a sulphonated azo dye was achieved. Achieving microbial degradation of azo dyes is a tedious and uneconomical task. *Paice* and *Jurasele* alternatively proposed the use of oxidative enzymes of lignin biosynthetic pathway from plants.

The development of stable, highly active microbial oxidative enzymes has long been an exclusive goal of biotechnology for a wide variety of commercially attractive processes, from polymer synthesis of bioremediation. Oxidative enzymes are a large class that function by oxidation-reduction reaction mechanisms to introduce oxygen into or remove electrons from organic compounds. Oxygen donors in the form of peroxidases (ROOH) or molecular oxygen (O_2), donate oxygen and a second substrate/co-factor, which is required to regenerate the resting state enzyme. Catalase reaction specificity has been shifted to peroxidase, the high pH, thermal and oxidative stability of a fungal peroxidase has been dramatically improved. Microbial oxidative enzymes are of particular commercial interest because of their potential to effectively substitute chemical methods currently used in a variety of applications: introduction of oxygen functionality into natural and synthetic chemicals, biosensor, environmental remediation of pollutants, polymer synthesis, degradation of coal and breakdown of woody pulps for fermentation and paper production.

Methane mono-oxygenase, toluene mono-oxygenase and toluene dioxygenase are involved in biochemical pathways responsible for the initial steps in the oxidation of methane, phenol and toluene, benzene respectively by a process called co-metabolism. Specifically, plant peroxidases and plant laccases/polyphenoloxidase are considered to be good biocatalysts. Use of plant peroxidases in removal of phenolic pollutants from aqueous solutions is well documented. It has been demonstrated that horseradish peroxidase can catalyse free radical formation followed by spontaneous polymerisation of a variety of aromatic compounds including phenol, chlorophenols and other substituted phenols. *Bhumai* studied degradation of dyes using horseradish peroxidase. Parathion hydrolase enzyme produced by *Flavobacterium* was reported to be degrading 97 per cent of diazinan in 24 hours and rapidly hydrolyse organophosphate insecticide coumphos to less toxic compounds/products.

The effectiveness of organophosphoric acid anhydrase enzymes and biocatalysts to detoxify phosphonfluoridate type agent has already been reported. The enzymatic biocatalysts have been shown to have superior rates and efficiencies in comparison with chemical decontamination approaches. *Coprinus macrorhizus* peroxidase, like many other peroxidases, catalyses the oxidation of a wide variety of aromatic compounds by reduction of hydrogen peroxidase or other organic hydroperoxides. Normally enzyme activity as catalyst and its persistence in system will depend on its origin. *Taylor* used fungal peroxidase for selective removal of aromatic compounds from waste-water.

The catalyst ability of peroxidase is a function of pH with a broad optimum over a pH range of 6 to 9. Normally aromatic removal is dependent on the amount of catalyst added since the catalyst had a finite lifetime and also conversion is dependent on the contact time. Extracellular fungal peroxidases oxidatively dechlorinate polychlorinated phenols at the fourth position *Valli* and *Gold* studied degradation pathway of 2,4-dichloro phenol by extracellular fungal peroxidase and showed to have several cycles of oxidation and other reactions in the degradation pathway. *Saunder* studied peroxidase enzyme conversion of phenols into polyphenols, which are commonly insoluble in water and can be removed by filtration. *Arseguel* studied horseradish peroxidase enzyme for degradation of phenols in the presence of inert supporting material such as talc and reported that phenolic waste-water can be depolluted successfully. Adsorption of the reaction products on the talc effectively protected biocatalyst against contamination by oxidative products, thereby prolonging the catalytic action and leading to almost complete elimination of phenols in aqueous media.

White rot fungus *Phanerochaete chrysosporium* which produce lignin peroxidases enzyme was studied for removal of phenol and found to enhance performance of the subsequent biological process. Laccases which are blue copper oxidases with broad specificity towards catalysis of four electron reduction reactions from oxygen to water are highly efficient in degrading a variety of aromatic compounds containing hydroxyl and amine groups. Moreover these biocatalysts, in presence of redox mediators, can also oxidise non-phenolic, halogenated aromatics and other xenobiotic compounds. The treatment of oils and fats has been recently investigated in order to evaluate the possibility of replacing traditional chemical processes with biotechnological processes. In anaerobic degradation of industrial waste-wasters containing both dissolved and insoluble solids, particle hydrolysis is considered as the rate limiting step. The soluble organics are usually methanised long before the suspended solids (SS) have been hydrolysed and the complete degradation of the solids requires extended hydraulic retention time (HRT). During the anaerobic treatment of slaughterhouse waste-water containing high concentration of suspended solids, especially fats, seriously impairs the performance of anaerobic reactors operated at high organic loading rates. The application of pre-treatment to hydrolyse particles particularly fats, may thus accelerate and stabilise the anaerobic digestion of waste-water. *Masse* tested the effect of four hydrolysis pre-treatment of fat particles in slaughterhouse waste-water—NaOH along with three lipases of plant, bacterial and pancreatic origin respectively and found that pancreatic lipase showed most efficient treatment by reducing the initial average particle size by up to 40 per cent after 4 hrs. of mixing at room temperature. Particle size reduction has been shown to increase the rate and extent of anaerobic treatment. The effect of enzymatic pre-treatment on to hydrolysis and size reduction of fat particles was studied by *Masse,* was stressed the need for future investigation on the effect of enzymatic pre-treatment on the rate of anaerobic digestion. *Cammaropta* studied enzymatic pre-hydrolysis with fermented basassu cake (*Penicillium restrictum* lipase) and a pre-treatment for anaerobic degradation of waste containing high fat content. Due to pre-treatment, COD removal upto 90 per cent and decrease of VSS upto 90 per cent was observed.

Enzyme Reactors

Enzyme reactor is described as the vessel in which a reaction, catalysed by enzymes or cells, takes place, together with all the devices necessary for control and sampling. Batch continuous flow, continuous, flow stirred tank (CFST), plug flow, fluidised bed reactor configurations have been extensively studied using enzymes. The performance of the different reactors depends on the physico-chemical properties of the reaction mixture. *Giorno* compared the efficiency of lipase to hydrolyse vegetable oil triglycerides into fatty acids and glycerol using different reactor configurations, viz.; a traditional emulsion stirred tank reactor (STR), an emulsified organic aqueous enzyme membrane reactor, where the reaction occurred in emulsion and the aqueous phase was ultrafiltered through a membrane, thus separating the product and a biphasic organic aqueous enzyme membrane reactor where the two phases are separated by the membrane that also contained the immobilised enzymes.

The results showed that compared to apparent volumetric reaction rate of the free enzyme, the catalytic activity of the immobilised enzyme was considerably more stable. Biocatalytic membrane reactors combined selective mass transport with chemical reactions and the selective removal of products from the reaction site increases the conversion of product inhibited or thermodynamically unfavourable reactions. Recent trend towards environmental friendly technologies make these membrane reactors particularly attractive because they do not require additives, are able to function at moderate temperature and pressure and reduce formation of by-products. *Al-Kassim* and *Taylor* used batch reactor for removing aromatic compounds from waste-water using peroxidase. *Lewandowski* studied degradation of 2-chlorophenol by *Penicillium chrysosporium* in packed bed reactor with silica based porous support and an alginate immobilised cell system.

Applications of Immobilised Enzymes

More recent development in the field of enzymes is their application in immobilised form in waste-water treatment. This aspect is of considerable interest owing to the low cost of the substrate, the production of re-usable material and the elimination of waste disposal. One of the most important developments appears to be the use of immobilised lactase for the hydrolysis of whey, waste material from cheese manufacture. Cellulose is a major constituent of domestic and commercial waste and it is hydrolysed by a complex of enzymes known as celluloses. The end product, glucose, can be used for single cell protein production, conversion to fuels, as sweetener and as the starting material. To treat complex effluents with immobilised enzymes which are capable to depending the polluting material with the pollutant as a substate, the waste-water would be brought in contact with the enzyme for a time sufficient for the pollutants to be converted to a less toxic form.

Munneche has shown that parathion hydrolase, covalently immobilised on glass can be used to detoxify organophosphate pesticide in industrial waste-water. *Talbot* described the use of immobilised parathion hydrolase to degrade parathion to *p*-nitrophenol and diethyl thiophosphoric acid. a-amylase has a potential use in waste-water treatment and soluble form of this enzyme has been used to clarify the starch clay-suspensions produced in large volumes by the paper industry. The advantage of using immobilised a-amylase include elimination of the enzyme denaturing step which is required with soluble a-amylase and it would enable large volume of water to be treated continuously. Other potential use of immobilised enzymes in waste-water treatment include the removal of carcinogenic aromatic amines such as benzidene, naphthylamine and aminophenyl found in waste-waters from the coal, resin, plastic and textile industries using immobilised peroxidase.

Thus enzymatic treatment has a promising future in waste-water treatment. Much work has to be done on application of enzymes as biocatalysts in degradating xenobiotic compounds to enhance the efficiency of subsequent biological process. The process monitoring with respect to optimisation pH, substrate specificity and inhibition towards enzymes and kinetic constants, etc. will immensely enhance the understanding in this respect. In order to fully establish the use of enzymes, as biocatalysts in waste-water treatment reactors, studies on the design of bioprocesses, particularly for large-scale production, the control of the reaction and kinetic mechanism and immobilisation procedures need to continue.

BIOENERGY

Energy conservation and energy production from cheaper resources are the need of the hour. Biotechnology can make a lot of contribution in this context, by increasing the acceptability of biomass, biogas, fuel alcohol as feasible alternatives. Advantages of bioenergy generation will be its ecofriendly nature, and the ready and cheap availability of raw materials. If bioenergy generation is coupled with the tapping of unutilised biomass, wasteland utilisation for biomass production can be achieved. Developments in reactor design and gene manipulation of micro-organisms has made the task of generating bio-energy from wastes a reality.

Biotreatment of wastes can fall into two categories: aerobic and anaerobic. Over the years, aerobic biotreatment (a more stable process) has been considered to be more efficient in BOD/COD reduction, for nitrogen and phosphorus removal, and has been applied on a wide range of wastes. Anaerobic biotreatment has been considered to be sensitive to toxicity and less reliable in performance in terms of COD reduction. However, in recent years this has been proved untrue, and anaerobic processes have come to be preferred as they produce less sludge, consume less energy, can operate at high organic loading rate and produce no odour or aerosol nuisance. Anaerobic biotreatment of wastes is now getting importance due to energy production that is possible along with pollution abatement.

It is not the energy production which is aimed for distribution of this energy to other places but this local energy production fulfils the energy needs of the particular unit which uses anaerobic biotreatment for waste disposal. Thus instead of any energy expenditure on the effluent treatment plant, there is energy generation from it to meet the overheads of treatment.

Bioenergy from Food Industry Wastes

Effluents from the food processing industry are most suitable for biogas production, as they typically have high BOD due to easily biodegradable organic matter. Effluents rich in carbohydrates are most suitable for methane production, but even those with high fat and protein contents are suitable. Breweries and other processing plants including dairy cooperatives, sugar plants, grain mills, gasohol plants, produce waste-water that is characterised by its content of complex organic matter, either in solution or as volatile, suspended solids, which can be treated anaerobically to generate methane. Maximisation of the efficiency of this conversion demands a focused engineering approach, depending in the specific nature of the effluents. Consequently data gathered from an on-site pilot plant is essential prior to the engineering of a full size plant. The price of the pilot plant operation will be recouped many times in design, construction, and operational savings of the industrial scale plant.

By-product wastes from vegetable processing originate in canneries in two main forms: discrete solids (leaves, trimmings, stems, peels, pods, husks, cobs, silk, and defective processed vegetables) and screened effluent (washing, husking, desilking, blanching, cutting, peeling, slicing, clipping, screening, grading, and inspection). Some waste solids from canneries processing vegetable crops are given away to farmers for use as animal feed. The remainder must be either land applied or placed in a landfill. The voluminous effluent streams from canneries represent high BOD sugar solutions, which cannot be released to surface waters and must therefore be spray irrigated on farmland or processed at high expense through private or municipal treatment plants. Costs of disposal vary depending on the type of waste and the method of treatment employed.

Ethanol can be produced from these wastes. Solids from vegetable wastes are pressed hydraulically, resulting in a liquid rich in soluble sugars. The pressed residue contains the bulk of the crude fibre carbohydrate (cellulose and hemicellulose). The soluble or extractable carbohydrates are present in the liquid portion of the screened effluent. It is estimated that the ethanol yield from the by-product carbohydrates will produce 172 gallons per ton from the starch fraction and 120 gallons per ton from the cellulose/hemicellulose fraction. The extractable carbohydrates are directly fermentable without enzymatic saccharification. The extractable carbohydrate concentrations in the liquid portion of these wastes are very low, though ideally, sugar concentrations of the liquid feedstock for fermentation should be in the range of 15–20 per cent.

Bioenergy Production from Paper and Pulp Industry Wastes

Major by-products of processing cellulosic biomass include, lignin, furfural, carbon dioxide and inorganic material. Lignin has a fuel value. Inorganic material like clay, titanium dioxide, calcium carbonate may be required to be removed as it may be harmful to saccharification process in ethanol production.

However, pulp and paper mill waste is rich in cellulose and has a potential for ethanol production. It was determined that pulp and paper mill sludge has an ethanol yield potential of 51 to 74 gallons of ethanol per dry ton. Cellulose must be hydrolysed to glucose before fermentation to ethanol. Conversion efficiencies of cellulose to glucose may be dependent on the extent of chemical and mechanical pretreatments to structurally and chemically alter the pulp and paper mill wastes.

The method of pulping, the type of wood, and the use of recycled pulp and paper products also could influence the accessibility of cellulose to cellulase enzymes. The chemical contaminants in the sludge from pulp and paper processing, have the potential to interfere with the hydrolysis of the cellulose material to glucose, or inhibit fermentation activity however printing inks do not have any adverse effects on enzyme hydrolysis or yeast fermentation. This may reduce ethanol yield, but these effects have not yet been quantified.

Biogas Production from Rubber Industry Effluents

Anaerobic digestion by the Biotim system in Malaysia is the first full-scale application for rubber industry effluents. These effluents have high sulphates and ammoniacal nitrogen which are inhibitory to digestion. Smell problem in the ponding system and high operation cost in the oxidation ditch make it necessary to set an alternative anaerobic system. Loading rate is 18.5 kg COD/m^3 day, hydraulic residence time (HRT) is 1 and 1.85 days respectively in Biological Conditioning Reactor (BOR) and Methane Upflow Reactor (MUR). 70 per cent COD is removed and biogas yield is 580 m^3/day.

Hydrogen Gas Production from Industrial Wastes

Hydrogen gas is a perfect renewable fuel. Hydrogen when burnt does not cause any pollution but regenerates water. Hydrogen production process operates at a normal temperature and no toxic materials are produced in the process. If cost of production is managed, hydrogen can become a common fuel. Recently some waste-water treatment processes are reported, which produce hydrogen gas along with effluent treatment. Hydrogen gas has 3 times calorific value per unit weight of petroleum and it does not generate CO_2 (so no green house effect gas) or other pollutants.

Microbial Production of Energy from Industrial Waste—New and Innovative Methods

Methane, a major gaseous fuel can be produced in abundant quantities from industrial wastes. This would substitute either partial or total energy needs of an industry along with stablilisation/disposal of effluent. Several process improvement approaches such as innovative fermentation modes, application of novel reactor designs, treatment of feeds, use of biostimulants may be used in order to optimise methane production. Methane, a major gaseous fuel is obtained by microbial transformation of substrates in a variety of natural or man-made anaerobic environments. The conditions necessary for microbial production of methane include:

1. The presence of degradable organic substrates.
2. The presence of suitable nutrients.
3. The presence of sufficient water to submerge or dissolve organic substances and to dilute soluble compounds to below toxic levels.
4. The presence of suitable type of bacteria.
5. Optimum pH 6.6–7.6 and temperature (mesophilic, 30 to 38°C and thermophilic, 49 to 57°C).

This process can accept pure as well as heterogeneous feeds including almost all types of organic wastes. The bioconversion of complex organic compounds to methane is achieved by anaerobic digestion, which is carried out in airtight reactors/digesters.

Due to rapid industrial growth, the problem of effluent disposal of a wide variety of waste-waters has been attracting global attention and it has been proved that it is possible to recover biogas (methane) from the waste which in turn can be stabilised. It is, therefore, possible to generate clean fuel, methane with simultaneous alleviation of difficult environmental problems.Enzymic Treatment of Waste Materials

BIOREMEDIATION

Bioremediation is a pollution treatment technology that uses biological systems to catalyse the destruction or transformation of various chemicals to less harmful forms. The general approaches to bioremediation are to (i) monitor the natural biodegradation process (intrinsic bioremediation); (ii) carry out environmental modification, such as through nutrient application and aeration (biostimulation); and (iii) add micro-organisms (bioaugmentation). The end products of effective bioremediation, such as water and carbon dioxide, are nontoxic and can be accommodated without harm to the environment and living organisms.

While conventional technologies call for moving large quantities of toxic waste-contaminated soil to incinerators, bioremediation typically can be performed on-site and requires only simple equipment. Bioremediation, though, is not the solution for all environmenatal pollution problems. Like other technologies, bioremediation is limited in the materials that it can treat, by conditions at the treatment site, and by the time available for treatment.

Where applicable, bioremediation is a cost-effective means of restoring environmental quality. Its cost effectiveness as compared to chemical and physical treatment technologies, especially for dilute contaminants, is the main driving force for the use of bioremediation.

Bioremediation of contaminants or pollutants is simply an extension of normal biological metabolism. As biocatalysts, enzymes can enhance the rate of feasible conversions. The characteristics of these processes are detoxification and mineralisation of the pollutants to biomass, carbondioxide and water.

Environmental contamination or pollution can be regarded as a malaise to the ecosystem, while, bioremediation can be regarded as a kind of environmental medicine. The nature of waste being generated has changed significantly with time, much historical pollution can be traced to either improper waste management practices that advocate disposal rather than treatment, or to both, accidental or incidental spillages, leaks, etc. Like for most diseases, suitable drug therapy is evolved on rational lines, there is a need for case specific evaluation and treatments. This investigation relates to the utilisation of aquatic macrophytes for pollution abatement. The macrophytes described herein are water hyacinth and water lettuce which are very commonly found in our Indian environments, hence readily available for application.

Water is an important commodity which is recently becoming highly polluted. As a result of this, several living forms which are better suited to the polluted waters have outbursted in our aquatic environments causing ecological imbalance.

Every water body has its own capacity to maintain its natural state through self purification process, however, when the discharge of pollutants is heavy, the process of self purification of water bodies is adversely affected and the water remains polluted. The EPA US has listed 129 pollutants occurring in industrial waste-waters of which 13 are metals. Attention has been focused on heavy metals as environmental pollutants since the occurrence of 'Itai-Itai' and 'Minimata' diseases caused by Cd and Hg poisoning. Chromium and Nickel are also toxic metals entering into the streams for various industrial effluents.

Aquatic plants are important components of lakes, ponds, rivers and streams ecosystems. Aquatic weeds like *Eichhornia crassipes, Pistia stratiotes, Lemna minor* etc., have drawn much attention for removal of heavy metals and organic pollutants.

Pollution abatement through efficient and cost-effective alternatives to conventional processes had been carried out by utilising nature's own technology for instance, aquatic plants like *Pistia, Eichhornia* and *Ceratophyllum* which help in the removal of toxic metals for aquatic environment. Methods so far developed for removing heavy metals from aquatic environments appear to be highly expensive hence available weeds which have the physiological capacity to absorb these toxic inorganics as well as organic pollutants were tried with some success. This biological system is a dynamic system, hence a study was undertaken to utilise these living systems to sponge the toxic metal contaminants from various effluents containing high concentrations of the soluble metals.

Bioaugmentation to Enhance Microbial Bioremediation

The two major approaches in bioremediation of hazardous/oily wastes and environmental pollutants are microbial inoculation (seeding) and the bioaugmentation of naturally occurring microbial activities. In most cases *bioaugmentation* consists of environmental modification to eliminate some limiting factor that is restricting the rates of microbial growth and metabolism of a polluting substance. For this approach to work the pollutant must not be *recalcitrant*—that is, micro-organisms must have the genetic and physiological capability to degrade the substance. The most common factors controlled to stimulate biodegradative activities by bioaugmentation are nutrient concentrations usually nitrogen and phosphorus concentrations, molecular oxygen concentration, redox potential, and moisture levels. Additionally, co-substrates can be provided as growth-supporting substances.

The potential use of bioaugmentation of bioremediation of hazardous and oily wastes and pollutants is an outgrowth of basic studies on microbial metabolism and ecology. Studies beginning in the 1950s focused on elucidating pathways of microbial metabolism of hydrocarbons with the aim of developing single-cell protein bioreactors using petroleum-based substrates. During this period the basic understanding of metabolism of hydrocarbons was developed. These studies expanded to examine the metabolism of chlorinated hydrocarbons, many of which are used as pesticides or otherwise wind up as persistent and/or toxic environmental contaminants. These basic physiological studies on microbial metabolism were supplemented beginning in the late 1960s with an examination of the ecological factors controlling the distributions and activities of micro-organisms in the environment. Some of these studies aimed at elucidating the factors controlling the rates of microbial degradation of pollutants in the environment. During this period it was recognised that micro-organisms had limitations and sometimes fail to degrade wastes and pollutants at rates fast enough to preclude adverse environmental impacts. The fallibility of micro-organisms was revealed, particularly in terms of their ability to degrade xenobiotics and to exhibit rapid degradative activities under all conditions.

While bioaugmentation could do little to overcome the limitations of microbial evolution to develop enzymatic degradation pathways, it could overcome limitations imposed by environmental restrictions on microbial degradation. Conditions could be optimised in bioreactors, as they had been for decades in industrial fermentors and sewage and other waste treatment facilities, to achieve microbial degradation of wastes and pollutants. *In situ* treatments could also be designed to overcome factors limiting microbial degradative activities in nature.

Usefulness of Micro-organisms in Bioremediation

With the advent of heavy industrialisation and manufacturing of different chemicals, introduction of potentially hazardous and toxic organic compounds in the environment can not be prevented, and slowly the influx of such compounds in soil, air and water is increasing.

The global environment provides us with infinite opportunities and challenges. As in agriculture, we need 'green revolution' in environment, and this can be achieved only by a 'gene revolution.'

The workhorses of biotechnology are microbes, more precisely bacteria and today they seem to provide many solutions to man made problems. Biotechnological answers are emerging rapidly not only for detoxifying the environment through bioremediations, but also to enhance our resource utilisation by bio-conversion and application of probiotics.

Bioremediation is the use of the entire organism or enzymes, to transform toxic substances to benign or less toxic or more preferably to CO_2 and water mineralisation. Detection, however, does not ensure cure, and hence the bioremediation processes have to be instantly upgraded, using bugs with novel catalytic capabilities coupled with the use of biosensors. This requires strategies to evolve and adapt micro-organisms that can oxidise, modify and breakdown toxic materials by their ability to use them as a carbon and nitrogen sources.

The selection of micro-organism, that have the specific ability to detoxify unfavourable chemicals, development of enzymes and proteins that can catalytically convert these wastes, and bioprocess technology to harness these abilities are the major steps to be taken in bioremediation and bioaugmentation.

Biological treatment is one of the largest and most important controlled application of micro-organisms in manufacturing and service industries. Micro-organisms, especially bacteria, are capable of mediating biodegradation of a wide variety of simple to complex and recalcitrant organic compound owing to their ability to adaptation and genetic manipulation under the prevailing environmental conditions.

LIGNIN AND CELLULOSE BIODEGRADATION

Lignin and cellulose, which constitute the structural components of vascular land plants, are the most important renewable sources of organic carbon on earth. *Bassham* valued the carbon incorporated into the lignocellulose of arboraceous plants at half of the global amount annually fixed by photosynthesis. This is equivalent to 1.8×10^{12} tonnes of biological substances. Cellulose and lignin are also present in quantity in industrial and municipal effluents and residues. The development of a process for their biodegradation could offer a solution to a major problem of waste disposal and environmental impairment caused by both man and industry. In fact, municipal waste could be converted by biological composting into fertilisers and humus, so achieving several important benefits. Agricultural utilisation would reduce the consumption of inorganic fertilisers, not only because of the direct nutrient contribution, but also because they would improve the capacity of soil to retain chemicals; the pollution of rivers and the sea caused by emission of inorganic fertilisers would be reduced; the quality of soil would be improved by the addition of organic materials; and the emission of polluting and toxic gases generated during municipal waste incineration would be avoided.

Large quantities of lignocellulose residues must also be handled in food factories and plants for primary agricultural production. These materials could be treated biologically. They are dispersed in water and are relatively rich in cellulose and lignin, which could be converted into chemicals, fuel and protein, thus recovering, at least partially, the costs of waste treatment and disposal. The recovery of cellulose from waste paper, mainly newsprint and its conversion to fermentable sugar, is another potentially attractive process. Lignin, lignin-degradation products and chlorinated lignins must be recovered from pulp and paper-mill effluents in order to reduce both colour and toxicity. This is presently done with difficulty and then only partially, by expensive processes such as physicochemical treatment, ultrafiltration and activated-carbon adsorption. The successful biological treatment of these waste-waters would solve waste-disposal problems and allow the recovery of process waters which could then be recycled within a closed-circuit paper-mill.

Lignin and related aromatics are relatively inert biologically and especially in anaerobic environments, can cause severe pollution by the accumulation of toxic and recalcitrant organic materials. Biomass in the form of agricultural and forest wastes is most readily available in large quantities. Conversion of a fraction of this material into fuel, chemical, gas and feed protein would make a significant contribution to the overall problems of resource recycle and conversion. The conversion of cellulose and lignin is difficult and costly whether by biological or chemical processes. The production of gaseous and liquid feedstocks by direct thermal methods (e.g., pyrolysis) is technically feasible but, since the moisture content of biomass is generally high, pyrolytic and chemical treatment seem to be less attractive than biological processes. A large capital investment is required for a plant converting lignin to phenolic compounds in a two-stage process consisting of hydro-cracking followed by hydrodealkylation. Whether the overall process would be cost-effective is not yet clear.

Biochemical processes for the conversion of lignocellulose materials are based on studies of lignin and cellulose degradation in various natural environments. The body of experimental evidence suggest that the lignin degradation system is oxidative, non-specific and extracellular. Lignin, like other high molecular weight and insoluble polymers, cannot enter the microbial cell by diffusion or active transport, therefore biodegradation must involve extracellular catalysts. However, the first step in lignin decay might be non-enzymatic and could be caused by 'activated oxygen'.

The recent discovery of an extracellular enzyme produced by lignin cultures of *Phanerochaete chrysosporium* brings into question the activated-oxygen concept. Nevertheless, the presence of oxygen seems to be essential. No significant mineralisation of lignin under anaerobic conditions has been observed, with the exception of natural and synthetic lignin incubated anaerobically with anoxic sediments of aquatic ecosystems. Several micro-organisms, fungi imperfecti, soil bacterial and basidiomycetes can degrade lignin and related aromatic compounds, but their activity depends on the lignin substrate.

Much less obscure is the enzymology of cellulose degradation. Bacteria and fungi degrade cellulose and synthesise new microbial cell and metabolic products such as carbon dioxide and methane, but very few of these offer cellulase systems capable of *in vitro* conversion of cellulose into useful products, such as soluble sugars. Cellulase enzymes consist of a multiplicity of different components which can be classified into three major types: endoglucanases, exoglucanases and cellobiase.

The mode of action and the kinetics of cellulase enzymes has been studied mostly using fungal cellulases isolated from *Trichoderma* and *Aspergillus* species bacterial cellulases from *Clostridium* species. Bacterial and fungal systems work at different pH optima and exhibit different extracellular endo- and exo-activity ratios.

Enzymology of Lignin and Cellulose Degradation

Lignin is a interspersed with hemicellulose and forms protective seals of cellulose microfibrils which prevent access of degradative enzymes to the cellulose and hemicellulose in woody plant tissues in nature. Different fungi, bacteria and microflora in the soil have developed systems able to modify and efficiently degrade this heterogeneous aromatic polymers, since no long-term accumulation of lignocellulose residues takes place in nature. Among the micro-organisms involved in lignin biodegradation, fungi seem to play the dominant role and have been most studied, in particular the white-rot basidiomycete *Phanerochaete chrysosporium* and formerly its imperfect stage *Sporotrichum pulverulentum*. Currently there is much research effort being devoted to the biochemistry of lignin degradation. The identification of individual reactions of the lignin degradation process is essential for defining the biochemical mechanism. In turn, the biochemical information is necessary for optimum design of industrial bioprocesses. To meet this end, different experimental approaches have been developed. They consist of studies on fungal degradation of extractive lignin at the macromolecular level, on the biodegradation of lignin model compounds, on the structure identification of the low molecular weight compounds and, most recently, on the isolation, purification and characterisation of lignolytic enzymes.

It may be assumed that a very complicated biodegradation takes place, with many enzymes working together. Also, the possibility of an interaction between the degradation of lignin and that of carbohydrates with a mutual effect of the enzymes involved cannot be excluded.

Lignin has no hydrolysable repeating units as occur in polysaccharides, proteins and nucleic acids and since it is unlikely that lignin is first decomposed to monomers and then the monomers disintegrated, many complex reactions must occur during its fungal degradation. The polymer is non-stereoregular and, therefore, biodegradation is probably heavily oxidative and non-specific. Reaction patterns include: (i) cleavage of the C–C bonds in the side chains; (ii) the reductive cleavage of –O–4 bonds; (iii) 3–O– and 5–0–demethylation and 4–0– methylation, which could occur either at the macromolecular level or on lignin fragments; (iv) oxidative cleavage of aromatic rings; and (v) other oxidative, reducing and hydroxylation reactions.

BIODEGRADATION OF XENOBIOTIC COMPOUNDS

Xenobiotic compounds are man-made chemicals, but although foreign to the biosphere they need not necessarily constitute an environmental problem. On the other hand, since most of them have to be chemically inert for their particular application, many xenobiotic compounds have been found to resist biological degradation.

Regarding the release of synthetic chemicals into the biosphere, two types can broadly be distinguished: deliberate area contamination (e.g., by application of herbicides and pesticides) and involuntary emissions (e.g., such as caused by waste-waters, the dumping of wastes and local contamination of waters or soil by accidents). Special attention must be applied to industrial waste-waters which are permanent emission sources of xenobiotic compounds often being recalcitrant or hazardous to the environment.

The proven ability of the biosphere to evolve in changing world and the extraordinary evolutionary potential of micro-organisms, gave reason to hope that new degradative capabilities for xenobiotic compounds could be developed readily. In recent years, many synthetic chemicals were found to be biodegradable under laboratory conditions by specially enriched, adapted, or constructed micro-organisms. Under *in situ* conditions and in communal clarification plants, however, mineralisation of recalcitrant xenobiotics is still difficult or not attainable at all for a large number of possible reasons. It should be noted here that synergistic interrelations within microbial communities and threshold concentrations below which growth, adaptation or induction of catabolic activity will not occur. Therefore, even micro-organisms harbouring specific catabolic pathways often cannot be established in communal or large industrial clarification plants which collect a variety of waste-water streams.

To enable degradation of rather recalcitrant waste-water components, particular treatment systems should be constructed and integrated as a final stage into each production process. This concept, termed 'decentral waste-water treatment', allows the use of specially enriched and adapted micro-organisms under selective conditions and suitable substrate concentrations.

Enobiotic compounds, toxic chemicals, hazardous wastes also come from chemical industry, mining, metal processing, canning and tanning industry. The list of such compounds includes; solvents, wood preservative chemicals, coal tar wastes, plasticisers, refrigerants, biocides, pesticides, insecticides, halocarbon propellants, polychlorinated and polybrominated biphenyls, synthetic fibres, detergents, polyethylene, polystyrene, PVC, oils, surfactants etc. These chemicals form a significant percentage of recalcitrant compounds.

Biodegradation will occur successfully if conditions (pH, temperature, oxygen availability, redox potential, carbon/energy source, nutrient availability etc.) are suitable for growth of relevant micro-organisms. Many a times, industrial wastes may contain one or more unfavourable factors (e.g., effluent from pesticide industry may have alkaline pH, high salt content, very low carbon source of easily degradable nature, high ammoniacal concentration etc.). In such circumstances it is illogical to expect that potential micro-organisms will attack the xenobiotics. Here an integrated approach will be required to eliminate problematic conditions like pH, high ammonia, high salt etc. (using methods like neutralisation, ion exchange or membrane techniques). Complex organics can then be degraded by potential micro-organisms. If mixing of effluents from different point sources is responsible for such conditions, then due care can be taken to avoid the mixing of different waste-streams. In a few cases the pollutant is used by micro-organisms as an energy or carbon source. Biological detoxification may be carried out using pure cultures or mixed cultures, with the latter having a potential advantage in the degradation of toxic compounds in hazardous wastes. Mixed cultures are particularly useful when complete degradation of toxic organics to CO_2, CH_4, H_2S, N_2 etc., is required. Pure cultures, on the contrary, face limitation of nutrients, limitation of adaptive enzymes and produce toxic intermediates which may accumulate. Enrichment and selection procedures are useful in selecting mixed cultures carrying out degradation. But use of genetically engineered organisms is also increasing which have specific capacity to degrade such tough chemical substances. For example, genetically engineered organisms have been developed to specially degrade aromatic and highly chlorinated compounds.

Genetically engineered micro-organisms (GEMs) should be used for small volume of wastes and preferably for attached growth systems to avoid their release in environment in general. Multistage treatment should be used so that other biodegradables are removed by conventional mixed cultures and only a second stage is used for recalcitrant wastes.

BIOMETHANATION: A NEW TECHNIQUE FOR PRODUCTION OF ELECTRIC POWER

Biomethanation is one of the most important techniques for the biodegradation of organic substances. It can be economically efficient in the field of environmental protection, *inter alia*, stabilising sewage sludge and treating high-strength organic industrial waste-waters.

In the present world, large amount of biodegradable waste is generated everyday. Many techniques have been adopted to make use of the wastes for various purposes to satisfy human needs. Biodegradable waste can be disposed by several means like incineration, land fills, composting, dumping in the sea or other water bodies etc.

These methods have their own hazards and demerits. So instead of disposing biodegradable wastes, many projects are being investigated to make use of it in order to produce various products. This section deals with the production of electricity from biodegradable solid vegetable waste by means of biomethanation technique.

Biomethanation technique is adopted to treat biodegradable solid wastes and also to produce electricity. The various sub-processes, which have to be performed to produce biogas by biomethanation process are as follows:

1. Waste receipt and segregation of non-biodegradables.
2. Shredder.
3. Food preparation.
4. Maceration.
5. High rate anaerobic digestion through BIMA digester.
6. Biological desulphurisation.
7. Biogas storage.
8. Power generation.
9. Dewatering of digested substrate.
10. Odour control system.

Fig. 17.5 shows the schematic representation of the biomethanation process and production of electricity. The vegetable waste is received in bunkers and later on they are loaded in conveyers. The non-biodegradable wastes are manually removed from a slow moving conveyer.

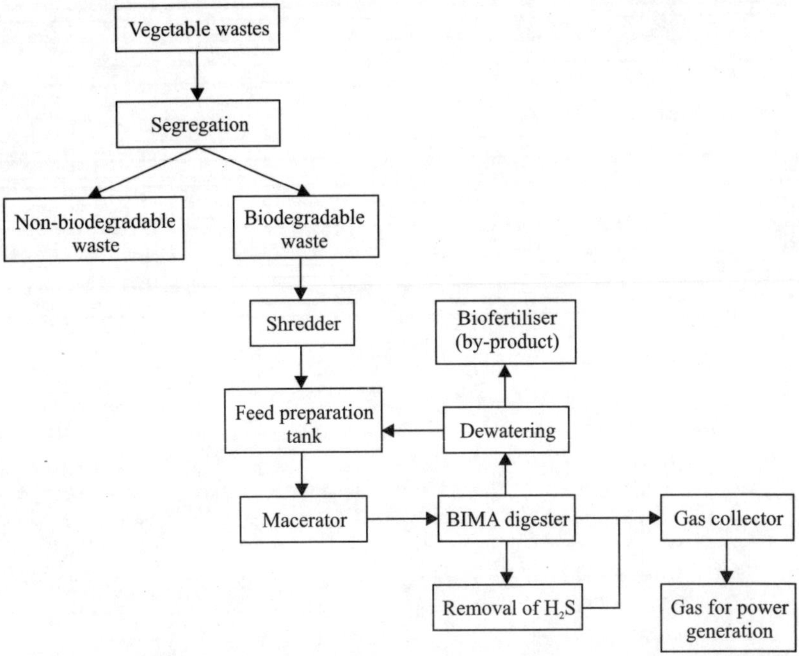

Fig. 17.5. Schematic diagram for the production of electric power.

The biodegradable wastes are then transferred into a shredder for size reduction. The shredded wastes are collected in a feed preparation tanks in which water is added from dewatering equipment. A mixer is provided in each tank to prepare feed solution.

The feed solution is pumped through the macerator for further size reduction and to cut the fibrous particles, which escapes from the shredder. The waste solution from macerator is pumped to a patented high-rate anaerobic digester, called BIMA digester, for anaerobic decomposition of organic wastes to produce biogas. The generated biogas is stored in a gasholder. The gas may contain hydrogen sulphide, which has to be removed. A biological system has been provided to remove H_2S gas from the biogas.

The biogas is used as a fuel in the gas engine to produce electricity. The digested substrates leaving the digester are mechanically dewatered in screw press. The dewatered cakes are used as biofertiliser and the water is sent back to the feed preparation tank for recycling. Biogas is produced and sent to gas engine for the production of mechanical energy; later on it is transformed into electrical energy by coupling a generator with the gas engine shaft.

Glossary

Activated sludge. Active biological solids in an activated sludge waste-water treatment plant.

Activated sludge process. Biological waste-water treatment process in which a mixture of the waste-water and activated sludge is aerated in a reactor basin or aeration tank. Active biological solids bio-oxidise the waste matter and the biological solids are removed by secondary clarification or final settling.

Advanced waste treatment. The use of physical, chemical, and biological means to upgrade the quality of a secondary effluent.

Aerated lagoon. Waste-water treatment pond in which mechanical or diffused air aeration is used to supplement oxygen supply.

Aeration tank. A tank or basin in which mixed liquor, waste-water, sludge, or other liquid is aerated.

Aerobacter aerogenes. One of the species of bacteria in the coliform group.

Aerobic. Presence of free molecular oxygen is required.

Aerobic bacteria. Bacteria requiring free molecular oxygen for their life processes.

Aerobic digestion. Digestion of suspended organic matter by aerobic microbes.

Algae. Primitive plant-like organisms, single or multicellular, usually aquatic, and capable of utilising food materials through photosynthesis.

Alkalinity. Ability of a water to neutralise an acid. Bicarbonate, carbonate, and hydroxide ions will augment it. Expressed as mg/l of equivalent $CaCO_3$ or meq/l.

Allowable breakthrough concentration. Maximum acceptable concentration of a solute in the effluent from carbon adsorption or ion exchange column.

Anaerobic. Requiring combined oxygen, such as SO_4^{-2}, PO_4^{-3}, NO_3^{-1} and the absence of free molecular oxygen.

Anaerobic bacteria. Bacteria that require combined oxygen and the absence of free molecular oxygen.

Anaerobic digestion. Digestion of suspended organic matter by anaerobic microbial action.

Anion. A negatively charged ion in a gaseous medium or in an electrolyte solution that migrates to the anode when an electrical potential is applied.

Bacilli. Rod-shaped or cylindrical bacterial cells.

Bacteria. Universally distributed, rigid, essentially unicellular microscopic organisms lacking chlorophyll, usually having a spheroid, or spiral shape. Some use organic matter as a foodstuff, while others use inorganic matter.

Binary fission. The manner in which most bacteria multiply. The parent cell divides, usually into two daughter cells.

Biochemical action. A chemical change resulting from the metabolism of living organisms.

Biochemical oxidation. Oxidation caused by biological activity resulting in a chemical combination of oxygen with organic matter to produce relatively stable end-products.

Biochemical oxygen demand. Oxygen required by microbes in the stabilisation of a decomposable waste under aerobic conditions.

Biodegradation. Biological oxidation of natural or synthetic organic materials by soil micro-organisms, either in soils, water bodies, or waste-water treatment plants.

Biological oxidation. An oxidation caused by biological activity resulting in a chemical combination of oxygen with organic matter to produce stable end-products known as both biochemical oxidation and bio-oxidation.

Biomass. Living biological matter.

Brackish water. Water having a dissolved solids content between freshwater and seawater.

Buffer action. Action of certain ions in solution to oppose a change in pH.

Bulking sludge. Activated sludge that settles poorly because of a floc with a low-bulk density.

Cake. Sludge that has been dewatered.

Carbon adsorption. Use of activated carbon to remove dissolved organic matter from a water or waste-water.

Centrate. Liquid flow leaving a centrifuge.

Centrifuge. A mechanical device that uses a centrifugal force to separate solids from a liquid.

Chemical coagulation. The destabilisation and initial aggregation of colloidal and finely suspended matter by the addition of a floc-forming chemical coagulant.

Chemical oxygen demand (COD). The amount of oxygen required to chemically oxidise the organic and sometimes inorganic matter in a water or waste-water. Usually expressed in mg/l. COD test does not measure the oxygen required to convert ammonia to nitrites and nitrites to nitrates, COD is frequently assumed to be equal to the ultimate first-stage biochemical oxygen demand.

Chemical sludge. Sludge produced by chemical coagulation or chemical precipitation.

Coagulant. A compound that causes coagulation or a floc-forming agent.

Coagulation. In water or waste-water treatment, the destabilisation and initial aggregation of colloidal and finely divided suspended solids by the addition of floc-forming chemicals.

Coarse rack. A rack with relatively wide spaces between the bars, usually one inch or more.

Cocci. Spherical bacterial cells.

Coliphage. A virus pathogenic to coliforms.

Complete treatment. Waste-water treatment that uses both primary and secondary treatment.

Conventional digester. A low-rate anaerobic digester.

Culture. Microbial growth that has been developed by provision of suitable nutrients and environment.

Degradation. Breakdown of substances by biological oxidation.

Desalting. Removal of salts from a brackish water or a tertiary effluent.

Dewater. Remove part of the water present in a sludge.

Dewatered sludge. Sludge that has had some of its water content removed.

Diatomaceous-earth filter. A filter usually used in water treatment that utilises a buildup layer of diatomaceous earth as a filter medium.

Diffused air aeration. Aeration produced in a liquid by the use of compressed air passed through air diffusers.

Digester. Tank used for sludge digestion. See sludge digestion.

Dissolved oxygen. Oxygen dissolved in a liquid, usually expressed in mg/l. Abbreviated as DO.

Dissolved solids. Dissolved substances in a water or waste-water.

Dry suspended solids. The suspended matter in water and, in particular, waste-water, which is removed by laboratory filtration and is dried for one hour at 103°C.

Effluent. Waste-water of other liquid, partially or completely treated, or in its natural state, flowing out of a basin, reservoir, treatment plant, or industrial treatment plant or parts thereof.

Enzymes. Organic catalysts that are proteins and are produced by living cells.

Excess activated sludge. Waste-activated sludge.

Facultative anaerobic bacteria. Bacteria that use either free molecular oxygen, if available, or combined oxygen. Also known as facultative bacteria.

Fermentation. Biochemical change caused by a ferment, such as yeast enzymes. Biochemical change in organic matter or organic wastes caused by anaerobic biological action.

Final effluent. Effluent from the final clarifier, final sedimentation basin, or final settling tank at a waste-water treatment plant.

Fixed-bed. In carbon adsorption or ion exchange treatments using columns or open beds, this refers to a bed that is stationary in the column or in the structure for the open bed.

Floc. The small, gelatinous masses formed in the water by the adding of coagulant. In waste-water treatment, the small, gelatinous biological solids formed at an activated sludge treatment plant.

Flocculation. Slow stirring of a coagulated water or waste-water to aggregate the destabilised particles and form a rapid-settling floc. In biological waste-water treatment where a coagulant is not used, aggregation may be accomplished biologically.

Flocculator. Basin in which flocculation is done or a mechanical device to enhance the formation of floc in a liquid.

Flotation. Raising of suspended matter to the surface of a liquid, where it is removed by skimming.

Fresh sludge. Undigested organic sludge.

Fungi. Small, multicellular, non-photosynthetic, plant-like organisms lacking chlorophyll, roots, stems, or leaves that feed on organic matter. Their decomposition after death may cause disagreeable tastes and odours in a water. They are found in water, waste-water, waste-water effluents, and soil.

Gravity filters. Filters that have gravity flow of the water through the filter bed.

Grit. Dense, mineral, suspended matter present in a water or waste-water, such as silt and sand.

Helminth. Intestinal worm.

Herbicide. Agent used to inhibit or kill plant growth.

Heterotophs. Bacteria that feed on preformed organic matter.

Influent. Water, waste-water, or other liquid flowing into a reservoir of a treatment plant.

Inoculate. Introduce viable cells in a medium.

Microbe. Micro-organism.

Micro-organism. Minute organisms, some being plant-like or animal-like, visible only by means of a microscope: microbe.

Mixed culture. Microbial culture consisting of two or more species.

Nematode. Worm that feeds on minute organisms.

Organic sludges. Sludges that have a high organic content. Usually, these are primary or secondary sludges at a waste-water treatment plant.

Pathogen. Pathogenic or disease-producing organism.

pH. The reciprocal of the logarithm of the hydrogen ion concentration in gram moles per litre. Neutral water has a pH of 7.

Primary sludge. Sludge from a primary clarifier.

Proteins. Complex nitrogenous compounds formed in living organisms. The compounds consist of amino acids bound together by the peptide linkage.

Raw sludge. Settled sludge that is undigested. Also called undigested sludge and fresh sludge.

Raw waste-water. Waste-water before it receives any treatments.

Returned sludge. Settled activated sludge that is returned to mix with the raw or primary settled waste-water.

Sanitary landfill. Landfill for disposing of solid wastes.

Secondary effluent. Effluent leaving the secondary or final clarifier at a waste-water treatment plant.

Secondary sludge. Sludge from the final clarifier at waste-water treatment plant. For the activated sludge process, it is the sludge to be recycled. For the trickling filter process, it is the trickle filter growths that have sloughed off–that is, the trickling filter humus.

Secondary treatment. Treatment of a waste-water by biological oxidation after primary treatment by sedimentation.

Sedimentation. Removal of settleable suspended solids from a water or waste-water by gravity settling in a quiescent tank or basin. Also called clarification or settling.

Sedimentation basin. Basin or tank through which a water or waste-water is passed to remove settleable suspended solids by gravity settling. Also called sedimentation tank, settling tank, settling basin, or clarifier.

Sludge. Accumulated solids removed from a sedimentation basin, settling tank, or clarifier in a water or waste-water treatment plant. Precipitate resulting from the chemical coagulation, flocculation, and sedimentation of a water or waste-water.

Sludge cake. Sludge that has been dewatered to increase the solids content to 15 to 45 per cent based on dry solids.

Sludge conditioning. Treatment of sludge, usually by chemical means, to enhance its dewatering characteristics.

Sludge digestion. Biological oxidation of organic or volatile matter in sludges to produce more stable substances.

Sludge moisture content. Amount of water in a sludge, usually expressed by percentage of dry weight.

Sludge processing. Collection, treatment, and disposal of sludge.

Sludge rakes. Rakes used in the collection of settled sludge from a clarifier used in water or waste-water treatment.

Sludge thickener. Tank or piece of equipment that concentrates the solids in a sludge.

Sludge thickening. To increase the solids content of a sludge.

Sludge treatment. Processing of waste-water sludges to render them innocuous. Common methods are anaerobic or aerobic digestion followed by sludge dewatering.

Sludge Volume Index (SVI). Volume in ml occupied by one gram of mixed liquor after settling for 30 minutes in one-litre graduated cylinder. Expressed as ml/gm.

Sludge loan. Sudden load, either organic or hydraulic, to a waste-water treatment plant. Same as shock load.

Substrate. Food substances for micro-organisms in a liquid solution.

Sulphur bacteria. Bacteria that use dissolved sulphur or dissolved sulphur compounds for their metabolism or growth.

Suspended matter. Solids in suspension in a water or waste-water that can be removed by laboratory filtration techniques, such as membrane filtration.

Suspended solids. Solids in suspension in a water or waste-water that can be removed by laboratory filtration techniques, such as membrane filtration. Same as suspended matter.

Tapered aeration activated sludge process. Activated sludge plant with a rectangular reactor basin and air diffusers or aerators spaced along the reactor length in accordance to the oxygen demand.

Tertiary treatment. Use of physical, chemical, or biological means to upgrade a secondary effluent.

Undigested sludge. Untreated, settled sludge that has been removed from a sedimentation basin or tank. Also called fresh or raw sludge.

Viable cells. Living cells.

Virus. Smallest form capable of producing disease in humans, animals, and plants, being 10 to 300 m in diameter.

Volatile solids. Quantity of solids, either suspended or dissolved, lost in ignition at 600°C.

Waste activated sludge. Excess activated sludge produced by the microbial solids in an activated sludge plant. This amount has to be wasted from the system at the rate it is produced. Same as excess activated sludge.

Waste-water. Spent water that consists of combination of liquid and water-carried wastes from dwellings, business buildings, industrial plants, and institutions, along with any surface or groundwater infiltration. The term has taken precedence over the word sewage.

Water softening. Process of removing calcium and magnesium ions from a water. The removal may be partial or total.

Zeolite. Natural minerals or synthetic resins having ion exchange capabilities.

Zooglea. The gelatinous material resulting from the attrition of bacterial slime layers. An important constituent of activated sludge floc and trickling filter growths.

References

Adams, J.A., and **Rogers, D.F.**, *Basic Biotechnology*, McGraw-Hill, New York.

Alliger, G., *Biotechnology Principles and Applications,* Interscience, New York.

Baum, B., and **Parker, C.H.**, *Industrial Biotechnology*, Reinhold, New York.

Berenson, C., *Genetically Engineering*, Wiley-Interscience, New York.

Billmeyer, W.F., *Biotechnology challenges,* John Willey and Sons, New York.

Blake, W.P., *Biotransformation,* McGraw-Hill, New York.

Brydson, J.A., *Food Biotechnology,* Elsevier Applied Science, London.

Clauser, H.R., *Handbook of Biotechnology*, Reinhold, USA.

DeGarmo, E.P., *Microbial Ecology*, John Wiley-Interscience, New York.

Donnet, J.B., *Mixed Culture in Biotechnology,* Marcel Dekker, New York.

Ewing, G.W., *Biochemical Engineering and Biotechnology*, McGraw-Hill, New York.

Faulk, B.F., *Plant Biotechnology*, Van Nostrand Reinhold, New York.

Gardon, J.L., *Principles of Biotechnology,* Interscience, New York.

Hampel, B.H., *Biotechnological Methods of Pollution Control*, Reinhold, New York.

Harvey, B.G., *Handbook of Biotechnology*, Prentice-Hall, UK.

Heckert, W.W., *Principles of Gene Manipulation*, Willey-Interscience, New York.

Jacobs, M.B., *Biodegradation in Effluents*, Wiley-Interscience, New York.

James, A.N., *Biotechnology and Environment*, Noyes, UK.

Jolles, Z.E., *Separation Processing Biotechnology*, Noyes, UK.

Lewis, F.W., *Membrane Separation in Biotechnology,* John Wiley and Sons, New York.

Morton, M., *Genetic Engineering*, Van Nostrand, Reinhold, USA.

Ostwald, W.J., *Ethics and Safety in Biotechnology*, Academic Press, New York.

Penn, W.S., *Biocatalyst and Bioseparation,* Maclaren, London.

Roger, S.W., *Microbiology,* Academic Press, New York.

Siting, M., *Methods in Enzymology*, Noyes, UK.

Stephenson, R.N., *Bioprocess Principles*, Reinhold, USA.

Swern, W., *Fundamentals of Biotechnology,* Reinhold Publishing Company, USA.

Van Wazer, J.R., *Biodegradation,* Wiley, New York.

Walker, E.E., *Recent Trends in Biotechnology*, Wiley-Interscience, New York.

Wright, P., *Principles of Biodegradation,* Maclaren, London.

Yaffe, L., *Biotechnology of Plant and Animal*, McGraw-Hill, New York.

Alfrey, T., *Biotransformation and Biodegradation,* Interscience, New York, USA.

Bauer, E., *Manual of Biotechnology*, Van Nostrand Reinhold, USA.

Culbertson, B. and **Pittman, C.,** *Principles of Fermentation Technology*, Noyes, UK.

Dym, J.B., *Industrial Biotechnology*, Van Nostrand Reinhold, USA.

Gacher, R., *Environmental Biotechnology*, Henser, USA.

Gaylord, M.W., *Bioseparation Processes—Theory and Practice*, Noyes, UK.

Griff, A.A., *Principles of Separation*, Van Nostrand Reinhold, USA.

Hertzberg, R.W. and **Manson, J.A.,** *Biocatalysts*, Academic Press, UK.

Irwin, C., *Biotechnology Today and Tomorrow*, Wiley-Interscience, New York.

Iwen, E.D., *Environmental Science and Technology*, Elsevier, USA.

Kusy, P.F., *Biotechnology of Medicines*, Noyes, UK.

Mark-Bikales-Overberger-Menges, *Industrial Microbiology*, Wiley Interscience, New York.

Matthews, G.A.R., *Industrial Biotechnology*, Elsevier, USA.

Parker, S.P., *Biotechnology and Waste-Water Treatment*, McGraw-Hill, USA.

Richardson, P.A., *Techniques of Biotechnology*, Noyes, UK.

Rosato, D.V., *Recycling of Animal and Industrial Wastes*, Van Nostrand Reinhold, USA.

Tobin, W.J., *Energy from Waste*, Noyes, UK.

Index